Measurement and Analysis of Mortality

The International Union for the Scientific Study of Population Problems was set up in 1928, with Dr Raymond Pearl as President. At that time the Union's main purpose was to promote international scientific co-operation to study the various aspects of population problems, through national committees and through its members themselves. In 1947 the International Union for the Scientific Study of Population (IUSSP) was reconstituted into its present form. It expanded its activities to:

- stimulate research on population
- develop interest in demographic matters among governments, national and international organizations, scientific bodies, and the general public
- foster relations between people involved in population studies
- disseminate scientific knowledge on population.

The principal ways through which the IUSSP currently achieves its aims are:

- organization of worldwide or regional conferences
- operations of Scientific Committees under the auspices of the Council
- organization of training courses
- publication of conference proceedings and committee reports.

Demography can be defined by its field of study and its analytical methods. Accordingly, it can be regarded as the scientific study of human populations primarily with respect to their size, their structure, and their development. For reasons which are related to the history of the discipline, the demographic method is essentially inductive: progress in the knowledge results from the improvement of observation, the sophistication of measurement methods, the search for regularities and stable factors leading to the formulation of explanatory models. In conclusion, the three objectives of demographic analysis are to describe, measure, and analyse.

International Studies in Demography is the outcome of an agreement concluded by the IUSSP and the Oxford University Press. The joint series is expected to reflect the broad range of the Union's activities and, in the first instance, will be based on the seminars organized by the Union. The Editorial Board of the series is comprised of:

Measurement and Analysis of Mortality

New Approaches

Edited by
JACQUES VALLIN
STAN D'SOUZA
ALBERTO PALLONI

CLARENDON PRESS · OXFORD
1990

Oxford University Press, Walton Street, Oxford OX2 6DP
Oxford New York Toronto
Delhi Bombay Calcutta Madras Karachi
Petaling Jaya Singapore Hong Kong Tokyo
Nairobi Dar es Salaam Cape Town
Melbourne Auckland
and associated companies in
Berlin Ibadan

Oxford is a trade mark of Oxford University Press

Published in the United States
by Oxford University Press, New York

British Library Cataloguing in Publication Data
Measurement and analysis of mortality: new approaches. —
(International Studies in demography)
1. Man. Mortality rate. Measurement & Statistical analysis
I. Vallin, Jacques II. D'Souza, Stan III. Palloni, Alberto IV. Series
304.64072
ISBN 0-19-828723-2

Library of Congress Cataloging in Publication Data
Mesure et analyse de la mortalité. English
Measurement and analysis of mortality : new approaches / edited by Jacques Vallin, Stan D'Souza, and Alberto Palloni.
p. cm. — (International studies in demography)
First published in French.
Includes bibliographical references.
1. Mortality—Statistical methods. 2. Mortality—Developing countries—Statistical methods—Case studies. 3. Developing countries—Statistics, Vital. I. Vallin, Jacques. II. D'Souza, Stan. III. Palloni, Alberto. IV. Title. V. Series.
HB1321.M4713 1990 304.6'4—dc20 90-30500
ISBN 0-19-828723-2

Typeset by Colset Private Limited, Singapore

Printed and bound in
Great Britain by Courier International Ltd
Tiptree, Essex

Contents

Contributors

Nicolas Brouard	Institut National d'Études Démographiques (INED), Paris, France
Riccardo Capocaccia	Istituto Superiore di Sanità, Rome, Italy
Graziella Caselli	Dipartimento di Scienze Demografiche, Università di Roma 'La Sapienza', Rome, Italy
Juan Chackiel	Centro Latinoamericano de Demografia (CELADE), Santiago, Chile
Stan D'Souza	United Nations Development Programme (UNDP), Cotonou, Benin
Viviana Egidi	Dipartimento di Scienze Demografiche, Università di Roma 'La Sapienza', Rome, Italy
Douglas C. Ewbank	Population Studies Center, University of Pennsylvania, Philadelphia, Pa., USA
Philippe Fargues	INED, Paris, France
Olivier Fontaine	Institut Français de Recherche Scientifique pour le Développement en Coopération (ORSTOM), Dakar, Senegal
Michel Garenne	ORSTOM, Dakar, Senegal
Antonio Golini	Istituto di Richerche sulla Popolazione (IRP), Rome, Italy
Ronald H. Gray	School of Hygiene and Public Health, Johns Hopkins University, Baltimore, Md., USA
M. Karim	International Centre for Diarrhoeal Disease Research (ICDDR, B), Dhaka, Bangladesh

Robert D. Mare	Center for Demography and Ecology, University of Wisconsin, Madison, Wis., USA
France Meslé	INED, Paris, France
Charles B. Nam	Center for the Study of Populations, Florida State University, Tallahassee, Fla., USA
Alberto Palloni	Center for Demography and Ecology, University of Wisconsin, Madison, Wis., USA
John H. Pollard	School of Economic and Financial Studies, Macquarie University, Sydney, Australia
A. Razzaque	ICDDR, B, Dhaka, Bangladesh
Patience W. Stephens	Office of Population Research, Princeton University, Princeton, NJ, USA
Jeremiah M. Sullivan	Westinghouse Institute for Resource Development, Columbia, Md., USA
Chukwudum Uche	Department of Sociology and Anthropology, University of Benin, Benin City, Nigeria
Jacques Vallin	INED, Paris, France
Étienne van de Walle	Population Studies Center, University of Pennsylvania, Philadelphia, Pa., USA
Arduino Verdecchia	Istituto Superiore di Sanità, Rome, Italy
Susan Zimicki	Population Studies Center, University of Pennsylvania, Philadelphia, Pa., USA

Abbreviations

ACME	Automated Classification of Medical Entities (US computer system)
AID	US Agency for International Development
CBR	Crude birth rate
CDR	Crude death rate
CELADE	Centro Latinoamericano de Demografia (Santiago, Chile)
CNR	Consilio Nazionale della Ricerca (National Research Council, Rome)
DHS	Demographic and Health Survey (conducted by Westinghouse)
DMB	Data Management Branch (Dhaka, Bangladesh)
DSS	Demographic Surveillance System (Matlab, Bangladesh)
EMW	Ever-married woman
FHI	Family Health International
FRG	Federal Republic of Germany
GDR	German Democratic Republic
GNP	Gross National Product
GRR	Gross reproduction rate
ICD	International Classification of Diseases
ICDDR, B	International Centre for Diarrhoeal Disease Research, Bangladesh (Dhaka)
ID	Identification number

IFORD	Institut pour la Formation et la Recherche en Démographie (Institute for Demographic Training and Research Yaoundé, Cameroon)
IHD	Ischaemic heart disease
IMIAL	Infant Mortality Research in Latin America
IMR	Infant mortality rate
INED	Institut National d'Études Démographiques (Paris)
INSEE	Institut National de la Statistique et des Études Économiques (National Institute for Statistics and Economic Studies, Paris)
INSP	Institut National de la Santé Publique (National Institute for Public Health, Abidjan, Ivory Coast)
IRP	Istituto di Ricerche sulla Popolazione (Institute for Population Research, Rome)
ISI	International Statistical Institute
ISS	Istituto Superiore di Sanità (Rome)
ISTAT	Istituto de Statistiche (Central Statistical Institute, Rome)
IUSSP	International Union for the Scientific Study of Population
LDC	Less developed context
MCH/FP	Maternal child health and family planning (Matlab, Bangladesh)
NCHS	National Centre for Health Statistics (Bangladesh)
NLS	National Longitudinal Survey of Labor Market Experience of Mature Men (USA)
NRR	Net reproduction rate
NTA	New trial area (Matlab, Bangladesh)
ORS	Oral rehydration solution

ORSTOM	Institut Français de Recherche Scientifique pour le Développement en Coopération (Paris)
OTA	Old trial area (Matlab, Bangladesh)
SEER	Program for Cancer Surveillance Epidemiology and End Results (USA)
SES	Socio-economic Status
SGF	Statistique Générale de la France (Paris)
SRS	Sample Registration Survey (India)
TFR	Total fertility rate
UN	United Nations
UNDP	United Nations Development Programme (Cotonou, Benin)
UNFPA	UN Fund for Population Activities
WFS	World Fertility Survey
WHO	World Health Organization

1 Introduction

JACQUES VALLIN

Institut National d'Études Démographiques (INED) Paris, France

STAN D'SOUZA

United Nations Development Programme (UNDP) Cotonou, Benin

ALBERTO PALLONI

Center for Demography and Ecology, University of Wisconsin Madison, Wisconsin, USA

Created by the International Union for the Scientific Study of Population (IUSSP) in 1985 at its General Congress in Florence, the committee on Comparative Mortality Changes, chaired by Stan D'Souza, is the third of its type to be set up since the International Union decided to give mortality studies a prominent position in its programme.

The first of these three committees, chaired by John Pollard, published the proceedings of three international seminars held in Fiuggi (Italy) in 1980, Dakar (Senegal) in 1981, and Paris (France) in 1983. The first of these publications deals with the main biological and social aspects of mortality (Preston 1982); the second is devoted entirely to the discussion of optimal study designs and methodologies for the collection and analysis of mortality data as these stood in the early 1980s (Vallin *et al.* 1984); and the third publication reviews the research in various fields regarding the influence of social and health policies on the levels and patterns of mortality (Vallin and Lopez 1985*a*, *b*).

The second committee, chaired by Lado Ruzicka, organized a meeting in Tokyo in 1984 on the Biological and Social Determinants of Mortality (Ruzicka *et al.*, in press). The present publication is thus the fifth volume of a series devoted to the study of mortality.

Pursuing a course already well explored was no easy task for the third committee. It was decided to assign greater importance to mortality differentials and consequently to comparative studies. Prior to embarking on two promising projects, one on mortality in sub-Saharan Africa, and the other on patterns of mortality in China and its neighbouring countries, it was considered necessary to take stock of the methodologies now available for the collection and analysis of mortality data. This was the aim of the international seminar organized by the third committee in Siena, Italy, on 7–12 July 1986. Although the Siena seminar shares common ground with the one held in Dakar (i.e. methodologies), it is by no means a duplication. In fact, the sheer volume of new ideas and the accelerated development of old ones proved to be too much ground to cover in a single seminar. The reader of *Methodologies for the Collection and Analysis of Mortality Data* may find that some of the themes treated in the present book are familiar, but it will also be evident not only that the materials combined under similar subjects have been deeply modified, but that new themes have emerged, the importance of which was not even suspected five years ago.

Forty researchers from 20 countries and all five continents, working in a variety of fields (demographers, sociologists, physicians, anthropologists, epidemiologists, statisticians, actuaries), came to Siena to compare their points of view through 26 papers, most of them solicited by the committee. This material and the discussions that emerged from it form the backbone of the present publication. This book is not, however, a straightforward reproduction of the papers presented at the seminar. Some of these have been omitted, either because they have already been published elsewhere, or because they duplicated other papers, or because they were somewhat removed from the main methodological preoccupation of the seminar. The remaining papers that now form the different chapters of this book are somewhat different from the original ones and the changes represent, in most cases, suggestions raised in debates that took place during the course of the seminar. Fifteen of the 19 chapters were originally written in English and four in French, which meant that their publication in both languages required a substantial amount of translation, in which several specialists from various disciplines participated. This is the second time, after *Health Policy, Social Policy and Mortality Prospects* that the International Union has launched a dual publication, the present book being the twin of a French version published by INED (Vallin *et al.* 1988).

To combine a set of papers covering quite diverse subjects proved to be a very difficult task. Further complications were created as it was necessary to consider highly specialized topics representing the state of the art in the field, without following a rigid, 'cook-book' approach. We decided on three major parts, which differ considerably from those used in the organization and discussion in Siena:

1. new and conventional approaches for collecting mortality data;
2. models and methods of analysis applied to mortality;
3. methods for studying mortality differentials and causes of mortality.

1. New and Conventional Approaches for the Collection of Mortality Data

The first three chapters of this first section deal with recent developments in collection procedures which have now become classic in their general outlines: population laboratories, multi-round surveys, and detailed maternity histories. The Matlab population laboratory in Bangladesh is a mine of information for epidemiologists and demographers. However, the many improvements added over time present various methodological problems that are somewhat difficult to resolve. This is the case with the 1982 update of the population register, which is dealt with in Chapter 2 by Stan D'Souza. In Chapter 3, Étienne van de Walle presents the particularly interesting case of the Bobo-Dioulasso survey (Burkina Faso) and discusses the advantages and constraints of the methodology elaborated by the *Yaoundé Institut pour la Formation et la Recherche en Démographie* (IFORD) for measuring infant mortality and its determinants. Van de Walle's paper could serve to tone down the quarrel between the advocates of indirect measures and those of multi-round surveys. In Chapter 4, Jeremiah Sullivan compares the principles which ensured the success of the World Fertility Survey (WFS) regarding the measurement of infant and child mortality to those which are being implemented by the Demographic and Health Survey (DHS) conducted by Westinghouse in some 40 countries.

The next two chapters (5 and 6) explore from different angles the potential results of more intense co-operation between demographers and practitioners of related disciplines. Over-concerned with how representative and generalizable their results are, demographers sometimes become dependent on methods of investigation which are not only cumbersome and expensive, but opaque to the complexity of the mechanisms underlying the phenomena under study. In the case of mortality, two disciplines close to demography may be particularly useful for filling in gaps in our knowledge of these mechanisms. In Chapter 5, Ronald Gray underlines the relevance of epidemiologic methods of investigation. In particular, he singles out case-control studies which involve particularly small and consequently very economic samples, but allow quite rigorous testing of many types of working hypothesis. Problems of observation and selection bias naturally exist, but they are known and in some cases their impact can be controlled. In Chapter 6, Chukwudum Uche discusses the possibilities offered by medical anthropology. Although generally more preoccupied with atypical situations, medical anthropologists concentrate on an array of intensive forms of investigation focusing on patterns of behaviour potentially related to the phenomenon of interest. It is up to the demographer to determine the conditions under which these behavioural patterns can be invoked as explanatory factors in situations different from those that helped to detect them.

The last three chapters in Part I (7 to 9) deal with a strategic issue in mortality analysis which, although somewhat neglected, is of central importance in developing countries: the medical causes of death. Even more so than for measures of total mortality, problems of defective data collection and classification of causes of death are seemingly insurmountable. In Africa, few deaths are officially reported and even fewer have a medical diagnosis of the cause of death. It is consequently impossible to obtain reliable medical cause-of-death statistics comparable to those existing in developed countries. Can this important gap be filled by surveys conducted by laymen? The recent findings of Susan Zimicki in Bangladesh (Chapter 7) and Michel Garenne and Olivier Fontaine in Senegal (Chapter 8) indicate that this is possible after preliminary operations to simplify nosology and identify symptoms easily recognizable by informants. In Chapter 9, Patience Stephens deals with the concrete problems presented by the treatment of measles reporting.

2. Models and Methods of Analysis Applied to Mortality

In Part II, three chapters deal with problems specific to countries in which statistical data are scant and of low quality. The other three chapters of the section focus on developed countries where data are more complete and accurate. In the 1960s, the scarcity and poor quality of data available in Africa led William Brass and others to develop relatively powerful models to produce estimates of mortality and fertility from simple statistics of survivorship and reproductivity. These advances led to the productive use of data from national retrospective population surveys conducted mostly in French-speaking countries. Similar and perhaps more serious problems created by the paucity of data remain crucial today, as national vital statistics have progressed only slowly while the enthusiasm for large national population surveys seems to have withered away. Thus the need to explore the viability of alternative sources of information to generate estimates and establish their consistency is quite current. In Chapter 10, Philippe Fargues presents several methods to estimate infant and child mortality from data collected in maternity hospitals on survival of preceding offspring. To avoid some of the drawbacks associated with the nature of the data, Fargues identifies the solution proposed by Brass and Macrae, and develops two new approaches that appear to yield more accurate estimates. In Chapter 11, Douglas Ewbank discusses the selection of the optimal model of mortality for the application of indirect techniques to measure infant and adult mortality in East Africa. In Chapter 12, Alberto Palloni reviews procedures measuring the levels of mortality. He shows that, with the exception of continuous registers, all tools for the measurement of excess mortality give inaccurate and long-delayed signals while the crisis runs its course. He also discusses various procedures for assessing the ultimate population impact of a crisis.

More accurate and very complete vital statistics systems in developed countries have stimulated preoccupation with very different theoretical and methodological issues and given rise to quite diverse research traditions. Two areas of study that have emerged (or re-emerged) rely strongly on the use of mortality models: the study of cohort effects and cause-of-death analysis. By and large, the adult mortality effects of a cohort's past history are a factor of secondary importance, being largely concealed by strong period changes due to the medical and sanitary progress of the last two centuries. It may, however, play a role in the explanation of certain aspects of recent mortality trends. Yet, to uncover the full extent of these effects requires not only fine tuning of the methods of analysis that are available, but also the development of new ones. In Chapter 13, Graziella Caselli investigates three types of approach and applies them to the study of cohort mortality in France and Italy. Using period data, Caselli first proposes the measurement of relative mortality variations in successive age groups. She then introduces a contour-line representation of the mortality surfaces by age and cohort since the end of the nineteenth century. Finally, she applies a logit model that enables her to assess the role played by cohort effects in the evolution of adult mortality.

The use of models is equally valuable for analysing cause-specific mortality. Using a classification based on a matrix of distances and focusing on trajectories of individual countries over a period of some 15 years, Nicolas Brouard shows in Chapter 14 how the cause profiles of mortality observed in low-mortality countries can be grouped together using a typology which is remarkably robust in the short to medium run. In Chapter 15, John Pollard proposes a formal method for measuring the contribution of single causes of death on the gain (or loss) in life expectancy experienced over a period of time. This method can also be extended to assess the relative contribution of changes in causes of death to observed changes in life expectancy between two countries, between the sexes within the same country, or between two periods within the same country.

3. Methods for Studying Differential Mortality and Causes of Death

The papers included in Part III deal more directly with the issues raised by the study of changes in cause-of-death structures and the methodological problems of and (significant) returns to the study of socio-economic differentials in mortality. Three chapters (16 to 18) consider from different angles appropriate ways of handling the medical causes of death. The next chapter (19) stresses the interest of longitudinal studies for measuring social differences in mortality. Finally, Chapter 20 offers health planners a method for estimating morbidity indirectly from mortality data.

The cause-of-death approach to studying mortality trends meets two major obstacles, statistical discontinuity due to the different revisions of the International Classification of Diseases and the often erratic variations in the proportion of deaths due to ill-defined causes. In Chapter 16, Jacques Vallin and France Meslé present a procedure for reconstructing a continuous series of death by cause, sex, and age as they applied it to France during the period 1925–78. In Chapter 17, Charles Nam deals with an important aspect of cause-specific mortality analysis. By and large, the researcher is constrained to use the so-called 'primary cause of death' data, despite the complexity of the morbid processes leading to death. Taking the example of cancer in the United States, Nam proposes a framework for analysing multiple causes of death that makes nearly full use of the information available on death certificates. In Chapter 18, Juan Chackiel discusses the virtues and defects of cause-of-death studies in Latin America and proposes a research agenda. He suggests that, even if the data were to remain far from satisfactory, one can still use them to assess the contribution of changes in the cause structure of mortality to trends in overall mortality and to identify some of the biological and social factors affecting mortality.

The quantitative assessment of socio-economic differentials in mortality is limited by well-known methodological problems which have been amply discussed elsewhere. For the last ten years, a network of specialists, working with various international and governmental sponsorship, have given much thought to this subject, and the Dakar Seminar revealed some of their contributions. Without pretending to exhaust the subject here, Chapter 19 emphasizes the importance of longitudinal data sets, even when these have not been explicitly created for the purpose of studying mortality. Robert Mare uses data from a longitudinal survey on socio-economic careers of adult men aged 45–60 (NLS) conducted in the United States to assess the mortality impact of various background characteristics and on certain aspects of a man's occupational career history.

The last chapter of the book turns the subject of mortality by cause of death on its head. The question is no longer how to measure the contribution of causes of death, but how to use the observed cause-of-death structure to infer a population's state of health, its 'morbidity structure'. Egidi, Golini, Capocaccia, and Verdecchia attempt this difficult exercise in Chapter 20. They estimate the incidence and prevalence of neoplasms among Italian males, using a dynamic model integrating mortality by neoplasms and various assumptions concerning the relationship between morbidity and mortality.

The diversity of themes and authors contributing to this volume gives to it an unquestionably heterogeneous nature. This occurred by design since from the onset our intention was to be as exhaustive as possible, even at the expense of overall cohesiveness. We have purposely sacrificed harmony for the sake of completeness and produced dissonance whenever that appeared to be the only way to take stock of the most recent methodological preoccupations in the field of collection and analysis of mortality data. Each chapter stands on its own and can, of course, be read independently of the others. Our aim has been to offer the specialist an opportunity for reference and reflection on various methodological aspects which are evolving rapidly.

We must conclude this introduction by thanking all those who participated in the Siena seminar. They all contributed, with helpful suggestions and critical remarks. We also wish to acknowledge the financial and technical assistance of several institutions which permitted the seminar to take place in excellent working conditions and facilitated the production of the present volume: the Italian National Research Council (CNR), Rome; the Italian Central Statistical Institute (ISTAT), Rome; Banca Monte dei Paschi, Siena; the University of Siena for hosting the seminar and INED for permitting us to publish a French version of this book; and, last but not least, Luciano Petrioli, who, as Director of the Istituto di Statistica of the University of Siena, organized the seminar. Finally, thanks are due to Lucetta Moelwyn-Hughes and Linda Sergent for translating the French texts into English, to Martine Deville Velloz for organizing the Index, and to Valerie Bower, Mary Ellen Eberhard, and Janice Deneen for typing the manuscript and preparing it for submission to the publisher. To all of them our most sincere acknowledgements.

References

Preston, S. (1982), *Biological and Social Aspects of Mortality and the Length of Life* (Liège: Ordina Éditions).

Ruzicka, L., Wunsch, G., and Kane, D. (eds.) (1989), *Biological and Social Correlates of Mortality* (Oxford: Oxford University Press).

Vallin, J., and Lopez, A. (eds.) (1985*a*), *Health Policy, Social Policy and Mortality Prospects* (Liège: Ordina Éditions).

—— —— (1985*b*), *La Lutte contre la mort: Influence des politiques sociales et des politiques de santé sur l'évolution de la mortalité*, Travaux et Documents, Cahier no. 108 (Paris: INED, PUF).

—— Pollard, J., and Heligman, L. (1984), *Methodologies for the Collection and Analysis of Mortality Data* (Liège: Ordina Éditions).

—— D'Souza, S., and Palloni, A. (eds.) (1988), *Mesure et analyse de la mortalité: Nouvelles approches*, Travaux et Documents, Cahier no. 119 (Paris: INED, PUF).

Part I

Classic Approaches and New Possibilities for Collecting Mortality Data

2 The 1982 Data Base Update of the Population Laboratory at Matlab, Bangladesh: Implications for Mortality Analysis

STAN D'SOUZA

United Nations Development Programme (UNDP), Cotonou, Benin

and M. KARIM and A. RAZZAQUE

International Centre for Diarrhoeal Disease Research (ICDDR,B), Dhaka, Bangladesh

---

1. Introduction

A baseline regarding the state of the art of methodologies for the collection and analysis of mortality data was set at the IUSSP meeting held in Dakar in 1981 (Vallin *et al*. 1984). Increasing interest from a variety of quarters—social statisticians, health planners, demographers—in an integrated approach to the study of population, health, and development problems, coupled with rapid progress in the field of microprocessors, justifies an overall review of the methodologies now available for mortality analysis and data collection. In a preparatory report on social statistics prepared for the UN Commission on Statistics, the need for radical changes in the manner of collecting and processing has been expressed. The importance of the use of computerized data bases that are tractable and geared to the needs of users has been stressed (United Nations 1984*a*). The recent World Conference on Population held in Mexico passed a similar resolution (United Nations 1984*b*). Restricting attention to mortality, the frameworks proposed require both rich data sets rarely found in a single survey or study and elaborate linkage procedures (Mosley and Chen 1984). These requirements mean that most Third World countries have problems using the frameworks suggested and do not have the data-processing capacity to handle large data sets or the linkages necessary for in-depth analysis. The recent advances in both data base techniques applicable to microcomputers and the improvements in hard-disk capacity—including removable cartridges—has brought a new range of tools to the researcher, particularly in a Third World setting.

A paper at the Dakar seminar presented the operation of the population laboratory at Matlab, Bangladesh run by the International Centre for Diarrhoeal Disease Research, Bangladesh (ICDDR,B) (D'Souza 1984). The capacity of such laboratories to study morbidity and mortality processes was explained. In particular, the advantages accruing from the existence of longitudinal records of fairly accurate data was stressed. Among the problems raised was the backlog created in data processing. Data was processed outside Bangladesh, and interest centred mainly on scientific topics concerned with cholera. The use of the data for population studies was initiated in the late seventies, and plans to have the data cleaned and computerized in the United States at Johns Hopkins University were set up. Due to problems related to distance and lack of experience with Third World data, on the one hand, and difficult linkage problems, on the other, this project has not been an unqualified success. However, the longitudinal data assembled under this project is of great value.

Details of the Matlab Demographic Surveillance System (DSS), in existence since 1966, will not be presented here. However, a brief summary is appropriate. The DSS collects information on births, deaths, immigration, emigration, internal migration, and marital status. In its present form covering 149 villages, a three-tier data collection system is used. At the village level 110 female community health workers record vital events, visiting each household once every fortnight. A family register is maintained in each home. The work of the health workers is supervised by 12 to 16 health assistants, who in turn report to three or four senior health assistants. The vital events, recorded on special forms, are checked at the field station of Matlab before being sent on to Dhaka for data processing. An illustration of the death registration form is displayed in Figure 2.1.

The Matlab study area was divided in 1978 into an intervention and comparison area, in order to assess the impact of health interventions, particularly maternal and child health and family planning (MCH/FP) programme. A description of the rationale behind the design and content of the programme has been reported elsewhere (Bhatia *et al.* 1980). Of the 149 villages 70 are selected for specific health intervention studies, the remaining villages being utilized as 'controls'.

The first major attempt to computerize the system was made while updating records of events till 1978 (Becker *et al.* 1982). The chief problem with this work was that it was done on a retrospective basis in 1981 and was out of date before the files were set on to the limited capacity of an IBM System 34. However, the printed output from these data files was utilized for a 1982 census and socio-economic survey (SES). The last SES data for the area had been collected in 1974. In the intervening years several crises due to floods and famine as well as political instability had ravaged Bangladesh. It was therefore felt that a new SES baseline was necessary.

A pilot study of five of the 149 villages of the Matlab surveillance area was done prior to the census update in 1982. During February to April 1981 an

MDR 85

DEATH REGISTRATION FORM

DSS-MATLAB, ICDDR, B

1. Sl. No. [| |] 2. Village (Code) ________ 3. Date of death [| | | | |] D D M M Y Y

2. Place of death: Village ____________ Upazila ____________ [ ]

	Name	Current Id No.	Registration No.	Date of birth
5.	Deceased			D D M M Y Y
6.	Mother's			D D M M Y Y

7. Marital status at the time of death:
- 1 Never married
- 2 Married
- 3 Widowed
- 4 Separated
- 5 Divorced

8. Type of doctor consulted:
- 1 Allopath
- 2 Allopath quack
- 4 Homeopath
- 8 Kabiraj
- 16 Others
- 32 Doctor not consulted

9. Education at death ________ [|]

10. Occupation at death ________ [|]

11. Detail description of events and symptoms leading up to death:

12. Specific causes of death (HA's report):
- 02 Measles
- 03 Tetanus
- 04 Drowning
- 05 Murder
- 06 Suicide
- 07 Diarrhoea acute
- 08 Diarrhoea chronic
- 09 Dysentery acute
- 10 Dysentery chronic
- 11 Child birth
- 12 Jaundice
- 13 Others not covered above

13. Reported by: ____________ (Code) [|]

14. Date entered into:

Field Vol. ____________

Matlab Vol. ____________

15. Date of reporting [| | | | |] D D M M Y Y

16. Remarks: ____________

Fig. 2.1. Sample of a death registration form

update on these villages and a socio-economic survey was done. A nutrition and morbidity study followed from June to September in the same year.

This chapter is intended to serve neither as a review of the Matlab system nor of the state of mortality, fertility, or migration in the area. Instead, it briefly documents field procedures utilized for the 1982 update of the Matlab data base and the changes introduced in the system in order to ensure appropriate capture of events due to movement of the population. These new procedures facilitated the establishment of a computerized data base that allows researchers to effect appropriate linkages of records and to carry out analysis within the ICDDR,B.

The resulting data sets provide answers to three types of question, each of which has considerable importance for mortality studies and the evaluation of health interventions in the Matlab area:

(*a*) Does a comparison of the 1974 and 1982 SES data base show that the economic condition of the households in the study area has changed substantially over the period 1974 to 1982?

(*b*) Has the impact of the health interventions through the service component of the ICDDR,B significantly reduced mortality over the same period?

(*c*) Is the rationale for the division of the Matlab area into treatment and comparison zones still valid for the assessment of the health intervention studies undertaken there?

This chapter attempts to illustrate the power of an integrated data base within a Third World context of mortality and morbidity studies, where data on explanatory variables are rarely available.

2. Methods and Materials

The boundaries of the study area have changed over time. Previous censuses had been done in 1966 in the old trial area (OTA) consisting of 133 villages; in 1968 in the new trial area (NTA) covering 101 new villages; in 1970 in the OTA; and in 1974 covering both the OTA and NTA. A 1978 update of registered events was also undertaken. In order to cut costs only 149 villages in the area covered in 1974, with a population of about 160 000, were retained. Each of these census exercises provided a new individual registration number, which, although theoretically usable for linkage of events and individuals, did not prove useful without high-powered computer facilities. Further, the registration number, originally intended to provide identification of household and individual, tended to become quickly out of date, due to movement of the population or other changes of status. Through timely data processing it is possible to obtain a fairly up-to-date idea of the population resident in a particular year. However, some changes in information cannot be updated from existing records. The following are a few examples:

(*a*) When a family splits into separate households there are problems related to identification of the new household and the updating of relationships of members to the new household head.

(*b*) A similar problem arises through the death of the household head.

(*c*) Marital status changes also occur that are not updated through the usual registration data. Death of an individual cannot be easily linked to the spouse.

In addition, to perform computer updating of the 1978 computerized file, the 1982 census update was designed to adjust for these problems through on-site visits. An important error source in computer updating arising from linkage problems was also corrected through these visits. The principal innovation for the 1982 census update was the introduction of a dual numbering system. A current identification number (ID) was set to designate present residence. This number would change with a new address. A permanent registration number was also designated. Additional information was introduced on the family registration form to facilitate computer linkage between vital-event files and the register. An illustration of the family forms appears in Figure 2.2. Details of the field operation as well as changes introduced are provided in the Appendix.

The five-village pilot study enabled the research team to check the forms used as well as to assess the costs and streamline logistic operations for the intended census update. The anthropometric measures taken in the

Table 2.1. *No. and % of household heads, by sex, religion, and family size, 1982 and 1974*

	1982		1974	
	No.	%	No.	%
Sex				
Male	27805	87.0	25285	88.4
Female	4168	13.0	3315	11.6
TOTAL	31793	100.0	28600	100.0
Missing	2			
Religion				
Muslim	27345	86.0	24063	84.1
Hindu	4628	14.0	4533	15.8
TOTAL	31973	100.0	28596	100.0
Missing	2		4	
Family size				
< 4	9908	31.0	8901	31.1
5–6	10448	32.7	9306	32.5
7+	11617	36.3	10393	36.4
TOTAL	31973	100.0	28600	100.0
Missing	2			
Average	5.85		5.87	

UPDATE DATE 31/05/82 | INTERNATIONAL CENTRE FOR DIARRHOEAL DISEASE RESEARCH, BANGLADESH | PAGE 1

DEMOGRAPHIC SURVEILLANCE SYSTEM : MATLAB

FAMILY REGISTRATION BOOK

VILLAGE : V 47 | NAME : TULATALI | BARI : 007 | NAME : MIZI BARI | FAM-RES-NO: 0044 | SIZE : 06 | RELGN : MUSLIM

IND. NO	N A M E	SEX	RELN HHH	MOTHER'S REGN. NO	MAR STA	SPOUSE'S REGN. NO	ENT YER	MEMBER'S REGN. NO	DATE OF BIRTH	DATE OF DEATH	DATE OF MIGRN. IN	DATE OF MIGR. OUT	INTER-VIL MIGRTN.	REMARK
01	SHAIZUDDIN	M	HEAD		M	IV47003602	74	IV47003601	0/00/20					
02	HALEMA KHA	F	WIFE		M	IV47003601	74	IV47003602	0/00/35					
03	ANNA	F	DAUG	IV47003602	S		75	IV47003640	20/09/75					
04	TAZMAHAL	F	DAUG	IV47003602	S		74	IV47003607	23/10/71					
05	HELENA	F	DAUG	IV47003602	S		74	IV47003604	0/00/64					
06	RASHID	M	SON	IV47003602	S		74	IV47003605	6/07/66					

INTERNATIONAL CENTRE FOR DIARRHOEAL DISEASE RESEARCH, BANGLADESH
DEMOGRAPHIC SURVEILLANCE SYSTEM : MATLAB
FAMILY REGISTER

DATE 31/08/82

VILLAGE CODE : V 47 NAME : TULATALI BARI CODE : 007

BARI NAME : MIZI BARI FAM.RES.NO : 0044 UPDATE DATE 31/05/82

IND. NO.	N A M E	MO NO	SP NO	DATE OF BIRTH	SEX	MEMBER'S REGN. NO	R E P O R T SERIAL NO	R E M A R K
01	SHAIZUDDIN		02	0/00/20	M	IV47003601		
02	HALEMA KHA		01	0/00/35	F	IV47003602		
03	ANNA	02		20/09/75	F	IV47003640		
04	TAZMAHAL	02		23/10/71	F	IV47003607		
05	HELENA	02		0/00/64	F	IV47003604		
06	RASHID	02		6/07/66	M	IV47003605		

Fig. 2.2. Sample of a family registration form
(*a*) Page 1
(*b*) Page 2

nutrition survey were height, weight, and arm circumference on all available children aged 2 to 60 months. Weight for age, height for age, and weight for height, as percentages, were calculated using the National Centre for Health Statistics (NCHS) standard growth curves. Information on the health status of the children was collected from the mothers of the children. On the day that the anthropometric measures were taken they were asked whether the child had been ill during the preceding seven days.

3. Results

3.1. The Data Base

The comparative results presented in this section relate the SES data of 1982 to the previous census of 1974. Table 2.1 presents data on heads of households by sex, religion, and family size for 1982 and 1974. There is an increase of female heads of households from 11.6% in 1974 to 13% in 1982. The religious breakdown shows a diminution in Hindu household heads from 15.8% to 14%. Family size per household remains virtually the same at 5.9.

The increase in female heads may be due to an artifact of reporting resulting from a greater facility in 1982 for a woman to be declared household head when her husband dies or when she is divorced. Emigration to other parts of Bangladesh and perhaps India may explain the diminution of Hindu household heads.

Overall, there were 31 973 separate households in 1982 compared with 28 600 in 1974—an increase of 11.2%. Since family size has not changed very much, this figure approximates the percentage population increase over eight years within the Matlab area and yields an annual increase of about 1.4%. These data are for the 149 retained villages, the initial 1974 population being about 160 000 and the 1982 population about 187 500.

Several questions in the SES questionnaire relate to ownership of property or assets (type of house, items owned, land owned). For a comparison with 1974, two aspects of ownership are considered here, the number of cows owned and area utilized as residence (see Table 2.2). Income data were not collected since in a rural setting it is exceedingly difficult to quantify earnings, not only because of the usual reticence of respondents in revealing their income levels but also, more importantly, due to the variety of sources that contribute to it. The ownership questions were designed as a proxy for assessing economic and income status.

The group with no cows represents 62.7% of household heads in 1982 as against 58.2% in 1974, while heads of households who have three cows or more in 1982 are 15% of the total as compared to 17.1% in 1974. The total bovine population as estimated from the average per household was just under 30 000 in 1974 and just over 30 000 in 1982. It thus appears that

Table 2.2. *No. and % of household heads, by no. of cows owned and area of residence, 1982 and 1974*

	1982		1974	
	No.	%	No.	%
Cows owned				
0	20052	62.7	16649	58.2
1–2	7122	22.3	7050	24.7
3+	4799	15.0	4897	17.1
TOTAL	31973	100.0	28596	100.0
Missing	2		4	
Average	0.93		1.05	
Area of residence (sq. ft.)				
< 169	9637	30.3	9540	33.4
170–272	7692	24.2	7387	25.8
273+	14518	45.6	11669	40.8
TOTAL	31847	100.0	28596	100.0
Missing	128		4	
Average	299.5		245.0	

constant land availability and increasing population pressure leads to difficulties in expanding cattle ownership.

The data on area of residence show little improvement in housing conditions. Families with an average of six members are still confined to a single room. The average residential area of 299.5 square feet in 1982 corresponds to a room 17 × 17 feet—not very different from the earlier 15.7 × 15.7-foot room size obtained from the 245 square-foot average in 1974. If one considers that in 1982, 30.3% of the families were still living in rooms under 13 × 13 feet, it appears clear that economic progress has been minimal. Other items of ownership confirm the same picture that, while a worsening of the economic situation has not taken place, there are few signs of optimism regarding the present state of affairs.

Classification of household heads by occupation is shown in Table 2.3. Little change in occupational status has taken place since 1974. Perhaps an element worthy of attention is the sharp reduction of household heads who have declared themselves able to support themselves through ownership of land without having to work themselves: 4% of heads of households belonged to this category in 1974, while in 1982 this percentage was practically zero. There is an increase in the percentage of those reported as housewife or houseworker in 1982. This is probably due to differences in the manner in which the question was asked in the two censuses. Persons classified as 'others' in 1974 may have been classified in other groups in 1982. The category 'others' accounted for 17% of household heads in 1974

Table 2.3. *No. of household heads, by occupation, 1982 and 1974*

	1982		1974	
	No.	%	No.	%
Occupation				
Landowner	153	0.0	1177	4.0
Farmer/worker	9877	31.0	8891	31.0
Daily agricultural labourer	6771	21.0	5157	18.0
Fisherman	1592	5.0	1404	5.0
Skilled labourer	1477	5.0	1368	5.0
Service	2032	6.4	1019	4.0
Business	2321	7.3	1154	4.0
Housewife/worker	3523	11.0	1757	6.0
Mill/factory worker	1714	5.4	1786	6.0
Others	2515	7.9	4887	17.0
TOTAL	31975	100.0	28600	100.0

and only 7.9% in 1982. The percentage of household heads shown as daily agricultural labourers, the group which experiences the most precarious conditions, rises from 18% in 1974 to 21% in 1982. The percentage of household heads classified as farmer/workers and fishermen remains unchanged. Overall, this table presents a bleak picture of occupation status in a rural area of Bangladesh that, in many ways, is better off than some other regions of the country.

The overall socio-economic picture of the Matlab area has not appreciably worsened but there are few if any signs of improvement of the lot of the people. In fact, with the recorded increase in population pressure, survival will be possible only through external resources to the community including remittances of residents who migrate out of the Matlab area for work.

3.2. The Mortality Conditions

Studies on mortality in the Matlab area have been numerous. A collation of results from a variety of studies appears in a case-study prepared for the UN/WHO (D'Souza 1985). The 'cause-of-death' structure assessed by lay-reporting methods has been shown to be stable (Zimicki *et al.* 1985). In this section mortality results related to the census update and the pilot five-villages study are presented.

Figure 2.3 presents results from the Matlab demographic surveillance system for the period 1966–81. The level of the overall death rate has been fairly constant at just under 17 per 1000. This rate is about 2.4 points below the national level estimated through indirect techniques. However, the fluctuations of the infant and child mortality rates indicate a fragile mortality balance where infants and children are the most vulnerable to crises. Peaks in 1971 and 1974–5 reflect the unsettled conditions during the liberation

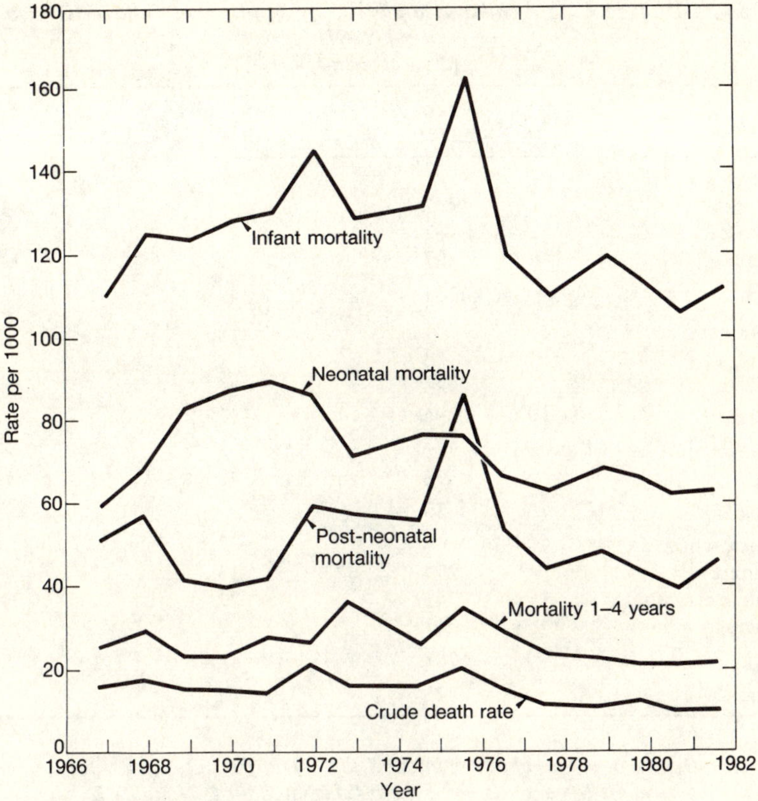

Fig. 2.3. Infant (neonatal and post-neonatal) mortality rate, children aged 1–4 years, death rate and crude death rate, Matlab, Bangladesh, 1966–1981

struggle and the famine respectively. A programme of health intervention in the treatment area was begun in 1978 and reductions in mortality in the treatment area have been recorded (Chen *et al*. 1983).

The existence of higher mortality among households with lower socio-economic levels has been documented in the Matlab area, using 1974 SES data (D'Souza and Bhuiya 1980). Confirmatory results are obtained from the five-village study and the latest SES data. Table 2.4 shows that children from households with the highest socio-economic level have death rates of about a third of those of the lowest level for the three characteristics of ownership tabulated—land, cows, and articles. Education levels of head of household and of mother (tables not shown) also suggest an inverse relationship with mortality rates.

Table 2.5 shows that the anthropometric measures discriminate neatly the groups of highest mortality risk. In particular, the severe and moderate categories for the 'weight for height' and severe category for the 'weight for

Table 2.4. *Deaths amongst children under 5, by land, cows, and articles owned, 1981–1982 (5-village study)*

	No. of children	No. of deaths	Rate (per 1000)
Land owned			
No land	442	15	33.9
< 1 acre	762	21	27.6
1 acre +	497	6	12.1
TOTAL	1701	42	
Missing	21		
Cows owned			
0	986	28	28.4
1–2	394	11	27.9
3+	312	3	9.6
TOTAL	1692	42	
Missing	30		
Articles owned			
Nothing	299	13	43.5
Quilt/hurricane	741	22	29.7
Others	661	7	10.6
TOTAL	1701	42	
Missing	21		

Table 2.5. *Deaths amongst children under 5, by height for age, weight for height, and weight for age, as % of the NCHS standard, 1981–1982 (5-village study)*

	No. of children	No. of deaths	Rate (per 1000)
Height for age (%)			
Severe (< 85)	312	17	54.5
Moderate (85–9)	618	14	22.7
Mild (90–4)	581	7	12.1
Normal (> 95)	211	4	19.0
TOTAL	1722	42	24.4
Weight for height (%)			
Severe (< 70)	32	5	156.3
Moderate (70–9)	325	17	52.3
Mild (80–9)	919		18.5
Normal (> 90)	446	3	6.7
TOTAL	1722	42	24.4
Weight for age (%)			
Severe (< 60)	233	24	103.3
Moderate (60–74)	996	13	13.1
Normal (> 75)	493	5	10.1
TOTAL	1722	42	24.4

age' measures characterize the groups of less than 25% of the child population experiencing more than half of the observed deaths.

Data on reported illness were collected on children in the five-villages study. The results showed confirmation of the links between poor socio-economic status, low nutritional status, and reported illness. However, the expected direct link between high socio-economic status and lower reported morbidity was not confirmed. It can be hypothesized that levels of tolerance of illness are lower among the richer groups, who thus tend to report illness when less fortunate groups would tend to ignore it. Details of this component of the study can be found elsewhere (Bhuiya *et al.* 1986).

3.3. Structural Variables

Information on village structures such as markets, schools, financial institutions, etc., were obtained in 69 of the 70 villages in the intervention area

Table 2.6. *Distribution of villages according to availability of selected public institutions, social and economic services, and other developmental programmes, Matlab, 1982 (%)*

	MCH/FP area (no. = 69)	Comparison area (no. = 79)
Public institutions		
Madrasa[a]	49.3	19.0
Primary school[a]	69.6	50.6
Junior high school[b]	24.7	10.1
High school[b]	62.2	68.5
College[b]	17.4	3.8
Rural health centre[c]	31.9	32.9
Government dispensary[c]	31.9	57.0
Government family welfare centre[c]	39.1	48.0
Services		
Post office[b]	84.0	88.6
Market with bank[b]	56.5	35.5
Service personnel		
Government health and FP workers	42.0	31.6
Midwife	100.0	84.8
Village practitioners	82.4	78.3
Other development programmes		
Union council	59.4	51.9
Youth club	47.8	54.4
Women's co-operative	7.3	7.6
Fishermen's co-operative	17.4	2.5
Farmers' co-operative	60.9	58.2
Adult education programme	5.8	21.5

[a] In or adjacent to village.
[b] Within $\frac{1}{2}$ hr.
[c] Within 1 hr.

Table 2.7. *Accessibility of villages to nearest transport point, and types of transport to Matlab Thana Headquarters (%)*

	MCH/FP area (no. = 69)	Comparison area (no. = 79)
Access to nearest transport point		
Paved/metalled road[a]	17.4	6.3
Kacha district council road[a]	26.1	39.3
Launch ghat[a]	66.6	50.6
Type of transport to Matlab		
Foot/boat		
Monsoon	87.0	59.5
Dry season	84.0	58.2
Launch		
Monsoon	5.8	16.5
Dry season	7.3	16.5
Foot/boat + launch		
Monsoon	7.3	24.0
Dry season	8.7	25.3

[a] Within $\frac{1}{2}$ hr.

Table 2.8. *Mean distance from and time of travel to Matlab Thana Headquarters, by area and season*

	MCH/FP area (no. = 69)	Comparison area (no. = 79)
Distance (miles)		
Mean	3.7	5.9
Range	0.5–8.0	0.5–18.0
Time (hours)		
Mean		
Monsoon	1.5	2.4
Dry season	1.4	2.6
Range		
Monsoon	0.25–3.50	0.25–4.50
Dry season	0.25–3.00	0.25–5.00

and in all 79 villages in the comparison area. There appear to be important structural differences between the two areas. Table 2.6 shows that the treatment area is better served with schools, markets, banks, and service personnel. Government dispensaries and family welfare centres appear to be more numerous in the comparison area. Tables 2.7 and 2.8 show that access to Matlab is substantially easier from the treatment area than from the comparison area. This fact has important implications with regard to the use of family planning and diarrhoea treatment facilities in Matlab (Rahman *et al.* 1982; Zimicki *et al.* 1984).

Table 2.9 shows that the religious composition of the populations and the

Table 2.9. *No. and % of household heads, by religion and family size, by area, 1982*

	MCH/FP area		Comparison area	
	No.	%	No.	%
Religion				
Muslim	13237	81.0	14108	90.2
Hindu	3099	19.0	1529	9.8
TOTAL	16338	100.0	15637	100.0
Family size				
< 4	5128	31.4	4780	30.6
5–6	5512	33.7	4936	31.6
7+	5696	34.9	5921	37.9
TOTAL	16336	100.0	15637	100.0
Missing	2			

Table 2.10. *No. and % of household heads, by years of schooling of household head, and highest education attained by household member, by area, 1982*

	MCH/FP area		Comparison area	
	No.	%	No.	%
Years of schooling (household head)				
0	8611	52.7	8774	56.1
1–6	5342	32.7	5063	32.4
7+	2383	14.6	1800	11.5
TOTAL	16336	100.0	15637	100.0
Missing	2			
Highest education (household member)				
0	5398	33.0	5598	35.8
1–6	6645	40.7	6503	41.6
7+	4295	26.3	3536	22.6
TOTAL	16338	100.0	15637	100.0

average family sizes in both areas are quite different. Hindus represent 19% of the household heads in the treatment area but only 9.8% in the comparison area. Also, 37.9% of the household heads in the comparison area have families with seven members and more, compared with 34.9% in the treatment area. Such a situation is not unexpected since the integration of a family planning component (MCH/FP) appears to have enjoyed considerable acceptance.

Table 2.10 provides data on education. Again, the two areas show differences. 56.1% of household heads have no education in the

Table 2.11. *No. and % of household heads, by no. of cows owned, floor space, and use of latrine, by area, 1982*

	MCH/FP area		Comparison area	
	No.	%	No.	%
Cows owned				
0	10111	61.9	9941	63.6
1–2	3823	23.4	3299	21.1
3+	2402	14.7	2397	15.3
TOTAL	16336	100.0	15637	100.0
Missing	2			
Floor space (sq. ft.)				
< 169	4876	30.0	4761	30.5
170–272	3651	22.5	4041	25.9
273+	7705	47.5	6813	43.6
TOTAL	16232	100.0	15615	100.0
Missing	106		22	
Use of latrine				
Both sexes	14507	88.9	11561	74.3
One or none	1805	11.1	4008	25.8
TOTAL	16312	100.0	15569	100.0
Missing	26		68	

comparison area, while in the treatment area the corresponding percentage is 52.7. The highest-level group (7+years) represents 14.6% of all households in the treatment area but only 11.5% in the comparison area. Data on the highest education attained by any household member confirms the same picture: in the treatment area 26.3% of households have at least one member who has completed seven or more years of schooling. The corresponding percentage in the comparison area is only 22.6. On the other hand, at the lower end of the education scale, 35.8% of households in the comparison area have no schooling while 33% have no schooling in the treatment area.

Table 2.11 considers the matter of ownership. Differences appear to be small between the two areas: 63.6% of household heads have no cows in the comparison area versus 61.9% in the treatment area. Ownership of other property differs little for the lowest groups in the treatment and comparison areas.

Finally, a striking difference exists between the two areas with regard to the use of fixed latrine: 88.9% of heads of households in the treatment area claim that they have access to a fixed latrine, while only 74.3% in the comparison area do so. This result could be due to the presence of the MCH/FP programme in the treatment area.

4. Discussion

In Bangladesh an adequate civil registration system has yet to take shape, and mortality information is based on small studies carried out in various parts of the country. The Matlab data base carries the longest continuous record of vital events. The additional information provided by a host of studies makes it probably the most useful resource for studies on mortality in Bangladesh and possibly throughout the Third World. Unfortunately longitudinal studies have suffered due to difficulties inherent in the linkage of events. This chapter has provided some details of the census update in the Matlab area in order to highlight areas where computerization and information from the field increase analytic efficiency. The introduction of the dual numbering system ensures that a permanent record is maintained while current identification continues to be available. The latter is necessary for field-workers to find individuals who have moved from their residence since the time of the last census. The permanent record ensures linkage of events—births, deaths, etc.—to the family register. This linkage has been further facilitated by the addition of spouse's and mother's registration number to the family registration book. The forms for vital events have also been modified to note current identification and registration numbers, while previous census numbers have been dropped (see Appendix).

The information collected points towards economic stagnation of the area during a ten-year period which may lead one to suspect that mortality differentials related to socio-economic status have been reproduced (Antonovsky 1979). The five-village study shows that, with relatively little extra cost and logistic difficulty, it is possible to graft studies relating to differentials in mortality and morbidity onto a longitudinal vital-registration system. The study confirms that simple anthropometric measures can help classify children at high mortality risk, a result that has been documented in many other studies (Martorell and Ho 1984). Due to the small number of deaths (42) in the five-village sample, a breakdown by sex was not presented. However, the higher female child mortality (1–4 years) found in previous cross-sectional studies (D'Souza and Chen 1980) has been confirmed by a longitudinal study using the Matlab data base and a follow-up from birth to 1978 of the cohort of children born during the period 1973–4 (Koenig and D'Souza 1984). These findings are important since they document the higher female mortality using registration data carefully collected, when indirect methods based on retrospective information show that male mortality was higher (Census Commission 1977). While indirect methods are undoubtedly useful, they should not be considered a substitute for careful registration, particularly in contexts where they frequently mask cultural biases that exist in the recall of selected events (for example, deaths of female children).

A review of the division of the Matlab area in 1978 into treatment and comparison areas is now possible. After the division was implemented, the SES data from the 1974 census were not easily accessible in computerized form. Further, important details about village structure and strategic variables for delivery and use of health facilities were not available either. The present census update and information collected both at the individual and household levels make it possible to sharpen the criteria for the proper distinction of the study area. The new results provide substantial information suggesting that an evaluation of the present design should be undertaken. The fact that the census update also collected data on structural variables, such as the existence of schools, markets, etc., has made it possible for the first time to base selection of villages for small studies on computerized criteria. The latter ensures that scientific sampling techniques can be undertaken in the design of health intervention studies. Thus, a scientifically designed vaccine trial was initiated in the area in 1984 utilizing the computerized data base (Clemens 1983). Using randomized allocation the comparability of the vaccine and the placebo groups was ascertained through use of the 1982 data. Not only are significant cost savings introduced by the use of sampling, but greater depth can be introduced into the comparison between the various groups involved in the trial.

A broader question can be raised at this stage regarding feasible designs for the study of mortality. Health intervention areas, particularly if they are successful, create differences in mortality within the study area. Once the validity of safe, simple, and relatively cheap interventions to reduce mortality or morbidity has been demonstrated, ethical considerations arise regarding the lack of provision of these interventions to the population in the comparative area. One example is the use of measles vaccine in the Matlab intervention area. The broadening of health intervention in the area to items other than diarrhoea and cholera control is of recent origin, and there is strong pressure to restrict the focus of intervention to diarrhoeal diseases alone. While there are some valid scientific and logistic reasons for such restrictions, administrators of study areas should provide convincing evidence to justify what would appear to be a lack of human concern.

Recent developments in data base technology have also called into question some linkage procedures carried out earlier. A project developed jointly with Johns Hopkins University has attempted to create a large file linking every event involving each individual. Unfortunately, apart from requiring very powerful computer capacity to handle this data set, the costs involved in using the file are particularly high. Relational data bases—the successors to hierarchical and network data bases—allow the relevant data to be stored in simple tables with appropriate indexes. The role of the registration number is clear since it would appear on any of the event files as well as the corresponding family registration books. Earlier computer technology suggested the building of large linked data files to be prepared for all future analysis. In a 'data base' environment, data should be maintained in small files that can be 'selected' and 'joined' specifically for the analysis in question. Since relational data bases are now available to microcomputers, the researcher can do much of this work without recourse to a group of skilled programmers. The need for programmers to know which fields should be at the head of the hierarchy or which ones should be linked by pointer chains when using earlier data base technology has been eliminated. Large data files will still require the use of mainframe computers, and an IBM 4330 has been acquired by the ICDDR,B. But the direction of microprocessor technology, including the ability to communicate with large computers as an emulated terminal, will clearly have an increasing impact in demographic data collection and analysis. These new trends are of particular importance for Third World countries, which have great difficulty in maintaining mainframe computers. While there is still a long way to go before these countries can have the basic facilities for research, the wider use of microprocessors will ensure that increasing amounts of analysis will be done in the country rather than in specialized centres located in the developed world. Cost considerations prevent population laboratories like that at Matlab from being replicated in other developing countries, but regional centres are a feasible alternative that should be explained. Although the burgeoning costs at the ICDDR,B have been in large part due to administrative reasons, it is hoped that recent budget streamlining will keep in place this outstanding resource for research in a variety of health fields in the Third World.

References

Antonovsky, A. (1979), 'Implications of Socio-economic Differentials in Mortality for Health Systems', in *Proceedings of WHO/UN Conference on Socio-economic Determinants and Consequences of Mortality*, Mexico City, 19–25 June.

Becker S., Razzak, A., and Sardar, A.M. (1982) 'Census Update 1978', *Demographic Surveillance System, Matlab*, viii, ICDDR,B Scientific Report no. 55.

Bhatia, S., Mosley, W.H., Faruque, A.S.G., Chakraborty, J. (1980), 'The Matlab Family Planning Health Services Project', *Studies in Family Planning*, 11, 202–12.

Bhuiya, A., Zimicki, S., Nahar, L., D'Souza, S. (1986), 'Socioeconomic Differentials in Child Malnutrition and Reported Morbidity in a Rural Area of Bangladesh', *Journal of Tropical Pediatrics*, Feb.

Census Commission (1977), *Report on the 1974 Bangladesh Retrospective Survey on Fertility and Mortality* (Dhaka: Census Commission).

Chen, L.C., Rahman, M., D'Souza, S., Chakraborty, J., Sardar, A.M., and Yunus, M. (1983), 'Mortality Impact of an MCH–FP Program in Matlab, Bangladesh', *Studies in Family Planning*, 14(89).

Clemens, J. (1983), *Cholera Vaccine Trial*, Research Protocol submitted to ICDDR,B Research Review Committee, Dec.

D'Souza, S. (1984), 'Population Laboratories for Studying Disease Processes and

Mortality: The Demographic Surveillance System, Matlab', in Vallin *et al.* (1984).
D'Souza, S. (1985), 'Mortality Structure in Matlab (Bangladesh) and the Effect of Selected Health Interventions', in UN, *Determinants of Mortality Change and Differentials in Developing Countries: The Five Country Case Study Project*, ST/ESA/SER. A/94 (New York: UN).
—— and Bhuiya, A. (1980), 'Socioeconomic Mortality Differentials in a Rural Area of Bangladesh', *Population and Development Review*, 8(4), June.
—— and Chen, L.C. (1980), 'Sex Differentials in Mortality', *Population and Development Review*, 6(2).
Koenig, M., and D'Souza, S. (1984), 'Sex Differences in Childhood Mortality in Rural Bangladesh', *Journal of Social Science and Medicine*, 22(1), 15–22.
Martorell, R., and Ho, T.J. (1984), 'Malnutrition, Morbidity and Mortality', in W.H. Mosley and L.C. Chen (eds.), *Child Survival: Strategies for Research*, supplement to *Population and Development Review*, 10 (New York: The Population Council).
Mosley, W.H., and Chen, L.C. (1984), 'An Analytic Framework for the Study of Child Survival in Developing Countries', in W.H. Mosley and L.C. Chen (eds.), *Child Survival: Strategies for Research*, supplement to *Population and Development Review*, 10 (New York: The Population Council).
Rahman, M., D'Souza, S., and Karim, R. (1982), 'Village Characteristics and Contraceptive Practices', Report presented at the *Seventh Annual Contributors' Conference, Bangladesh Fertility Research Program*.
UN (1984*a*), *Special Questions: Future Orientations of Work Related to Social Indicators*, E/CN.3/1985/3 (New York: UN).
—— (1984*b*), *Report on the International Conference on Population 1984*, E/CONF.76/19 (New York: UN).
Vallin, J., Pollard, J., Heligman, L. (eds.) (1984), *Methodologies for the Collection and Analysis of Mortality Data* (Liège: Ordina Éditions).
Zimicki, S., Yunus, M., Chakraborty, J., and D'Souza, S. (1984), *A Field Trial of Home-Prepared Oral Rehydration Solution in Rural Bangladesh*, Center for Population Studies Discussion Paper, no. 84(4) (Cambridge, Mass: Harvard University).
—— Nahar, L., Sardar, A.M., and D'Souza, S. (1985), 'Cause of Death Reporting in Matlab', *Demographic Surveillance System, Matlab*, xiii, ICDDR,B Scientific Report no. 63.

Appendix 2.1. The Nature of the 1982 Census Update

1. The 1982 *De Jure* Census Definition

It was decided that the resident population of the 149 villages in the DSS area in October 1978 would constitute the target population of the 1982 census update. In the *de jure* census a resident was defined as:

(*a*) A person resident in the DSS area on a regular basis, including persons who return to the area at least once a month and stay at least one night.
(*b*) A person present in the DSS for at least six months per year (usually

migratory workers). Temporary visitors and guests who were present at the time of the census were excluded from the count.

The mid-year population of 1982 was the target.

2. Registration Number

For individuals present at the 1974 census, this number would be the 1974 census number. For individuals entering after that date, the registration number assigned at the time of entry into the system was designated the permanent registration number. Forms would cease to carry earlier census numbers. However, a current ID corresponding to actual location would be noted. Thus an individual living in village V47 and having family residence number 0044 and being the head of the household would have current ID V47-0044-01. The last two digits (-01) constitute his individual number and would be assigned by the computer based on relation to head of household. A check code (1 or 2) was also inserted in the file to indicate if the person entered before or after the 1982 census. This was to avoid possible duplication of registration numbers for new entries. The registration number of this individual would be his 1974 census number: V47-0036-01 (see Figure 2.1).

3. Family Register

A family register is maintained in each household of the DSS area. During the 1982 census update a computerized register replaced the previous handwritten forms. In addition to the introduction of the dual numbering system, the spouse's and the mother's registration number were added. Family registration books, consisting of computerized output by village, have been made for quick reference both in the field and in Dhaka. These procedures ensure that additions, changes, or corrections recorded in the field can be easily transferred to the computer at Dhaka and a new family registration form which is identical to the computerized records can be provided (see Figure 2.2).

4. Execution Procedures

The Demographic Surveillance System (DSS) workers collected the vital registration data, and workers specially recruited under an ICDDR grant obtained data for the SES. Eighteen teams were formed, all consisting of a staff member of the DSS and a newly recruited worker. The DSS workers were responsible for the census update and assisted in supervision of SES work done by the new workers. This procedure also ensured easy identification and access to homes since they form part of an ongoing longitudinal data collection system.

It was decided to collect the SES information at the same time as the field visits for the census update. This decision was based on efforts to economize on transport costs as well as to reduce inconvenience to Matlab inhabitants caused by renewed visits for sociological enquiry purposes.

Four stages in the execution can be noted: first, the 1978 census file was updated by computer by adding births and immigration events and deleting deaths and emigration events from 1 January 1979 to 31 December 1981. A printout by village and by family was obtained for this partially updated census file. In order to reduce linkage problems that had proved severe in work done at Johns Hopkins University,

a.

SES 1982 Questionnaire

SES 1982
ICDDR, B

a. Household Information (1 - 19)

b. Individual Information (20)

1. Card No. 1 (1), 2. Study No. ____ (2-4), 3. Village : ____ ____ (5-7) 4. Bari : ____ ____ (8-10)

5. Family No. ____ (11-14), 6. (74/) Family registration No. ____ (15-18) 7. Religion ____ ____ (19) 8. Date of Interview ____ (20-25)

9. Main sources of water during season (from to) :

Drinking : ____ ____ (26), Cooking : ____ ____ (27), Bathe : ____ ____ (28), Washing : ____ ____ (29)

Codes : Tube well (1), Tank (2), River (3), Ditch/Canal (4), Others (5)

10. Structure of the largest room : Wall : ____, Roof : ____ ____ (30)

Codes : Wall tin + Roof tin (1), Wall pucca + Roof tin (2), Wall tin + Roof others, Wall others + Roof tin (3), Others (4)

11. Items Owned : Lep (01), Hurricane (02), Bi-cycle (04), Any watch (08), Radio (16), Remittance (32) ____ (31-32)

12. No. of cows owned : ____ (33), 13. No. of boats owned: ____ (34)

14. Total land owned (excluding homestead) in decimals : ____ ____ (35-38)

If own land : Self cultivated (1)
Renting out (2) ____ (39)
Share cropping (4)

If no land : Takes rent (1)
Shares crop (2) ____ (40)
Others (specify) (4)

15. Use of fixed latrine : Male Yes/No, Female Yes/No ____ (41), Within 15 yards from used water sources : Yes/No ____ (42)

16. Highest education of former member now outside DSS/Abroad : ____ (43), Type : ____ (44), Years of schooling : ____ (45-46)

17. Highest education in the family : Type ____ (47), Years of schooling : ____ (48-49)

18. Occupation of household Head : ____ (50-51), 19. Family size : ____ (52-53)

Fig. 2.A1. Sample of questionnaire of household and individual information

b

Card No. 2 (1), Study No. ___ (2 4), Village : ___ ___ (5 7), Bari : ___ ___ (8 10), Family No. ___ (11 14), Religion : ___ ___ (1)

20. Individual Information :

Sl. No.	Year of Reg.	Registration No.	Name	Sex	Relation to Head	Marital Status	Sl. No. of spouse	Date of birth	Sl.No. of Mother	Education		Occupation	
										Type	Years of schooling	Primary	Secondary
16-17	18-19	20 - 28		29	30-31	32	33-34	35 - 40	41-42	43	44 - 45	46 - 47	48 - 49
1.													
2.													
3.													
4.													
5.													
6.													
7.													
8.													
9.													
10.													
11.													
12.													
13.													
14.													
15.													

Name of the interviewer :

Time taken (in minutes) :

Fig. 2.A2. Sample of questionnaire of household and individual information

three new variables were added to the census file: year of entry into the DSS, spouse's registration number, and mother's registration number. In the second stage, the computer printout was updated by Matlab field staff for the items requiring field visits. The forms utilized for this exercise are shown in Figures 2.A1 and 2.A2. The third stage involved the updating of the computer file to accommodate these changes, including split households, changes in relation to head and marital status, as well as corrections registered in the field. The fourth and final stage involved updating the census file to 30 June 1982 by computer with the help of the event files.

Four days were devoted to training, two days at the Matlab treatment centre and two days in the field. An SES questionnaire which was developed during the pilot SES on five villages was utilized, and written instructions were prepared for field-workers (see Figure 2.5). The training was undertaken in Bengali by senior staff of Matlab, DSS, and the Data Management Branch (DMB) in Dhaka. Daily supervision of the work was done by a separate team of supervisory-level persons from the regular DSS staff at Matlab. In addition, senior staff from the DMB visited Matlab every week for a two-day period. During their stay problems occurring in the field were discussed. Joint meetings of supervisors and staff ensured uniform solutions to field problems. The field-work started on 1 April and ended on 25 June 1982. Forms were carefully checked for errors, omissions, or inconsistencies by supervisors in the Matlab office before being sent to Dhaka for data processing. Coding of forms was done in Dhaka between July and September 1982. Data entry on the computer was slow due to heavy pressure on available resources at the computer centre. The data entry was completed by December 1982. The computer output was subjected to usual range and consistency checks. Error lists were thus prepared and manually corrected. The small proportion of errors (0.5%) that could not be corrected in Dhaka were then sent back to Matlab for final correction and the new information was included in the computer file.

As a complement to the SES survey of households, a village survey was first undertaken during which information on a few structural variables such as the existence of markets, schools, post offices, etc. was sought. Apart from maintaining computerized records on diskettes, 'village books' were prepared for use in Matlab and in Dhaka. SES characteristics of individual heads in a village, together with pertinent structural variables collected, were collated in these books. Besides providing an important resource of immediate information on households in the Matlab area, these volumes are very useful for the design of smaller studies.

3 The IFORD Surveys

ÉTIENNE VAN DE WALLE

Population Studies Center, University of Pennsylvania, Philadelphia, Pennsylvania, USA

1. Introduction

In much of French-speaking sub-Saharan Africa, the multi-round survey has long been the method of choice for collecting demographic data. Paradoxically, after taking an outstanding series of single-round demographic surveys in the 1950s and 1960s, the francophone statisticians and demographers became disappointed with the results, at the very time when William Brass and the Princeton group were using the African surveys to devise methods of indirect estimation (Brass *et al.* 1968). A clear rift came into being, which is not yet completely closed. On the one hand, what the French labelled the 'Anglo-Saxon school' came increasingly to rely, for the estimation of fertility and mortality, on retrospective questions on children ever born and surviving. These are now being routinely included in censuses and surveys of the English-speaking countries, while an 'orphanhood' question based on the same principle has been serving to estimate adult mortality. Meanwhile, the French school has accepted as an article of faith that the only way to improve the quality of demographic estimation in Africa was to improve data collection. And the multi-round survey has appeared, perhaps mistakenly, as the best way to do this. It is only very recently that the extraordinary success of the single-round retrospective World Fertility Surveys in Africa in producing results in a fast and reasonably economical way has led to a reconsideration of the theoretical advantages of the two methods of data collection.

Most multi-round demographic surveys are based on the matching and reinterview of a whole population. They present obvious advantages for obtaining accurate person-years denominators and for dating events. After a baseline survey has accurately identified a population at risk, it is theoretically straightforward to proceed with a rigorous accounting of all entries and exits that have occurred in the population by the time of reinterview. For the measurement of mortality and migration, the procedure has in principle obvious advantages over retrospective surveys because the reference period is accurately delimited by the survey rounds, and because the persons who may have died or may have left during the intervening period had been interviewed, described, tagged as it were, before they exited from the sample; in practice, however, the problems of matching a large and changing population have often turned out to be formidable. Meanwhile, the methodology has no obvious advantages in the instance of the so-called 'double events' involving very young children, born and dead between rounds, about whom the only information that can be collected is retrospective information (Adlakha and Nizamuddin 1984).

To fill this gap, a special survey was designed that would concentrate precisely on these infant deaths. The basic idea consisted of following a panel of births identified at the time of delivery, including all the births in a selected population or a representative sample of them. In the absence of complete registration of birth, the choice was made of large cities where a substantial proportion of the births occur in hospitals and maternity clinics. Once a cohort of births had been identified in a baseline survey, it would be followed by reinterview of the mothers at regular intervals. IFORD, the United Nations demographic training and research centre in Yaoundé, Cameroon, pioneered the design, organization, and analysis of these urban surveys, and we shall call them 'IFORD surveys' for short (Courbage *et al.* 1979). Starting in 1978, the methodology was applied, under IFORD leadership, in a series of capitals: Yaoundé (Cameroon), Ouagadougou (Burkina Faso), Lomé (Togo), Cotonou (Benin), and Brazzaville (Congo). Later, the same type of questionnaires and methodology was transferred, under the sponsorship of the Sahel Institute, to a group of secondary urban centres of Burkina Faso (then Upper Volta), the largest being Bobo-Dioulasso, and to Bamako, the capital of Mali.

The author became involved in the Bobo-Dioulasso survey, and much of this article is derived from his experience in this single survey. It is difficult to generalize from this experience, since it is not always possible to sort out what was peculiar to Bobo-Dioulasso from what is a general characteristic of this type of operation. The attempt is justified, however, because there has been very little discussion of the methodology of the IFORD surveys, at least in publications in English.

The design of the surveys involved, in principle, the registration of all births occurring in the maternity hospitals of a city during a period of 12 months, and the interviewing of the mothers at the time of delivery; in several instances, a representative sample of these births was deemed sufficient to provide a significant number of cases. In Bobo-Dioulasso, an attempt was made to include some of the births occurring to residents of the city in their homes. The mother was requested to give an address and various references, so that it would be possible to find her again at home after she had left the maternity clinic. She was then normally reinterviewed one month later, four months later, and then every four months until the end of the second year of life of the child, if the latter had survived. A child would thus be subjected to a total of eight interviewing rounds (using a baseline questionnaire once, and seven times with an identical questionnaire for the rounds). We shall not enter here into details of the actual content of the interviews. Be it sufficient to say that the baseline survey and the seven subsequent rounds covered more than 100 variables: demographic, socio-economic, and cultural characteristics of the mother and the household; access to health facilities, including vaccination; determinants of the birth interval, with information on feeding of the child and on the resumption of menstruation and intercourse; health of mother and child, disease and causes of death, etc. Overall, a great wealth of interesting data was collected in a longitudinal fashion but space is not available here to review and criticize the phrasing of the questions in detail.

I will limit the discussion in this chapter to the principle of the multi-round survey, and its particular application in the IFORD surveys. I shall organize the comments in two parts: the weaknesses and strengths inherent in the study of a cohort of births, and those inherent in the particular IFORD application (i.e. the urban, hospital-based panel study). The chapter ends with a discussion of a few design changes that might be considered in future multi-round surveys.

2. Justification of a Longitudinal Study of Mortality

We suspect that the governments, who are the largest potential users of health and mortality statistics, were sold on the principle of a cohort study without really knowing what they were getting into. They thought they would be obtaining period estimates of infant and child mortality in their capital city; instead of that, they got a sophisticated instrument for the study of the determinants of mortality in the first two years of life. Government services want, as a first priority, national estimates of crude demographic rates. The infant mortality rate for a capital city is a welcome addition to knowledge, with clear policy implications. Since surveys are infrequent and vital statistics are unreliable, any indication of trends is out of reach, and the life of an estimate is very long. Cohort studies based on year of births are unsuitable, however, for estimating the average values for a period. They are extraordinarily sensitive, not only to yearly fluctuations, as the information about births and deaths in the last year might also be, but also to seasonal variations. This is illustrated here with results from the Bobo-Dioulasso survey, where the issue has been studied in some detail (van de Walle 1986).

The Bobo survey was initially planned to record all the births of the entire calendar year 1981. However, it took some time before the field-work had reached its cruising speed, and the organizers decided to treat the first three months as a practice run, and to let the constitution of the panel cohort run between 1 April 1981 and 31 March 1982. This apparently innocuous change of plans may have decreased the mortality recorded during the survey period by close to 10%. If, however, the survey had started three months later, it might have reduced infant mortality by a significant amount, but at the cost of a serious increase of mortality in the second year of life. This

Table 3.1. *Mortality due to measles in the Bobo-Dioulasso survey in relation to no. of persons at risk, by month*

Month	Measles deaths	No. at risk*	Rate (per 10 000)
1981			
December	1	211	47
1982			
January	4	748	35
February	8	1288	62
March	21	1683	125
April	36	2196	164
May	13	2693	48
June	11	3216	34
July	5	3766	13
August	3	4334	7
September	1	4393	2
October		5450	
November		6077	
December		6366	
1983			
January		6294	
February		6229	
March		6182	
April	1	5944	2
May	3	5438	6
June	2	4921	4
July		4525	
August		3994	
September	1	3508	6
October		3004	
November		2458	
December		1905	
1984			
January		1380	
February	1	845	12
March	3	288	104

* These numbers consist of all children observed at mid-month who were aged above 9 months.

paradox is explained by the fact that the city was affected by a measles epidemic, an occurrence which seems to strike every other year at the beginning of the dry season. The epidemic reached its peak in April of 1982, as shown in Table 3.1, which gives death rates above 9 months of age by month.

As far as we know, measles was killing children of all ages, but particularly those who were too young at the time of the previous epidemic, and those who were born since. In the survey, however, because of the cohort configuration, the survey was killing mostly children between 9 and 12 months of age. The probability of getting measles is greatly reduced before 7 to 9 months of age, since children are still protected by their mothers' antibodies. Thus, only the children born in the early months of interviewing who were old enough at the beginning of 1982 were vulnerable to measles. This is illustrated by Figure 3.1, based on the survey results on the basis of a question asked at every round on disease having affected the child since the previous round. The highest reported incidence of measles coincides with the time when mortality from measles is also reported. (The results are plotted at the middle of the four-month period between rounds.) It appears clearly that the children born in April, May, and June 1981 were

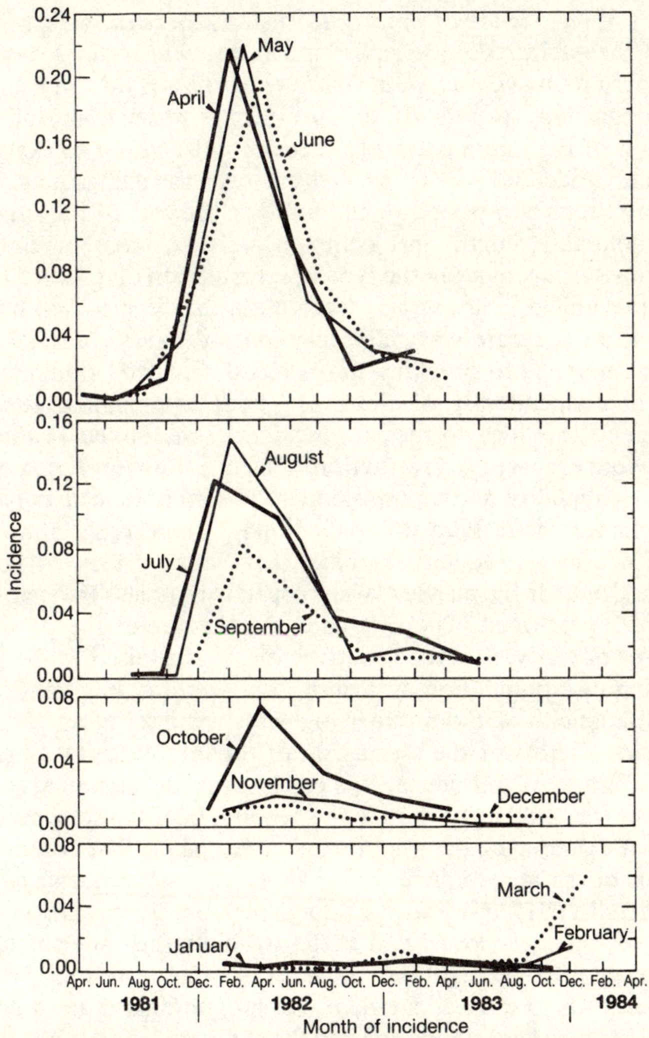

Fig. 3.1. Incidence of measles by month of birth

bearing the brunt of the 1982 epidemic. The younger a child was at the time of the epidemic, the less likely it was to be affected. Had the survey started three months earlier as planned, the population at risk would have been roughly doubled. In 1983, however, there was no epidemic, and the birth cohorts which had been spared the 1982 epidemic were struck with a vengeance in 1984, after the survey was mostly over. (There is some trace of the beginning of the new epidemic in Table 3.1 and Figure 3.1.) If the survey cohort had been constituted over the year 1982, not only would the level of the infant mortality rate have been much lower, but the structure of mortality during the two first years of life would have been quite different.

In other words, as long as there are strong seasonal fluctuations in death rates and the recurrence of epidemics such as measles, which are a basic feature of the African disease environment, the *multi-round survey should not be trusted to provide a reliable indication of the level and age pattern of mortality*. One way to remedy this problem is to lengthen the period of constitution of the initial cohort to three or more years, so that the effect of recurrent epidemics will be evened out over the age groups. However, lengthening the survey period would increase the cost of the survey.

The potential strength and comparative advantage of the IFORD surveys consists in providing the type of information that served to compile Figure 3.1: detailed, continuous information that is specified by date and duration in an accurate way. The phenomena under observation can be studied in reference to previous characteristics of the population at risk, established independently of the events that are studied. In a specific application of this idea to measles incidence, we looked at the probability of previous events (e.g. vaccination) affecting the contracting of measles, and at the probability of dying subsequently from other causes for children who had had measles (van de Walle 1986). There is an almost infinite variety of interesting sequential studies that can be made with this type of data that are outside the purview of retrospective studies. Two issues should, however, be mentioned. The first concerns the size of the panel required for this type of study; the second is the ethical issue raised by the prolonged interviewing of a population on health issues without intervention.

Most discussions of the number of births required in an IFORD-type survey were in terms of the estimation of the infant mortality rate within acceptable limits of confidence. It is on the basis of such an argument that the decision was reached to sample the hospital births in some cities such as Yaoundé and Bamako. Sampling experts estimated that somewhat more than 10 000 births were required to obtain sufficiently accurate estimates of infant mortality. If, as we are arguing, the surveys cannot give reliable estimates of mortality levels and patterns unless they are pursued over a much longer period, then the issue of sample size must be entirely reconsidered. On the other hand, if the longitudinal studies are used in analyses stressing causal inference and the study of differentials rather than the reliable measurement of mortality levels, then smaller samples might be quite acceptable. Two of the longitudinal studies described by Pebley (1985), the Narangwal and the El Progresso projects, involved birth cohorts of fewer than 3000 children.

The second concern is the ethical issue. The recurrent resurvey of a panel of children to scrutinize their health and their mortality can only be justified by a desire to improve their chances of surviving in the long run. In the mind of a population that sees no direct benefit from the insistent peeking into its life and suffering, questions are inevitably raised. Pebley (1984: 282) has noted that accurate data on diet and feeding practices, morbidity and treatment, immunization, and use of medical services cannot be collected through censuses and retrospective surveys:

> Since mortality in developing countries is usually the result of repeated insults to a child's system, such data should ideally be collected repeatedly over the course of a young child's life. Because of the expense of continuous data collection on a sufficiently large sample and the ethical problems associated with a surveillance program not accompanied by treatment of health problems detected, attempts to collect these data in nonintervention settings have been rare.

The marked advantages of longitudinal surveys and the drawbacks of the IFORD-type survey would together argue for a multiplication of population laboratories of the type illustrated in Bangladesh by the Matlab project and, in Africa, by the Keneba (Gambia), Sine-Saloum (Senegal), and Machakos (Kenya) projects. These operations have been continued over a number of years and have generally combined data collection with medical intervention. It remains to be seen whether such laboratories can function satisfactorily in the context of an African city, and whether vital registration systems cannot instead be upgraded to the point where they would produce the data that are required on a continuous and long-term basis.

3. The Specific Problems of the IFORD Design

In this section, we shall discuss the issues of losses and sample attrition that have been raised by the IFORD design, rather than by longitudinal surveys of a birth cohort as such. To some extent some of the problems mentioned below may be blamed on features of the survey that could have been remedied by better field procedures. It may well be, however, that large attrition is inevitable in longitudinal surveys of this kind. The essence of the IFORD design consists in registering a cohort of births in city maternity clinics and hospitals, in tracking the women and their children down at their place of residence, and in following the survivors for two years in a multi-round design of visits at home by an interviewer. It was assumed that maternity clinics would concentrate an overwhelming majority of the births in the city, and thus come closer to recording all births in a representative segment of the population than would ever be possible in a vital-registration system of the type that is commonly found in African urban areas. Figure

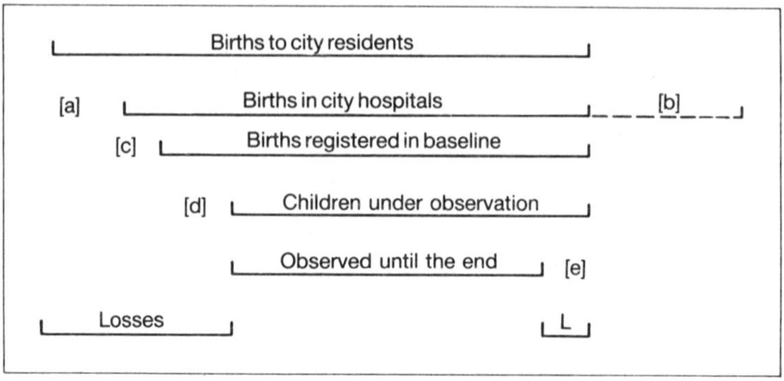

Fig. 3.2. Schematic representation of coverage in the IFORD surveys

3.2 presents the various types of problems of coverage that may be encountered in an IFORD survey. The top line represents the target population, the births to residents of the city. An unknown percentage of these births occur outside maternity wards: these are marked (*a*) in the diagram. But if the maternity wards of city hospitals do not cover all deliveries by residents, conversely they drain an unknown number of births (marked (*b*)) from a large hinterland outside the city, which do not belong to the universe covered by the survey. Unfortunately these cannot always be clearly identified from hospital or survey records as births to non-residents. It cannot be assumed that the survey will register all the births of residents born in hospitals; a subset (*c*) may be missed. Altogether a proportion of the panel recorded in the maternity clinics will not be identified in the field for reinterview ((*d*) in Figure 3.2). Finally, some children will be lost before the survey is over, and some will migrate; these are indicated as (*e*).

(*a*) Births of city children outside hospitals and maternity clinics must represent a proportion of all resident births that varies according to the particular city examined, but which is never totally negligible. This was recognized by the designers of the survey from the beginning, and they were prepared to accept the loss in view of the presumed small proportion of such births (Houehougbe 1982). The biases resulting from the use of an unrepresentative sample may, however, affect more than the level of mortality. It is possible that the structure of mortality by age and by socio-economic characteristics is quite different among the children who were born at home. For example, in Bobo-Dioulasso, where antitetanus injections were administered to all the children born under medical supervision, the risks of dying during the first weeks of life must have been significantly reduced among survey children. On the other hand, the loss of representativeness must be balanced against the benefits of having a well-identified population born in a controlled environment: trained medical personnel are able to ascertain some characteristics that would otherwise be likely to be unknown to the mother, such as the vitality of the child and its birth-weight. It can be argued that an accurate measurement of the frequency of stillbirths and deaths during delivery is out of reach of a retrospective survey and can only be achieved when records are kept by trained attendants, as in the IFORD surveys. Indeed, most mothers might not know and might not care whether or not the child had lived for a few seconds.

(*b*) In most city hospitals in Africa, there are births from the countryside which do not belong in a survey of residents. Their mothers should not be interviewed, but it would be helpful if the survey kept a list of all births born in the maternity hospital, with indication of the place of residence of the mother, for the purpose of checking the coverage of the survey. This is particularly needed because of the likelihood of the situation described in (*c*) below.

(*c*) In Bobo-Dioulasso, a comparison of the administrative records of the maternity clinics with the survey registers revealed that a surprisingly large number of hospital births to residents had been missed by the interviewers. After scrutiny, it turned out that 'difficult births', i.e. non-viable children or stillbirths, children whose mother had died during or immediately after delivery, and deliveries involving surgical procedure (such as a Caesarian) were much more likely to be missed. Omission of the latter category occurred because the care of surgery patients was the responsibility of another department of the hospital; but, in addition, it is likely that women exposed to the trauma of a difficult birth are not ideal subjects for a social-science survey. The solution here is to make liberal use of the administrative records of the hospital, after making sure that they mention the place of residence of the mother.

(*d*) Most of the IFORD surveys seem to have been plagued by excessive losses of panel members. For example, in Brazzaville the losses were almost 35% of the sample (Duboz 1984: 159). It is likely that the largest number of these losses represent failures to reidentify some panel members at their homes who had been registered in the hospital at the time of delivery. In the Yaoundé survey, 12.3% of the registered births could not be found again one month later. Brouard, who gives this figure, comments that the fate of these children is clouded in mystery. If all were alive at the end of the month, the risk of dying was only 17.9 per 1000; if all had died, it was 140.8 per 1000 (Brouard 1986). Such losses seem inseparable from the methodology of the IFORD surveys. Finding a woman on the basis of an address given from a hospital bed, in an African city where street names are often not known and where house numbering often does not exist, and where the population is very mobile, can be as difficult as finding the proverbial needle in a haystack.

(*e*) Once they have been identified in the field during a first round, the risk of losing a mother and her child appears to decrease considerably. There are moves within the city, but neighbours can often indicate the new address. Serious problems arise, however, in distinguishing between a move inside the city, a temporary departure followed by a return within the survey period, and a definitive migration. The IFORD surveys showed little interest in the accurate recording of migration. When a mother had migrated, no attempt was made to ascertain whether the child was still alive at the time of departure, so that observation had to cease at the time of the round prior to that when the exit was reported. (In principle, it would have been better to include either the entire time prior to departure or half of the period between rounds.) It should be noted, moreover, that migration of mothers and children belonging to the survey population represented a type of sample attrition that was not compensated for by new entries. Hence, there was a tendency for the survey population to become less and less representative of the population of the city as time elapsed and the most mobile part of the population left the sample and was not replaced.

The losses discussed under (*a*) to (*e*) are unfortunately not likely to be distributed randomly in the population. They are likely to be associated in some way with mortality, the phenomenon that the survey attempts to measure and to analyse. This is obvious in the case of (*b*), the difficult births. But there are biases lurking in all categories of deficient coverage, and the net result seems to be an underestimation of mortality (Ouaidou and van de Walle 1986). Urban births occurring outside hospitals will not be immunized against tetanus. Dead children are less likely to be subject to follow-up than surviving ones. Finally, migration may be associated with mortality in various ways: a death may lead to departure, or the unstable residents may belong to the most destitute segments of the population; an unmarried teenage girl, whose child has a high risk of dying, is likely to leave to join a new household. The initial results of the analysis of the mortality in Yaoundé by socio-economic characteristics indeed shows unexpected differentials, such as higher mortality among the children of the best-educated parents and in the central neighbourhood inhabited by professionals and civil servants (Dakam–Ngatchou 1986). It is tempting to attribute these unexpected results to defects in coverage.

With hindsight derived from the accumulated experience, a tentative conclusion can be suggested: matching births in maternity clinics with children and mothers at home is a daunting task in a large capital city, and significant bias is likely to accompany the almost unavoidable losses. Mobility will always be the bane of urban multi-round designs. A special effort must go into the measurement of the various biases attending this kind of survey before any more are attempted.

4. Suggestions for a Revision of the Longitudinal Survey

It is clear that we need more studies of infant and child mortality in Africa, and that retrospective techniques have serious limitations when it comes to studying mortality differentials and mortality determinants. The IFORD surveys have filled a void, and it is hoped that their analysis will be pursued vigorously after more awareness has developed of what they can and cannot achieve.

The following list of suggestions is meant as a basis for discussion of design changes that might be considered for the next generation of urban infant and child mortality surveys. The list incorporates some lessons drawn from the IFORD experience, but it is limited, as is the rest of this chapter, to the basic structure of the survey and makes no reference to other features of the survey methodology (e.g. questionnaire content, data base management, and analysis) that might also deserve revision.

(*a*) *Identification of the panel of births.* Perhaps panel studies should start with a representative sample of households or women identified at their home before the delivery of a child. Such a methodology was attempted in a rural multi-round survey in Senegal sponsored by the Sahel Institute, and it seems that the problems of coverage were considerably reduced. The success was perhaps due in part to the fact that rural populations are less mobile than urban ones.

(*b*) *Longer baseline period.* The observation of the initial cohort should be prolonged beyond a period of one year, at least for three years and preferably for more. Thus the impact of epidemics and exceptional-period mortality would be distributed more realistically, over more age groups.

(*c*) *Longer follow-up period.* There is no good a priori reason to interrupt the period of observation when a child reaches his second birthday. In populations with late weaning and high child mortality, the third year of life is still a period of high risk. The inclusion of an entire birth interval would enhance the interest of these surveys for the study of both mortality and fertility.

(*d*) *Medical intervention* should accompany demographic and epidemiological observation.

(*e*) *Observation of all mothers should continue* until the end of the follow-up period. The IFORD surveys were conceptualized as follow-up studies of survivors. There were no provisions made to interview mothers past the time when their child had been ascertained to be dead. This sometimes left out precious information. When the child had died in the maternity clinic soon after birth, for example, the information normally collected in the subsequent rounds (including some socio-economic information) was not available. The reporting of stillbirths was often badly neglected. An opportunity to study fertility and the determinants of the birth interval (including mortality) was partly wasted.

(*f*) *Special care for the recording of migration* should provide for observation of the mother and child until the time of departure. At the very least, the interviewers should ascertain whether the child was alive at the time of departure. There should be a mechanism of recruitment of immigrants to compensate for panel attrition by emigration.

(*g*) *Measurement of coverage* and of the biases resulting from panel losses should be built into the survey design.

5. Conclusion

An issue raised by the methodology of the IFORD survey is whether the large sample attrition that has been the rule rather than the exception so far was the result of faulty field procedures or whether the losses were a logical result of the approach. In his discussion of the respective advantages of multi-round and retrospective surveys, Tabutin warns that 'the application (in a broad sense, involving resources, controls, etc.) of a system is at least as important as the system itself' (Tabutin 1984: 138). It is possible, and even probable until the results of the available IFORD surveys clearly invalidate this conclusion, that the problems of matching children identified at birth within the city population are so intractable as to make very large losses almost inevitable in a typical African capital city. Certainly, the possibility of improving field procedure, for example by resorting to micro-computers and direct data entry during field-work and better integration of the administrative statistics of hospitals, could significantly improve the quality of the data. But it must be realized that the possibility of very large losses is looming large. In evaluating whether more IFORD-type surveys should be attempted, this issue will have to be carefully weighed. In addition, there should be more careful consideration of the nature of the expected results: the data, because they are collected longitudinally will be more enlightening for the study of differentials than as mere indications of the level and structure of mortality. Finally, there are serious issues of an ethical nature raised by a longitudinal survey unaccompanied by intervention, which ought to be brought out into the open.

References

Adlakha, A., and Nizamuddin, M. (1984), 'Mortality Data Collections: A Review of Integrated Multi-purpose Household Surveys and Multi-round Demographic Surveys', in UN, *Data Bases for Mortality Measurement*, Population Studies, no. 84 (New York: UN).

Brass *et al.* (1968), *The Demography of Tropical Africa* (Princeton, NJ: Princeton University Press).

Brouard, N. (1986), 'Une modélisation de l'enquête sur la mortalité infantile et juvénile à Yaoundé', in *Estimation de la mortalité du jeune enfant de (0–5 ans) pour guider les actions de santé dans les pays en développement* (Paris: Les Éditions de l'INSERM).

Courbage, Y *et al.* (1979), 'Enquête sur la mortalité infantile et juvenile à Yaoundé: Présentation méthodologique', in *La Mortalité des enfants dans le tiers-monde*, Chaire Quételet, Département de démographie, Université Catholique de Louvain (Liège: Ordina Éditions).

Dakam-Ngatchou, R. (1986), 'Niveau, tendance et déterminants de la mortalité à Yaoundé', in *Estimation de la mortalité du jeune enfant de (0–5 ans) pour guider les actions de santé dans les pays en développement* (Paris: Les Éditions de l'INSERM).

Duboz, P. (1984), 'Mortalité et morbidité infantile et juvénile en République Populaire du Congo', in *La mortalité des enfants dans les pays en développement*, Cahiers ORSTOM série Sciences Humaines, no. 2 (Paris: ORSTOM).

Houéhougbé, A. (1982), 'Examens de quelques problèmes liés à l'échantillonnage, aux déperditions et aux méthodes d'estimation dans les EMIJ', Paper presented at the *Yaoundé Colloquium on Infant and Child Mortality Surveys*.

Ouaidou, N., and van de Walle, E. (1987), 'Réflexions méthodologiques sur une enquête à passages répétés: L'EMIS de Bobo-Dioulasso', *Population*, 42(2), 249–66.

Pebley, A. (1984), 'Intervention Projects and the Study of Socioeconomic Determinants of Mortality', in W.H. Mosley and L.C. Chen (eds.), *Child Survival: Strategies for Research*, supplement to *Population and Development Review*, 10 (New York: The Population Council).

Tabutin, D. (1984), *La collecte des données en démographie: Méthodes, organisation et exploitation* (Liège: Ordina Éditions, and Département de démographie, Université Catholique de Louvain).

Van de Walle, E. (1986), 'Anatomie d'une épidémie de rougeole vue par la lorgnette d'une enquête à passages répétés', in *Estimation de la mortalité du jeune enfant de (0–5 ans) pour guider les actions de santé dans les pays en développement* (Paris: Les Éditions de l'INSERM).

4 The Collection of Mortality Data in WFS and DHS Surveys[1]

JEREMIAH M. SULLIVAN

Westinghouse Institute for Resource Development, Columbia, Maryland, USA

Prior to 1970, reliable information on infant and childhood mortality was available for only a few countries in the developed world. Since 1970, the availability of data on this area has expanded rapidly and will continue to do so until 1990. This change is due primarily to two large-scale survey programmes: the World Fertility Survey (WFS) and the Demographic and Health Survey (DHS). This chapter describes the mortality data collected in both.

The chapter contains five sections. In Section 1, the objectives of WFS and DHS are reviewed. In Section 2, the mortality data collected by WFS and DHS are compared. In Sections 3 and 4, the analytic potential of those data is discussed. Section 5 is a summary.

1. WFS and DHS Objectives

The WFS programme was implemented by the International Statistical Institute with funding primarily from the UN Fund for Population Activities (UNFPA) and the US Agency for International Development (AID). Between 1972 and 1984 WFS provided technical and financial support to conduct 42 fertility surveys at the national level in developing countries of Asia, Africa, and Latin America. The WFS programme had the objective of providing information on fertility and the factors affecting fertility (WFS 1975). This objective was translated into survey instruments designed to obtain data on fertility, contraceptive use, and mortality.

The DHS programme (1984–9) is being implemented by the Institute for Resource Development (of Westinghouse Electric Corporation) and the Population Council, with funding from AID. DHS provides technical and financial assistance for 35 surveys, primarily at the national level, throughout the developing world. Thirty-one surveys focus on data pertaining to fertility, contraceptive use, mortality, and health practices, while four are in-depth investigations of specific fertility, mortality, or health topics. For

[1] For their comments on an earlier draft of this paper, I wish to thank R. J. Lapham, C. F. Westoff, A. R. Cross, L. H. Ochoa, and R. Cornelius.

the former surveys, the main survey instrument is an individual woman's questionnaire, to be applied in its entirety in each survey. Only minor modifications to make the questions culturally meaningful or culturally acceptable are anticipated. Of course, additional questions of interest to particular countries may be added to the DHS questionnaire in any country.

In terms of understanding and analysing mortality and morbidity conditions, the DHS programme differs from WFS primarily by collecting data on maternal and child health practices and on health indicators (e.g. anthropometric measures for children under the age of 3). These data are useful in their own right for the design and evaluation of health programmes. In addition, they complement the other data on infant and childhood mortality and their socio-economic and demographic determinants collected by DHS.

Since the DHS programme began in September 1984, survey agreements have been finalized in 15 countries (Table 4.1). Discussions are under way in ten additional countries.

Table 4.1. *DHS surveys by country, executing agency, and target date for field-work**

Country	Executing agency	Field-work
El Salvador	Salvadorian Demographic Association	May–July 1985
Liberia	Ministry of Planning and Economic Affairs	March–July 1986
Senegal	Ministry of Economics and Finance	March–July 1986
Brazil	Brazilian Society for Family Welfare	June–August 1986
Peru	National Institute of Statistics	August–November 1986
Ondo State, Nigeria	Ministry of Health, Ondo State	September–November 1986
Sri Lanka	Department of Census and Statistics	November–December 1986
Dominican Republic	National Institute for Population and the Family	September–December 1986
Colombia	Colombian Regional Population Centre	October–December 1986
Mexico	Federal Ministry of Health	January–March 1987
Mali	Sahel Institute, Bamako	February–May 1987
Nepal	New ERA, Kathmandu	January–March 1987
Trinidad and Tobago	Family Planning Association of Trinidad and Tobago	April–May 1987
Burundi	Department of Population, Ministry of Interior	April–June 1987
Morocco	Ministry of Health	April–June 1987

* Status as of Sept. 1986.

2. WFS and DHS Survey Instruments

In the WFS programme, data pertinent to mortality analysis were collected at the community and household levels and from individual female respondents. DHS also collected data at these three levels.

2.1. Community-Level Information

WFS and DHS community-level data collection differ in terms of the types of data collected and the methods of data collection. The WFS community-level questionnaire typically included questions on transportation, communications, government institutions, health and family planning facilities, schools, electrification, and agricultural and non-agricultural activity (Casterline 1984). In contrast, the DHS community schedule primarily collects information on health and family planning services. The schedule includes questions about services within the community (family planning field-workers, traditional birth attendants, etc.), the proximity to health and family planning facilities (hospitals, clinics, etc.), and detailed information on facilities within 30 kilometres of a community (staff, operating schedule, date established, etc.). Table 4.2 shows the types of data collected in the WFS and DHS community questionnaire. The central point to note is that the DHS schedule is considerably more focused on health and family planning facilities, and on the conditions under which they offer services.

In WFS, community-level data were collected in 17 out of 42 surveys (in 15 of these 17 for rural areas only). The data were collected from village leaders, usually by the supervisory staff of the individual woman's survey. In DHS, except for a few of the early surveys, community-level data are considered an integral part of each survey and are obtained for all sample points in a survey. Special teams of enumerators visit both the community and nearby health and family planning facilities to obtain detailed information on those facilities.

Table 4.2. *Information collected in the WFS and DHS community-level questionnaires*

WFS community module	DHS community schedule
Transportation and community	General information
Health	Type of access road
	Availability of public transport
Family planning and prevalence education	Health and family planning service availability
Availability of electricity	Coverage by community contraceptive distribution system, FP mobile clinic, FP field-worker, traditional birth attendant, trained midwife
Agricultural activity	
Non-agricultural activity	
Access to government agencies	Proximity to hospital, clinic, health centre, pharmacy, private doctor
Ethnic/religious/language groups	
	Information on health facilities within 30 kilometres (travel time, staff, hours of operation, services available, cost, date service delivery began)

2.2. Household-Level Information

The WFS household schedule provides for collecting data to select respondents for the woman's survey, to calculate the denominator of fertility rates, to describe the characteristics of dwelling units, and to estimate vital rates. The model DHS schedule is designed only to collect data to select respondents for the woman's survey. The types of data collected in the WFS and DHS household schedules are shown in Table 4.3. While there are over 25 items listed for the WFS schedule, the DHS schedule only lists household members and visitors and records information on whether a listed person is a usual resident, whether he or she slept there last night, age, and sex. (Also shown in Table 4.3, for ease of comparison with the WFS data, are the data on characteristics of dwelling units collected in the DHS woman's questionnaire.)

There are several reasons for the relative austerity of the DHS household schedule. First, about half of the WFS surveys involved ever-married woman (EMW) samples, while most DHS surveys involve all-woman

Table 4.3. *Information collected in the WFS and DHS household questionnaires*

Type of data	WFS	DHS
Data for respondent selection		
Name	Yes	Yes
Relationship to head	Yes	No
Usual resident	Yes	Yes
Slept in house last night	Yes	Yes
Sex	Yes	Yes
Age	Yes	Yes
Marital status	Yes	No
Education	Yes	No
Data on characteristics of dwelling unit		
Possession of modern objects*	Yes	No
Source of drinking water*	No	No
Source of water for general use*	No	No
Toilet facilities*	Yes	No
Lighting	Yes	No
Premises electrified*	No	No
Building materials	Yes	No
Main material of floor*	No	No
Number of rooms	Yes	No
Soap on premises*	No	No
Data for vital rate estimation		
Children ever born data	Yes	No
Last live birth data	Yes	No
Household deaths, last 24 months	Yes	No
Orphanhood data	Yes	No
Spouse survival data	Yes	No

* A question concerning this information is included in the DHS woman's questionnaire.

samples. Thus, in WFS it was necessary to collect information on marital status and education (for the denominators of education specific rates), but this is not necessary in the DHS schedules. (Of course, for DHS surveys involving EMW samples, it will be necessary to include questions on marital status and education in the household schedule.) Secondly, it was decided that in DHS, data on the physical characteristics of dwelling units would be collected in the woman's questionnaire. This approach ensures that a minimum set of data is collected in each survey and that those data are processed and available for analysis with the individual woman's data. These are compelling reasons considering that, in WFS, only about half of the surveys collected any information on the characteristics of dwelling units (Singh 1984) and that delays of several years were involved in providing data tapes which merged household- and individual-level data. Thirdly, it was judged that the level of effort and the cost of obtaining vital-events data in the WFS household schedule was not commensurate with the results obtained and that vital-events data should not be collected at the household level in DHS, at least not by the methodologies employed in WFS.

The absence of mortality data collection in the DHS household schedule is a major departure from WFS and requires further explanation. In WFS, provision was made in the household schedule for collecting (*a*) childhood mortality data (survivorship of children ever born and of the last live birth) and (*b*) adult mortality data with the WFS general mortality module (household deaths in the last 24 months, orphanhood data, and spouse survival data). The rationale which led to this is interesting. In the early years of WFS, there was considerable concern about the accuracy with which fertility data would be collected in the woman's questionnaire, a questionnaire designed with more concern for collecting data on the proximate determinants of fertility than data for estimating fertility levels (Caldwell 1985). Also, it was thought that the size of the WFS samples (2000 to 10 000 women), while adequate for investigating the factors affecting fertility, would be too small for estimating fertility levels in the year preceding the survey at an acceptable level of sampling error. The idea of an expanded household survey which collected data on vital events provided a solution to these problems. In all, 13 WFS surveys employed expanded household surveys. The samples ranged in size from 8200 to 35 000 households, with a median size of 15 000 (Scott and Harpham 1984). In some countries, the cost of the expanded household surveys exceeded the cost of the woman's survey (Timaeus 1984).

As it turned out, the WFS woman's surveys provided fertility and childhood mortality data with sufficient accuracy to warrant estimating vital rates at the national and, usually, subnational levels. The problem of sample size was circumvented by pooling data for periods of three to five consecutive years. This result effectively eclipsed the usefulness of the fertility and child survival data of the household schedules. The usefulness of the data from the general mortality module is also questionable but, in this case, because of reliability problems. A recent evaluation of results from the general mortality module clearly indicates that credible adult mortality estimates were not produced with any degree of consistency (Timaeus 1984).

2.3. Individual Respondent Information

It is necessary at this point to describe the newly developed DHS woman's questionnaires. The DHS questionnaires attempt to strike a balance between flexibility and standardization while satisfying the different constituencies interested in DHS data. This has resulted in creating four core questionnaires: 'A core short' and 'A core long' (for countries with high levels of contraceptive use, the long version is expanded to obtain information for studying fertility differentials) and 'B core short' and 'B core long' (for countries with low levels of contraceptive use, the latter again obtaining information on non-proximate determinants of fertility). The four cores contain a large number of identical questions and each core is divided into eight sections: respondent background, reproduction, contraception, health, marriage, fertility preferences, husband's background, and height and weight measurement (children 3–36 months of age).

For purposes of mortality analysis, the most significant differences among the four DHS questionnaires occur between the long and the short cores. They differ in three ways. First, the long cores contain a full birth history, while the short cores contain a five-year truncated birth history. Secondly, the long cores collect breast-feeding and maternal-care information for all births in the last five years, while the short cores obtain these data only for the last live birth. Finally, the long cores obtain information for studying fertility differentials which may be useful for studying mortality differentials, i.e. co-residency with parents, whether or not the respondent has worked for wages, and what she did/does with her earnings.

The data useful for mortality analysis in the WFS and DHS woman's questionnaires are shown in Table 4.4. Both WFS and DHS collect a basic set of background information on the respondent (age, education, religion, ethnicity, marital status, etc.) and her husband (education and occupation). As might be expected, the differences are minimal; the principal difference stems from the fact that WFS obtained a full marital history while DHS only asked if the respondent was married more than once.

The data collected on the biological characteristics of each live birth (date of birth, sex, birth order, etc.) are also similar in WFS and DHS. The primary differences are that the WFS data permit the calculation of pregnancy intervals (although this writer is unaware of any analysis using such a variable) and the determination of mother's marital status at the time of birth.

A major difference between the WFS and DHS questionnaires is in the collection of data on health practices. Both questionnaires collect data on the duration of breast-feeding. However, the DHS questionnaires always include questions about various aspects of child and maternal health care while such questions were only infrequently included in WFS.

(*a*) In all four DHS cores, for each child under the age of 5, mothers are asked questions on *immunization* and *treatment given for diarrhoea*. The series of questions on immunization begins by asking the mother if she has a health card for the child. If a health card is available, the dates of each immunization are recorded on the questionnaire. If no card is available, the mother is asked if the child has ever been vaccinated. The questions on diarrhoea first determine if the child had diarrhoea in the last two weeks and in the last 24 hours. If yes, questions are asked to determine if the child was taken to a health professional or a health facility for treatment, if oral rehydration therapy was employed, and if any other treatment was given.

(*b*) Also, in all four DHS cores, information is obtained on maternal care: type of health practitioner seen for pre-natal care, type of attendant at delivery, and whether or not a tetanus injection was received during the pregnancy. (In the long cores this information is obtained for all births in the last five years, while in the short cores, for only the last live birth.)

(*c*) Finally, in the B cores (short and long), mothers are asked for each child under age 5 if the child had a fever or cough in the last four weeks. If yes, questions are asked to determine if the child was taken to a health professional or a health facility for treatment and what treatment was given.

The WFS and DHS woman's questionnaires also differ in that anthropometric measurement of children is only undertaken in the DHS questionnaires. In most of the DHS woman's survey, height and weight measurements are taken of the natural children under 3 years of age of female respondents. The objective is to obtain data for about 1500 children. In large DHS surveys, this may be accomplished by a subsample of female respondents. It should be understood that the universe for the anthropometric data is children living with their mothers, rather than all children in the population. However, in populations where fostering is an important practice, information on its prevalence can be obtained in both the household schedule and the birth history so that it will be possible to estimate the proportion of all children in the households which were included in the DHS anthropometric sample.

3. Mortality Levels and Time Trends

3.1. Bias in Indirect Childhood Mortality Estimates

Several recent studies have demonstrated that indirect childhood mortality estimates can be seriously biased (Garenne 1982; Pison 1984). As a result there is now a consensus among many demographers that, for the study of infant and child mortality, data suitable for direct estimation should be

Table 4.4. *Information for mortality analysis in the WFS and DHS woman's questionnaires*

Type of data	WFS	DHS
Data for background variables		
Age	Yes	Yes
Education	Yes	Yes
Religion/ethnicity/language	Yes	Yes
Marital status	Yes	Yes
Number of times married	Yes	No
Whether married more than once	Yes	Yes
Age at first marriage	Yes	Yes
Place of residence	Yes	Yes
Childhood place of residence	Yes	Yes
Children ever born questions	Yes	Yes
Husband's education	Yes	Yes
Husband's occupation[a]	Yes	Yes
Biological data on births[b]		
Date of birth	Yes	Yes
Sex	Yes	Yes
Birth order	Yes	Yes
Survival status	Yes	Yes
Age at death	Yes	Yes
Birth interval	Yes	Yes
Mother's age at birth	Yes	Yes
Pregnancy interval	Yes	No
Mother's marital status at birth	Yes	No
Data on health practices, births last 5 years		
Duration of breast-feeding[c]	Yes	Yes
Dates of immunizations	No	Yes
Treatment of diarrhoea	No	Yes
Tetanus injection for pregnancy[d]	No	Yes
Pre-natal care[d]	No	Yes
Attendant at delivery[d]	No	Yes
Treatment of fever[e]	No	Yes
Treatment of cough[e]	No	Yes
Anthropometric data, children under age 3		
Height	No	Yes
Weight	No	Yes

[a] In DHS, long cores only.
[b] In DHS, short cores restricted to births in last 5 years.
[c] In WFS, restricted to last two births; in DHS, long cores only.
[d] In DHS, short cores restricted to last birth.
[e] In DHS, B cores only.

collected (Preston 1985). The plan for the DHS programme is to use the birth history data for direct mortality estimation.

One study which documents the presence of bias in indirect childhood mortality estimates is based on the birth history data of the Indonesian Fertility Survey (Sullivan and Wilson 1982*a*, *b*). In that study, the birth history data were analysed so as to provide the statistics necessary for indirect estimation and for investigation time trends in mortality, age patterns of mortality, and differentials in mortality by mother's age at birth. Even when derived from the same data set, the indirect IMR estimates based on data reported by women aged 20–4 exceed the direct estimates for the four-year period preceding the survey by about 20%. It was clearly established that the difference was due to a positive bias in the indirect estimate arising from the relatively high mortality of births which occurred when the mother was under age 20. The indirect approach, when based on data reported by women aged 20–4, simply weights the births to young women more greatly than is warranted by their occurrence in a cross-section of time and, when their experience reflects unrepresentatively high mortality, this tends to positively bias the indirect estimates.

Not only is there potential for bias in indirect mortality estimates but there is evidence that the degree of bias can differ between population subgroups. The evidence comes from the East Java Population Survey in which rural IMR estimates exceed urban estimates by about twice as much when estimated by direct procedures (55–67%) than when estimated by indirect procedures (22–36%) (Table 4.5). The same result holds when analysing data from the Indonesian Fertility Survey; rural estimates exceed urban estimates by about twice as much when based on direct estimates (65%) as when based on indirect estimates (21–36%). This comes about because, for Indonesia, the indirect procedures overestimate mortality and they overestimate urban mortality to a greater extent than rural mortality. The differential overestimation is probably related to the fact that indirect

Table 4.5. *Urban/rural mortality differentials, East Java and Java–Bali, measured by direct and indirect procedures*

Survey and estimation technique	IMR estimates (per 1000)		Difference (rural over urban) (%)
	Rural	Urban	
East Java Population Survey			
Direct estimates			
Round 1	90	58	55
Round 2	74	44	67
Indirect estimates			
Women 20–4	101	83	22
Women 25–9	100	73	36
Women 30–4	98	78	26
Indonesian Fertility Survey			
Direct estimate	101	62	65
Indirect estimates			
Women 20–4	120	96	26
Women 25–9	102	75	36
Women 30–4	115	95	21

Sources: Sullivan and Wilson 1982*b*,*c*.

mortality estimates based on data reported by women below age 30 reflect, to a significant degree, the high mortality experience of births occurring to women under age 20. This mortality is influenced more by biological factors than by socio-economic factors. To the extent that urban/rural mortality differences are due to socio-economic factors, those differences are reflected in the mortality experience of births occurring to women over age 20. Since the indirect models employ data which disproportionately incorporate mortality of children born to younger mothers, the urban/rural mortality differentials estimated by indirect procedures understate the true differential.

3.2. Direct Mortality Estimation in WFS

Direct mortality estimates are as reliable as the data on which they are based. The quality of the mortality data in the WFS reproductive histories has been evaluated in several studies. Rutstein tested the internal consistency of the data by investigating mortality patterns by age, sex, birth order, etc. (Rutstein 1984*a*, *b*; 1985). It is well known that such tests are not sufficiently powerful to detect many kinds of data problem. Nevertheless, Rutstein's work has not revealed any major problems with the WFS data and has tended to establish their overall quality.

In addition to Rutstein's work, there have been studies which evaluate the WFS mortality estimates by comparison with other estimates. A number of these studies have recently been reviewed by Preston (1985). In some, the WFS mortality data for a particular country have been judged to be of high quality (Sri Lanka, Pakistan, and Costa Rica), while in others the data have been judged to be problematic (Bangladesh, Senegal, and Jordan). Preston concludes that, while the completeness of the reporting of neonatal deaths is questionable in some surveys, there is little evidence of serious problems with the WFS data, at least not for the 15-year period preceding a survey. Most analysts agree with this view (Chidambaram *et al.* 1984; Hobcraft 1984; Gille 1985).

In any case, the success of WFS in documenting infant and childhood mortality levels in developing countries does not mean that the reproductive-history approach is error-free. To place the potential for error in a quantitative framework, results from the WFS survey in Portugal will be cited. This is a particularly good survey to review because it has the best external standard for comparison of any WFS survey. For Portugal, the WFS infant mortality rate was 33 per 1000 (for the four years preceding the survey) as compared to the registered rate of 38 per 1000 (Chidambaram *et al.* 1984). This differential implies some omission of infant deaths. Considering the various pieces of evidence concerning the quality of WFS mortality data, it is probably safest to conclude that any effort to collect infant and childhood mortality data with a retrospective questionnaire will result in some data problems. The challenge for DHS is to minimize those problems and to maintain and improve upon the legacy of WFS.

3.3. WFS and DHS Reproductive Histories

In the WFS and DHS questionnaires, the reproductive-history section begins with a series of questions on children ever born (asked in terms of the standard Brass questions on sons and daughters living at home, living elsewhere, and who have died) and goes on to an event-by-event history of the respondent's reproductive experience. There are two major differences between WFS and DHS with respect to the event-by-event reproductive histories:

(*a*) WFS collected information for *pregnancies* (either in the form of a live-birth table plus any-other-pregnancies table or in the form of an integrated pregnancy history table) while DHS collects only information for live births and the current pregnancy.

(*b*) WFS collected a full pregnancy history while DHS collects either a full birth history or a truncated birth history (births in the last five years plus the birth prior to that period and the respondent's first birth).

The DHS rationale for collecting only a live-birth history is based on the WFS experience of collecting data on other pregnancies. In WFS, these data were collected to improve the completeness of infant death reporting. The idea was to ask respondents reporting a stillbirth after seven or more months of pregnancy if there were signs of life (crying or moving) at the time of pregnancy termination. Scott and Singh (1981) analysed the WFS data for 13 surveys and showed that the other pregnancy data increased the reporting of infant deaths hardly at all: in all cases by less than 1%.

The DHS rationale for including a truncated birth history in the short cores is based on economy. With WFS birth history data from Colombia and Kenya, it has been shown that a truncated birth history requires collecting about half as much data as does a full birth history and saves, on average, from three to five minutes of interviewing time per respondent. The truncated history may be preferable in a small number of countries where past demographic rates are well documented and where interest is in recent experience. However, the substantially greater analytic value of a full birth history data set will, for most countries, outweigh the economy realized by using a truncated history. This point is sufficiently important to cite examples of mortality analysis which can only be undertaken with a data set from a full birth history.

(*a*) *Analysis to document changing mortality conditions*. Data from a five-year truncated birth history are sufficient to calculate infant and child mortality rates but not for the study of mortality time trends or trends in the age pattern of mortality.

(*b*) *Analysis of the impact of short birth intervals on child survival*. One of the most important findings from WFS concerns the deleterious effect of short birth intervals (either the preceding or the subsequent birth interval) on child survivorship. The analysis on which these results are based requires data for a more extensive time period than is available from a truncated birth history.

(*c*) *Analysis to detect modelling bias in indirect childhood mortality estimates*. Since indirect estimation is an important source of mortality information in developing countries, frequently the only source at the provincial or state level, the lack of the data necessary to conduct this analysis can be a significant loss.

3.4. DHS Methodological Investigation

The completeness with which retrospective surveys collect infant and child mortality data is a controversial issue. DHS is addressing this issue and other methodological issues in surveys in Peru and the Dominican Republic in late 1986. The Peru survey uses the standard DHS questionnaire with 5000 respondents (full birth history) and an experimental questionnaire with another 3000 respondents (truncated birth history followed by a series of carefully crafted probes designed to detect unreported births followed by infant deaths). In the Dominican Republic, there are 8000 respondents to the standard DHS questionnaire and 4000 respondents to the experimental questionnaire. The usefulness of the probing questions in the experimental questionnaires can be assessed by the number of additional infant deaths that they detect.

Another procedure which DHS could employ to investigate the completeness of reporting of early infant deaths involves follow-up interviews with respondents who report themselves pregnant in a standard DHS survey. A follow-up interview conducted six months after the survey would provide data sufficient for calculating neonatal mortality rates, which can be compared to the neonatal estimates from the standard survey. In this approach, the issue of underreporting of early infant deaths might be investigated further by comparing distributions of deaths by days within the neonatal period.

4. Mortality Differentials and Determinants

WFS data have been used extensively for investigating mortality differentials and determinants. The availability of similar data sets for over 40 countries has been a great advantage in this work. Where consistent results are found across data sets, the credibility of those results is strengthened. On the other hand, a disadvantage of the WFS data is that they were not collected with the idea of testing specific hypotheses. Thus, much of the analysis of WFS data has documented associations between various factors and mortality, but has not demonstrated causal links or the mechanisms through which those factors, in particular socio-economic factors, affect mortality. The resulting atheoretical approach to the analysis of mortality differentials has been expressed elsewhere (Preston 1985).

The DHS data have the same strengths and weaknesses as the WFS data. Standardized core instruments generate data for comparative analysis but do not allow in-depth analysis of specific factors. However, taking advantage of the WFS experience, there are a number of ways in which DHS should be able to extend the analysis of determinants beyond what was possible with WFS data.

4.1. Availability of Health Data

DHS surveys collect health data at the community and individual levels. These data are useful for programme design purposes and may also advance the study of differentials and determinants. For example, differentials in the use of health services (maternal care, child inoculations, and treatment for childhood illness) by education might help clarify the causal links and mechanisms through which education affects mortality. Analysis of the anthropometric data (height and weight, for children under age 3) by socio-economic groups and by demographic characteristics might lead to a better understanding of the links between those factors and mortality differentials.

4.2. The Dependent Variable

Social scientists typically use the survival status of children as a dichotomous variable for the study of mortality differentials at the individual level. This will, no doubt, continue to be an important approach with DHS data. However, DHS anthropometric data make it possible to explore a new procedure proposed by Mosley and Chen (1984). They suggest using as the dependent variable a health status index which would allow classifying members of a birth cohort into a continuum of risk categories: living children by reference to their nutritional status (weight for age) and deceased children by assigning an extreme score on the health status index. This approach incorporates data on health status and mortality into a single model which may provide new insights on the relationship between various explanatory variables and child survival.

4.3. The Independent Variables

The DHS survey is collecting information on many variables that have proved useful for investigating mortality differentials and determinants. Of the five categories of proximate mortality determinants identified by Mosley and Chen (1984), variables representing four are collected by DHS:

maternal factors (age, parity, birth interval, etc.), environmental factors (source of drinking-water and material of the dwelling unit's floor), nutrient status (breast-feeding practices and anthropometric measurement), and personal illness control (circumstances surrounding childbirth, child immunizations, and illness treatment). The category for which DHS does not collect data concerns childhood accidents and injuries; however, while the breadth of factors covered by DHS is extensive, the information collected for any particular factor is frequently insufficient to draw definitive conclusions concerning its impact on mortality levels. For example, consider breast-feeding. DHS collects information on the duration of breast-feeding in the long cores in much the same way as did WFS. Analysis of such data has shown that breast-feeding has a positive effect on survivorship (Sullivan and Wilson 1982*c*; Adlakha and Suchindran 1985). With WFS data for Jordan, Egypt, Tunisia, and Yemen, multivariate analysis found that the risk of post-neonatal mortality was from three to eight times higher for children not breast-feeding at 1 month of age than it was for breast-feeders. The differentials are substantial but it is not certain that lack of breast-feeding is the cause of the elevated mortality. The problem arises because it is not known if the decision never to breast-feed or to stop breast-feeding was made by choice or if it was necessitated by illness of the mother or the child. In the latter case, the causation would run from illness to cessation of breast-feeding and death rather than vice versa. The data to resolve this issue were not collected by WFS and are not included in the DHS cores. However, the necessary data could be collected in a DHS survey by asking respondents the reasons for never or for stopping breast-feeding. These two questions have, in fact, been included in the DHS survey in Nigeria (Ondo State), Peru, and the Dominican Republic. Similarly, in other surveys, questions could be added to the DHS questionnaire to obtain data for a detailed analysis of the relationship between mortality and other factors (water supply, dwelling unit characteristics, child care practices, etc.). Some of these issues are dealt with in the DHS health module, an additional set of questions that can be incorporated into the survey questionnaire in cases where the country wants a greater focus on health.

5. Summary

There are important differences between WFS and DHS in the collection of data useful for mortality analysis. In terms of community-level data, DHS collects much more data on the proximity to health facilities and the services that they offer than did WFS. A second major difference between WFS and DHS is that DHS does not collect mortality data at the household level as was done with the WFS general mortality module. Finally and most significantly, at the level of individual female respondents, DHS obtains data on health practices (maternal and child care) and health indicators (height and weight of children under age 3). These data, in conjunction with the community data on health facilities, provide a set of health data not usually collected in large-scale fertility surveys. Significantly, they permit the joint analysis of population and health variables at the individual level.

For the individual woman's questionnaire, DHS has developed four model instruments: two incorporate a full birth history and two contain a five-year truncated birth history. Surveys that apply a truncated birth history have considerably less analytic potential than those that apply a full birth history. Restricting the collection of birth data to the five-year period preceding a survey essentially means forgoing the possibility of analysing time trends in mortality, the relationship between the length of birth intervals and infant and child survival, and the potential for bias when applying indirect childhood mortality estimation techniques for the country.

References

Adlakha, A., and Suchindran, C. (1985), 'Factors Affecting Infant and Child Mortality', *Journal of Biosocial Sciences*, 17: 481–96.

Caldwell, J.C. (1985), 'Strengths and Limitations of the Survey Approach for Measuring and Understanding Fertility Change: Alternative Possibilities', in J. Cleland and J. Hobcraft (eds.), *Reproductive Change in Developing Countries: Insights from the World Fertility Survey* (New York: Oxford University Press).

Casterline, J.B. (1984), 'The Collection and Analysis of Community Data: The WFS Experience', Paper presented to Finding Session 3, *World Fertility Survey Symposium*, London, Apr.

Chidambaram, V.C., McDonald, J., and Bracher, M.D. (1984), 'Contribution of WFS to the Study of Childhood Mortality', Paper presented to the Plenary Session, *World Fertility Survey Symposium*, London, Apr.

Garenne, M. (1982), 'Problems in Applying the Brass Method in Tropical Africa: A Case Study in Rural Senegal', *Genus*, 18(1–2): 119–34.

Gille, H. (1985), 'Policy Implications', in J. Cleland and J. Hobcraft (eds.), *Reproductive Change in Developing Countries: Insights from the World Fertility Survey* (New York: Oxford University Press).

Hobcraft, J. (1984), 'Use of Special Mortality Questions in Fertility Surveys: The World Fertility Survey Experience', in UN, *Data Bases for Mortality Measurement*, ST/ESA/SER. A/84 (New York: UN).

Mosley, W.H., and Chen, L.C. (1984), 'An Analytical Framework for the Study of Child Survival in Developing Countries', in W.H. Mosley and L.C. Chen (eds.), *Child Survival: Strategies for Research*, supplement to *Population and Development Review*, 10 (New York: The Population Council).

Pison, G. (1984), 'Age Patterns of Mortality in Eastern Senegal: Comparison of Micro and Survey Approaches', Paper presented at the *IUSSP Seminar on Micro-Approaches to Demographic Research*, Australian National University, Canberra, 3–7 Sept.

Preston, S.H. (1985), 'Mortality in Childhood: Lessons from WFS', in J. Cleland and J. Hobcraft (eds.), *Reproductive Change in Developing Countries:*

Insights from the World Fertility Survey (New York: Oxford University Press).

Rutstein, S.O. (1984*a*), *Infant and Child Mortality: Levels, Trends and Demographic Differentials*, WFS Comparative Studies, no. 43 (Voorburg, Netherlands: International Statistical Institute).

—— (1984*b*), *Socio-economic Differentials in Infant and Child Mortality*, WFS Comparative Studies (Cross-National Summaries: Preliminary Tables) (Voorburg, Netherlands: International Statistical Institute).

—— (1985), *Assessment of the Quality of WFS Data for Direct Estimation of Childhood Mortality*, WFS Comparative Studies, no. 44 (Voorburg, Netherlands: International Statistical Institute).

Scott, C., and Harpham, T. (1984), 'Major Issues of Survey and Sample Design', Paper presented at the Methodology Session 2, *World Fertility Survey Symposium*, London, Apr.

—— and Singh, S. (1981), 'Problems of Data Collection in the World Fertility Survey', *Proceedings of the World Fertility Survey Conference*, 1980 (Voorburg, Netherlands: International Statistical Institute).

Singh, S. (1984), *Comparability of Questionnaires: Forty-One WFS Countries*, WFS Comparative Studies, no. 32 (Voorburg, Netherlands: International Statistical Institute).

Sullivan, J., and Wilson, S. (1982*a*), *A Methodology for Detecting Bias in Indirect Mortality Estimates: The Case of Indonesia*, Population Paper no. 25 (Chapel Hill, NC: Carolina Population Center).

—— —— (1982*b*), *The 1980 Baseline Round of the East Java Population Survey: A Final Report* (Jakarta, Indonesia: Central Bureau of Statistics).

—— —— (1982*c*), *The Second Round (1981) of the East Java Population Survey: A Final Report* (Jakarta, Indonesia: Central Bureau of Statistics).

Timaeus, I. (1984), 'Estimation of Fertility and Mortality from WFS Household Surveys: Background Paper', Paper presented to the *World Fertility Survey Symposium*, London, Apr.

WFS (World Fertility Survey) (1975), *Survey Organization Manual*, World Fertility Survey Basic Documentation, no. 2 (Voorburg, Netherlands: International Statistical Institute).

5 Epidemiologic Methods and Case-Control Studies of Mortality and Morbidity

RONALD H. GRAY

Johns Hopkins University School of Hygiene and Public Health Baltimore, Maryland, USA

1. Introduction

The study of mortality has long been an interest of both demographers and epidemiologists, the two disciplines often using similar research methodologies and data. Historically epidemiologists have focused on morbidity as well as mortality with a primary interest in the measurement of disease frequency, aetiology, prevention, and therapeutic evaluation. However, more recently demographers have also turned to morbidity measurement as an adjunct to demographic surveys, particularly in developing countries (Tekce and Shorter 1984; Demographic and Health Surveys 1986). These studies serve both descriptive and analytic purposes, and, in the latter regard, the objective is to determine factors causing illness or death, or to evaluate interventions designed to reduce morbidity or mortality.

This chapter will briefly review the methodologies used by both disciplines to highlight similarities and differences, but the major emphasis will be on potential demographic applications of case-control methods, which have hitherto been the exclusive province of epidemiologists. The discussion will concentrate on studies of infant and childhood mortality or morbidity in developing countries, since this is the area in which an integration of disciplinary approaches may be most productive.

It is first necessary to explain some of the concepts of disease causation as they are applied in epidemiology. In general, epidemiology searches for a causal association between certain antecedent 'risk factors' or 'exposures' and subsequent outcomes of disease or death. Often this involves a complex web of multiple causation and requires careful attention to distinguish between direct and indirect causal associations. Thus the richness and detail of information particularly on the sequence of events is of critical importance.

An example of causal pathways is illustrated by using the framework for studying child survival proposed by Mosley and Chen (1984) shown in Figure 5.1. Outcomes such as child death or faltering growth can be related to antecedent diseases, and these in turn are associated with five proximate determinants of mortality (grouped into maternal, environmental, nutrition, injury, and illness control) which are influenced by underlying socio-

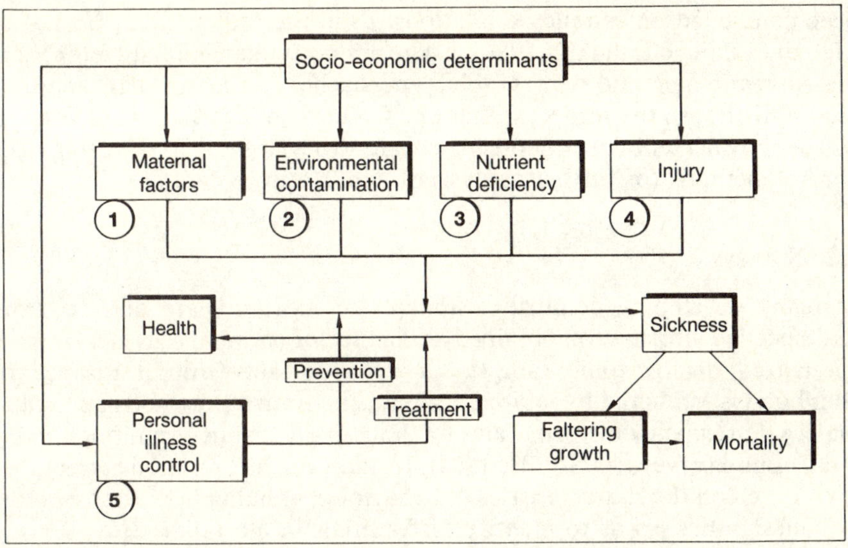

Fig. 5.1. Operation of the five groups of proximate determinants on the health dynamics of a population (proximate determinants are number 1-5). *Source*: Mosley and Chen 1984

economic characteristics. In epidemiologic terms, these proximate and underlying determinants would be considered as risk factors or exposures. For instance, in the well-established relationship between maternal education and child survival, the level of education would be regarded as a predisposing characteristic affecting exposure to the proximate determinants and thus modifying the risk of disease or death. An understanding of the proximate determinants might provide insight into the mechanisms through which education affects mortality and this is of critical importance in guiding policy and in planning or evaluating programmes.

2. Review of Demographic and Epidemiologic Methods

2.1. Vital Statistics

Vital statistics derived from registration data and population censuses have been widely employed to measure levels and patterns of mortality, causes of death, and differentials by socio-demographic characteristics, place of residence, and occupation, etc. (WHO 1977). Most studies are confined to industrialized countries or selected developing countries where registration data are of adequate quality and completeness. Also, in settings where national data are inadequate, special demographic surveillance studies have been conducted on sample populations, as in the Indian SRS, China, or Matlab in Bangladesh (UN 1984). There are no fundamental differences in the epidemiologic and demographic approaches to the analysis or use of vital statistics, so the subject will not be further explored here. However, as will be shown later in this chapter, there are strong arguments for using case-control methods for further analysis of vital-statistics data.

2.2. Surveys

In many developing countries estimates of mortality are derived from retrospective information obtained at the time of national censuses or from specialized demographic sample surveys. Mortality during infancy or childhood is estimated by indirect techniques, by direct methods applied to maternal pregnancy histories, and by 'household change techniques' using multi-round surveys (UN 1983, 1984). In addition, information is frequently collected on the characteristics or behaviours of individuals, households, or communities, so as to identify differentials in mortality. Such surveys have shown major differences in survival associated with birth spacing, maternal age or parity, breast-feeding, parental education, sanitary facilities, environmental ecology, etc.

These demographic surveys are conceptually similar to the cross-sectional or prevalence surveys used by epidemiologists to measure the prevalence of disease or disability and factors associated with differentials in health status. Although strictly speaking cross-sectional studies imply measurements of disease and exposure status at one point in time, most epidemiologic investigations also include some retrospective information (McMahon and Pugh 1970; Lilienfeld 1976). For example, morbidity surveys will frequently obtain information on the presence of disease at time of interview or over variable periods preceding interview. One disadvantage of this approach is that acute diseases with short duration or with short survival times may be under-represented, whereas more chronic illnesses of longer duration may be over-represented (Black 1984).

Information on current status at time of interview (e.g. illness) and retrospective information on past events (e.g. deaths or exposures) allows the estimation of disease or death rates by various categories of interest. There are numerous papers on infant mortality by maternal educational status derived from these types of survey, and this association will be used for illustrative purposes. Demographers conventionally analyse such data by proceeding forward from exposure (cause) to outcome (effect) and estimate death rates by maternal-education categories. However, an epidemiologist might follow the reverse logic and examine exposure frequencies in contrasting outcome groups. For example, he/she might estimate the proportions educated among mothers whose infants died compared with those whose infants survived. This apparently inverted logic, proceeding from outcome (effect) backwards to exposure (cause), is

Table 5.1. *Cohort or prevalence studies: relationship of outcome and exposure in a hypothetical population-based survey of maternal education and child survival*

		Outcome			Odds of outcome (death)
		Dead +	Alive −	Total	
Exposure	Low education +	A	B	m_1	A/B
	High education −	C	D	m_2	C/D
TOTAL		n_1	n_2	n	
Odds of exposure		A/C	B/D		

Notes: A = no. of deaths to uneducated mothers; B = no. of children surviving to uneducated mothers; C = no. of deaths to educated mothers; D = no. of children surviving to educated mothers.

the approach used in case-control studies to be discussed later in this chapter. Intuitively, this often appears to be a reversal of common sense, but in fact much observational science is based on such backward-looking methods. An example is the theory of evolution, whereby Darwin and Wallace reasoned backwards from present differences in species to draw inferences about the origin of observed differences (Sartwell 1979). In epidemiology, the backward-looking approach grew out of the need to study relatively rare chronic diseases such as cancer often caused by distant exposures such as previous smoking.

The forward and backward approaches are in fact symmetrical, as illustrated in Table 5.1, which arrays data from a population survey of maternal education and child survival in a two-by-two table. Also shown are the odds of outcome or exposure. The relative odds of child death (OR) among educated versus uneducated mothers is the ratio of the odds of death for uneducated (A/B) to educated (C/D):

$$OR=(A/B)/(C/D)=AD/BC. \quad (1)$$

Similarly, the relative odds of being an educated or an uneducated mother of dead versus surviving children is the ratio of the odds of exposures:

$$OR=(A/B)/(B/D)=AD/BC. \quad (2)$$

However, we are usually concerned with estimating mortality rates in exposed and non-exposed groups. From Table 5.1, the proportions of dead infants of mothers with no education (q_1), and educated mothers (q_2) are:

$$q_1=A/m_1=A/(A+B) \text{ and } q_2=C/m_2=C/(C+D). \quad (3)$$

The ratio of the proportions dead is the relative risk of death (R) associated with education which is:

$$R=\frac{(A/m_1)}{(C/m_2)}, \quad (4)$$

$$R = q_1/q_2 = \frac{(A/A+B)}{(C/C+D)} = \frac{\{A(C+D)\}}{\{C(A+B)\}}. \tag{5}$$

If the outcome of interest (e.g. death) is relatively rare then A and C are small compared to B and D, so $B \cong m_1$ and $D \cong m_2$ in equations (4) and (5), and the odds ratio approximates to the relative risk:

$$R \cong AD/BC. \tag{6}$$

This is identical to the relative odds of disease given exposure or the relative odds of exposure given disease in equations (1) and (2).

The major disadvantage of these surveys is that large samples are required in order to detect sufficient numbers of infrequent events. This precludes lengthy interviews, or, if long interview schedules are used, the cost and duration of field-work is often greatly increased and maintenance of quality control can be difficult. As a consequence one cannot obtain the detailed information often needed to identify causal pathways or important exposures illustrated in Figure 5.1 and only gross associations can be determined. As will be shown in Section 3, this problem can be overcome by the case-control study design.

2.3. Cohort Studies

In the classical epidemiologic concurrent prospective cohort study, individuals who are initially free of disease are followed over time to determine the incidence of disease in persons exposed (I_e) and non-exposed (I_o) to factors of interest. A variant is the historic cohort study in which a known exposure occurred in the distant past, and the incidence of subsequent disease or death is studied using available records or by special follow-up methods (Lilienfeld 1976; Feinleib and Detels 1985). The relative risk of disease or death associated with a given exposure is estimated from the ratio of the incidence rates for exposed and non-exposed as in equation (4) ($R = I_e/I_o$).

The closest demographic parallel is the longitudinal surveillance of births and death in which mortality rates are estimated according to subject characteristics (e.g. education) (D'Souza 1984).

Cohort studies have several advantages in epidemiologic terms. They theoretically allow a complete description of disease development and can examine a variety of outcomes in relation to a given exposure, while maintaining quality control in data collection. The main disadvantages are the large sample sizes required to detect differences in the incidence of rare diseases or events such as deaths, the long duration of follow-up required, the potential for biased losses to follow-up, and the high cost of prospective studies (Schlesselman 1982). In addition, changes in exposure factors may occur over time that are outside the control of the investigator, and this can undermine the original objectives of the study. Finally, in both developed- and developing-country settings, the populations enrolled in cohort studies are often atypical of the general population at large, frequently because the cohorts have better health status (Briscoe *et al.* 1985). Thus, a cohort study may have internal validity in the comparison between exposed and non-exposed, but limited external validity required for generalization of findings.

2.4. Experimental and Quasi-experimental Studies

Evaluations of health interventions are often based on an attempt to approximate to classical research designs which allow the testing of hypotheses by the random allocation of treatments. For example, in vaccine trials there is a random allocation of vaccines between non-immune individuals, and the efficacy of vaccination is assessed by subsequent disease incidence in experimental and control groups. Similar clinical trials are used to evaluate therapies such as oral rehydration for diarrhoea. The classical experimental approach can only be used when a treatment can be applied to individuals, but many health interventions are directed at community level and this requires a quasi-experimental approach whereby communities, often selected at random, are used as experimental and control populations. An alternative approach commonly used in primary health care evaluation is the comparison of disease or death rates before and after introduction of an innovation. Combinations of these two strategies may also be employed.

A drawback to both experimental and non-experimental studies is the large sample sizes needed for monitoring rare events, and a particular problem of the quasi-experimental method is possible lack of initial comparability between experimental and control groups. Moreover, in the quasi-experimental design, the comparison is made between populations rather than individuals, and this raises serious problems of misclassification, since one does not know whether individuals in the experimental group utilized the services provided, or whether individuals in the control group obtained services by other means. It is often difficult to maintain 'uncontaminated' comparison groups over the prolonged time needed for these studies, and there are ethical dilemmas if the control group is denied access to services.

The data required for experimental or quasi-experimental studies are usually derived from repeat surveys or cohort-type follow-up studies, but problems of sample size, logistics, expense, and misclassification bias have increasingly turned investigators towards the use of case-control methods (Gray 1984; Orenstein *et al.* 1984; Briscoe *et al.* 1985).

3. Case-Control Studies

3.1. General Features

In case-control studies individuals with an outcome of interest such as disease or death (cases) are selected for comparison with individuals without

Table 5.2. *Case-control studies: relationship of outcome and exposure in a hypothetical population-based survey of maternal education and child survival*

		Outcome			Odds of death
		Cases (dead) +	Controls (alive) −	Total	
Exposure	Low education +	a	b	m_1	a/b
	High education −	c	d	m_2	c/d
TOTAL		n_1	n_2	n	
Odds of exposure		a/c	b/d		

Notes: a = exposed cases; b = exposed controls; c = non-exposed cases; d = non-exposed controls.

these conditions (controls), and information is obtained on antecedent risk factors. The latter may consist of current or past attributes or exposures thought to be of relevance to the disease or condition under study. Cases are compared to controls to determine the relative frequency of exposures as shown in Table 5.2.

In this table, n_1 cases and n_2 controls are identified and the numbers exposed (a and b) or non-exposed (c and d) are determined by interview. The cases and controls represent different and often unknown fractions of the total cases or controls in the population, and the number of controls selected per case may vary. Thus, unlike the population-based data depicted in Table 5.1, one cannot directly estimate rates of disease or mortality from case-control studies.

In Table 5.2 the relative odds or odds ratio (OR) of child death for educated versus uneducated mothers is:

$$OR = (a/b)/(c/d) = ad/bc \tag{7}$$

This is identical to the relative odds estimated for a population-based study in equations (1) and (2), and approximately equal to the relative risk in equation (6) if the outcome is rare (see Appendix 5.1). The relationship between the odds ratio (OR) and relative risk (R) is complex, and the assumed approximation of equality is generally considered unsatisfactory if the event frequency is above 10%. However, even with common outcomes in which the rare-disease assumption does not hold, the relative risk can be estimated from the odds ratio (see Section 3.5).

It is generally assumed that the cases and controls are representative subsamples of all eligible diseased and non-diseased individuals, and that exposed individuals have the same eligibility for inclusion in the study as non-exposed individuals. Cases and controls are often selected from institutions such as hospitals, or identified from the community using population-based disease registers or reports of vital events. These sources of study subjects are not readily available in many developing-country settings, particularly in rural areas, and institution-based studies confront major problems of biased selection. Also, the lack of comprehensive disease registers or complete vital-statistics records limits the use of these population sources as sampling frames for subject selection.

In summary, case-control studies have not been widely applied in developing countries because of concerns over the inability to directly estimate rates, biased selection of study subjects, and because high disease frequencies (e.g. diarrhoea or malnutrition) and high death rates (e.g. infant mortality rates (IMRs) above 100 per 1000) invalidate the rare-disease assumption underlying the interpretation of the odds ratio (*OR*). Moreover, there is suspicion about the backward logic of proceeding from effect to cause and biases that may arise from retrospective data.

What then are the advantages of case-control studies? One advantage over cohort or prevalence studies is the smaller sample size requirements, particularly with relatively rare outcomes (see Section 3.3). A second major advantage is the ability to study multiple potential causes of disease or death (i.e. to trace the complex web of causation), since one can obtain more detailed information from individuals because the smaller sample sizes allow longer and more probing interviews, as well as the validation of interview data from medical or other records. The third major advantage is economy of time and expense, again because of smaller sample size requirements, and the lack of prospective follow up (Schlesselman 1982).

3.2. Linked Surveys and Case-Control Studies

Demographic or epidemiologic sample surveys provide estimates of mortality or morbidity but limited information on causal linkages. However, if these surveys were used to identify subjects for case-control studies it would be possible to obtain the additional data needed to trace causal pathways. Also, since subjects are identified from population samples, the problem of selection bias in institutional case-control studies would be overcome (see Section 3.4). The integration of surveys with case-control studies thus provides an optimum approach to field investigations, and the additional costs or logistic difficulties would be marginal since there could be economies in integrating personnel and support facilities for dual purposes. The only proviso is that the outcome of interest must be frequent enough or the sample sizes large enough to identify sufficient numbers of cases.

3.3. Sample Size Requirements for Case-Control Studies

The sample size issue is illustrated in Table 5.3, showing a hypothetical example of a study of IMR in relation to maternal education. It is assumed that 50% of the female population is educated, and that among educated women the IMR is set at either 25 or 50 per 1000. The objective is

Table 5.3. *Sample sizes required for cohort or prevalence studies and for case-control studies, using hypothetical study of infant mortality in relation to maternal education*[a]

Relative risk	Cohort or prevalence studies[b] outcome rate in non-exposed		Case-control study[c]
	25 per 1000	50 per 1000	
1.5	6074	2938	774
2.0	1808	868	272
3.0	594	280	116
4.0	266	150	76

[a] Total sample sizes required, estimated by methods given by Schlesselman (1982), assuming $\alpha = 0.05$, $\beta = 0.2$.

[b] Assumes 50% exposed and 50% not exposed.

[c] Assumes an equal number of cases and controls, i.e. the number of cases required is half the total shown.

to estimate the increased risk of infant death associated with no maternal education. The total samples required for a case-control study are substantially lower than for cohort or prevalence studies, particularly if the level of risk or the frequency of the outcomes are low. For example, to detect a twofold increased risk, the sample size for a case-control study is between 15% and 31% of that required for a cohort study, depending on the mortality levels, and the number of cases is only half the total case-control sample shown. Moreover, unlike cohort or prevalence studies, the sample sizes for case-control studies do not depend on the frequency of the outcome, since the ratio of cases to controls is fixed.

The sample size required for case-control studies depends on two parameters: the expected frequency of exposure among controls (p_o) and magnitude of the risk (R) to be detected. In unmatched studies with equal number of cases and controls, the required sample size (n) for each group is:

$$n = \frac{\{Z_\alpha \sqrt{(2\bar{p}\bar{q})} + Z_\beta \sqrt{(p_1 q_1 + p_o q_o)}\}^2}{(p_l - p_o)^2}, \tag{8}$$

where p_1 denotes the proportion of exposed cases which can be estimated from:

$$p_1 = \frac{p_o R}{[1 + p_o(R-1)]}, \tag{9}$$

and

$$\bar{p} = \tfrac{1}{2}(p_1 + p_o) \quad \bar{q} = 1 - \bar{p},$$
$$q_1 = 1 - p_1 \qquad q_o = 1 - p_o. \tag{10}$$

Z_α and Z_β are the values of the standardized normal deviates exceeded with a probability of α and β in the normal distribution. α is the probability of type I error and β the probability of type II error. The power is $1 - \beta$.

A simplified formula useful in most practical situations is:

$$n = \frac{2\bar{p}\bar{q}(Z\alpha + Z\beta)^2}{(p_1 p_o)^2}. \tag{11}$$

Sample size estimates for unequal numbers of cases and controls are given in Schlesselman (1982).

3.4. Bias in Observational Research

One major issue confronting case-control studies is the problem of bias, which is defined as 'any effect at any stage of investigation or inference tending to produce results that depart systematically from the true values' (Last 1983). Bias is common to all observational studies, but the main problems of relevance to case-control studies are biases in selection, subject classification, exposure information recall, and confounding due to indirect association.

3.4.1. Selection or ascertainment bias This form of bias arises when the likelihood of identification or selection of cases or controls is not independent of the exposure under study, as illustrated in Figure 5.2. If in a hospital case-control study there is selectively higher referral or admission rates among exposed cases (*ED*) due perhaps to socio-economic factors, multiple illnesses, or distance from the hospital, then the odds ratio will be seriously biased (Sacket 1979; Kleinbaum *et al.* 1981). In an institution-based study this form of admission or referral bias can be overcome by selecting controls with other illnesses of severity comparable to that of the cases, so theoretically the degree of selective pressure is the same for both groups. More importantly, if a demographic survey were used as a source of cases and controls, then these admission–referral biases would be eliminated.

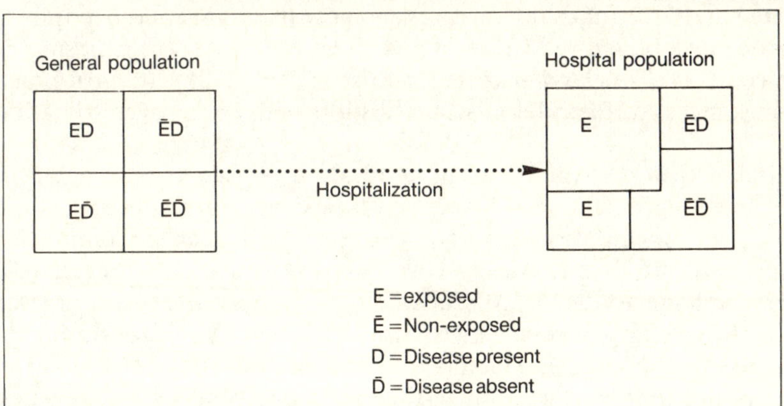

Fig. 5.2. Selection bias due to hospitalization. *Source*: Greenberg and Ibrahim 1985

Another source of selection bias is differential case detection in which cases with exposure are more likely to be identified than cases without exposure (or vice versa). This may arise if subjects with the exposure are more carefully screened than those without, or doctors are more likely to diagnose the disease in an exposed than in a non-exposed case due to prior suspicion of an association. Also, interviewers may tend to select cases or controls on the basis of their exposure status (Sacket 1979; Schlesselman 1982). These problems can be overcome by stringent criteria for subject selection, proper training of interviewers, or use of 'blinding' procedures.

In studies of living respondents bias may also arise as a result of differential survival in exposed and non-exposed. For example, if affluent mothers more rapidly seek treatment for their children's illnesses, they may improve survival relative to children of poorer mothers, and this will increase the apparent disease prevalence among the affluent though the disease incidence may be the same (Sacket 1979; Kleinbaum *et al.* 1981). This prevalence–incidence bias can be overcome by obtaining information on deaths or by including only new incident cases of disease.

Many of these problems of selection bias are not restricted to case-control studies, but also arise in cross-sectional or prospective investigations, and the selective admission–referral bias which particularly affects hospital-based case-control studies can be overcome by using population-based surveys as a source of case-control subjects.

3.4.2. Misclassification bias Misclassification bias occurs when there are errors in the determination of an individual's exposure or disease status. With non-differential misclassification the errors are similar in cases and controls, and this results in a dilution of the association between exposure and outcome (i.e. a spurious reduction in risk). Differential misclassification occurs when the likelihood of error is related both to disease status and exposure of interest. Differential misclassification may give a biased increase in risk if, for example, cases are more likely to report exposure than controls, or a biased reduction of risk if cases underreport exposure more than controls (Copeland *et al.* 1977; Briscoe *et al.* 1985). In most practical circumstances, differential misclassification leads to an overestimation of risk.

Misclassification can occur as a result of misdiagnosis, and this is particularly a problem when diagnosis is based on informant recall of illness rather than on clinical or laboratory tests or on a clear end-point such as survival or death. Even in the latter case, misclassification errors can arise if there is omission of deaths. Misclassification of exposures arises largely as a result of failure by respondents to recall or report exposures during interview (see Section 3.4.3). This may be due to inadvertent omission (recall lapse) or to deliberate omission with sensitive questions such as abortion or sexually transmitted diseases.

The degree to which a given variable results in misclassification depends

Table 5.4. *Sensitivity and specificity: a descriptive scheme*

Apparent status	True status	
	+	−
+	$1-\beta$	α
−	β	$1-\alpha$
TOTAL	1	1

Notes: $1-\beta$ = proportion of true positives detected (sensitivity); $1-\alpha$ = proportion of true negatives detected (specificity); β = proportion of false negatives; α = proportion of false positives.

upon the validity of the measure used, and the likelihood of misclassification is increased if indirect or proxy measures are employed. Validity implies the ability of a measure to correctly identify true positives (sensitivity) and true negatives (specificity). This is illustrated in Table 5.4.

The ability to detect true positives or the sensitivity of disease or exposure classification plays the most important role in influencing misclassification in case-control studies (Copeland *et al.* 1977). In general, bias is greatest with low sensitivities, low prevalence of exposure, and higher ranges of relative risk. In most practical situations the problem is one of omission leading to a reduction in sensitivity (i.e. reduced true positives detected) since respondents are unlikely to report a death, illnesses, or exposures which have not occurred.

The example in Table 5.5(*a*) gives the hypothetical 'true' distributions of exposures in cases and controls. It is assumed that the sensitivity of the history of exposure is 0.8 (i.e. 20% false negatives), and the observed distribution of exposures given this misclassification is shown in Table 5.5(*b*). Such non-differential misclassification reduces the *OR* from 3.5 in Table 5.5(*a*) to 2.9 in Table 5.5(*b*). Table 5.5(*c*) assumes differential misclassification with sensitivities of 0.8 for cases and 0.6 for controls (i.e. 20% more omission of exposure by controls) and the *OR* is biased upward to 4.2.

To avoid or minimize misclassification bias it is necessary, as far as possible, to validate diagnoses or exposures and to use clearly defined criteria for case or control classification. There needs to be a strict definition of disease diagnosis and it is desirable to obtain a homogeneous group or subgroup of cases. Validation studies may allow adjustment of the odds ratio if the specificity and sensitivity are known or can be estimated from other data (see Table 5.4). Also, the validity of responses can be improved by more detailed and intensive interviewing, and in this respect the smaller samples in case-control studies have a distinct advantage over larger samples required for cohort or prevalence surveys.

In summary, misclassification bias can be a serious problem, but it is a problem of all observational studies in general, rather than case-control studies in particular (Briscoe *et al.* 1985).

Table 5.5. *Study of several misclassification biases*
(*a*) Hypothetical 'true' classification of exposure

Exposure	Observed frequencies of disease		Total
	+	–	
+	60	30	90
–	40	70	110
TOTAL	100	100	200

OR = 3.5.

(*b*) Non-differential misclassification (assumes sensitivity of exposure $1-\beta = 0.8$)

Exposure	Observed frequencies of disease		Total
	+	–	
+	48	24	72
–	52	76	128
TOTAL	100	100	200

OR = 2.9.

(*c*) Differential misclassification (assumes sensitivity of 0.8 for cases and 0.6 for controls)

Exposure	Observed frequencies of disease		Total
	+	–	
+	48	18	66
–	52	82	134
TOTAL	100	100	200

OR = 4.2.

3.4.3. Information bias Information bias occurs when either the respondent provides inaccurate information or the interviewer elicits incorrect information on exposure status. For example, mothers of dead children (cases) may be more motivated than mothers of surviving children (controls) to provide a complete history of exposure; they may inadvertently over-report certain exposures or omit others because of anxiety or cultural proscriptions, and the interviewer may probe for information from cases more than controls (Sacket 1979; Schlesselman 1982). These errors in recall or interview can result in misclassification bias.

To avoid these problems it is necessary to use carefully structured questionnaires and to avoid open-ended interviews. The subjects should not be informed of the expected disease exposure association under study to avoid prompted responses, and ideally the interviewers should be kept blind with respect to a subject's case or control status to avoid unbalanced probing.

Where possible, information should be validated against available records such as death certificates, clinic attendance, vaccination cards, or growth monitoring charts, etc. In drug studies examples of drug packages can be used to facilitate accurate recall. Also, where the reporting of events is known to deteriorate over time, questioning may be restricted to recent events.

Although there is no satisfactory way of ensuring accuracy of recall information, the possibility of differential recall bias can in part be assessed during analysis. For example, the frequency of responses to 'neutral' variables not associated with the disease can be compared in cases and controls. If the distributions of 'neutral' variables are similar, this suggests that recall bias is not a major problem. If the distributions differ, estimates of the degree of bias can be obtained.

Finally, it must be noted that recall bias is a problem with all retrospective studies, and applies to both retrospective demographic or morbidity surveys as well as case-control studies.

3.4.4. Confounding Confounding refers to the effect of an extraneous variable which partly or wholly accounts for an apparent association between exposure and outcome, or masks such an association. A confounding variable is one which is independently associated both with the disease and with the exposure under study. Commonly a confounder is a risk factor for a disease which is also associated with the exposure of interest but not in a causal manner. This may lead to spurious associations, as illustrated in Figure 5.3(*a*). Alternatively, both the confounder and the exposure may be causally linked with the disease but non-causally with one another as in Figure 5.3(*b*).

In some situations potential confounding variables can be predicted a priori and the study designed to avoid problems. For example, age is closely linked to risk of death, so cases and controls can be matched for age. However, many confounders cannot be predicted in advance, and it would not be feasible to design a study which could cope with all potential confounders, but the existence of confounding can be detected and adjusted for during analysis using stratification (Schlesselman 1982). For example, maternal education might be closely linked to socio-economic status and to the chances of child survival. To assess and adjust for confounding, the effects of maternal education can be examined within strata of socio-economic status (SES). If education is confounded by SES, the effects should be reduced by stratification.

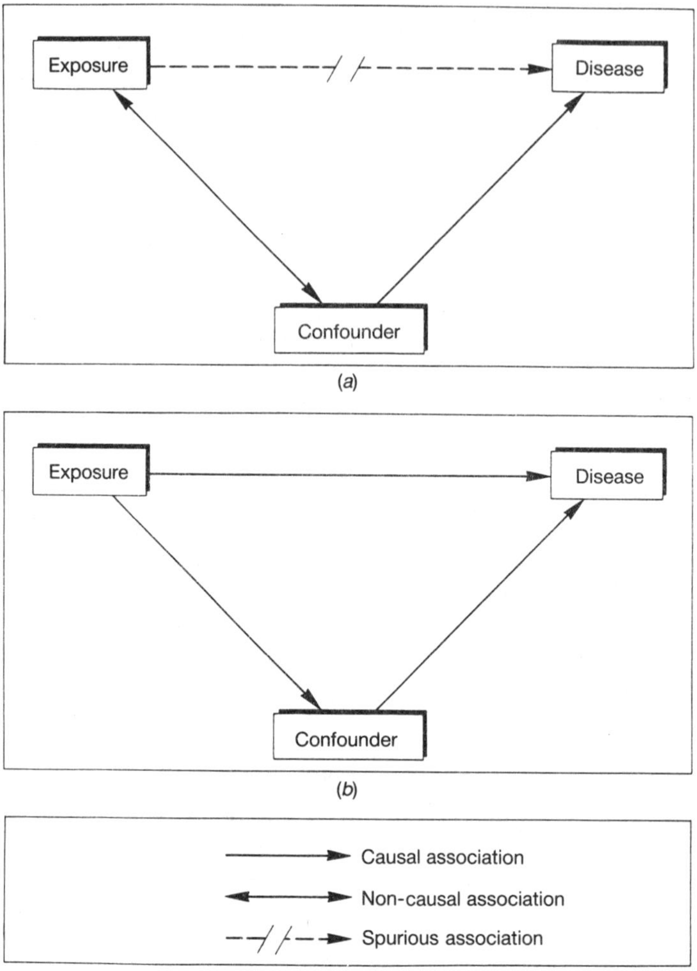

Fig. 5.3. Confounding
(*a*)
(*b*)

3.4.5. Multivariate analysis Methods have been developed for multivariate analysis of case-control data primarily using logistic regression (Breslow and Day 1980; Schlesselman 1982). These cannot be fully described here, but the models take the general form of:

$$OR\ (X{:}x) = \exp\{\beta_0 + \beta_1(X_1 - x_1) + \ldots, + \beta_i(X_i - x_i)\}, \qquad (12)$$

where OR represents the odds of disease given exposure to risk factors X, the betas *(β)* represent log odds of disease for exposure to each risk factor $x_1 \ldots x_i$, and X_i and x_i denote the values of variable x_i for exposed and non-exposed individuals, respectively, if these are binary variables.

Logistic models allow point estimates of the odds ratio for each variable derived from the exponent of the betas, $OR_i = \exp\{\beta_i(X_i - x_i)\}$, after adjustment for all other variables entered in the model. This implies adjustment for confounding. Also, estimates of interaction and effect modification can be obtained by specifying interaction terms $\Upsilon(X_1 x_i)$ in addition to the main effects variables. Confidence intervals for point estimates of risk and tests of statistical significance between models are also available (Breslow and Day 1980).

3.4.6. Use of case-control studies for common diseases or outcomes As shown in equations (5) and (6), the odds ratio *OR* is equivalent to the relative risk *R* only when the frequency of the outcome is 10% or less, and this approximation breaks down if the 'rare-disease assumption' is invalid. Many conditions such as infectious diseases, pregnancy outcomes, the prevalence of malnutrition, or even infant deaths occur with a frequency exceeding 10%. The violation of the rare-disease assumption is only a

Table 5.6. *Estimation of relative risk from odds ratio in relation to the prevalence of disease and exposure*

OR	P(D) = 5%					P(D) = 10%					P(D) = 15%				
	P(E) (%)					P(E) (%)					P(E) (%)				
	5	15	25	35	45	5	15	25	35	45	5	15	25	35	45
2	1.9	1.9	1.9	1.9	1.9	1.8	1.8	1.8	1.9	1.9	1.7	1.8	1.8	1.8	1.8
3	2.8	2.8	2.8	2.8	2.8	2.5	2.6	2.6	2.7	2.7	2.3	2.4	2.5	2.5	2.5
4	3.5	3.6	3.7	3.7	3.8	3.1	3.3	3.4	3.4	3.5	2.8	3.0	3.1	3.2	3.3
5	4.3	4.4	4.5	4.6	4.7	3.7	3.9	4.1	4.2	4.3	3.2	3.5	3.7	3.8	4.0
6	5.0	5.2	5.4	5.5	5.5	4.2	4.5	4.8	5.0	5.1	3.6	3.9	4.2	4.5	4.7
7	5.6	6.0	6.2	6.3	6.5	4.6	5.1	5.5	5.7	5.9	3.9	4.4	4.8	5.1	5.4
8	6.3	6.7	7.0	7.2	7.3	5.1	5.7	6.1	6.4	6.7	4.2	4.8	5.3	5.7	6.1
9	6.9	7.5	7.8	8.1	8.2	5.4	6.2	6.8	7.2	7.5	4.4	5.2	5.8	6.3	6.7
10	7.5	8.2	8.7	8.9	9.1	5.8	6.8	7.4	7.9	8.3	4.6	5.5	6.3	7.0	7.4

OR	P(D) = 20%					P(D) = 25%					P(D) = 30%				
	P(E) (%)					P(E) (%)					P(E) (%)				
	5	15	25	35	45	5	15	25	35	45	5	15	25	35	45
2	1.7	1.7	1.7	1.7	1.7	1.6	1.6	1.6	1.7	1.7	1.5	1.6	1.6	1.6	1.6
3	2.2	2.2	2.3	2.4	2.4	2.0	2.1	2.2	2.2	2.3	1.9	2.0	2.0	2.1	2.1
4	2.6	2.7	2.8	2.9	3.0	2.3	2.5	2.6	2.7	2.8	2.2	2.3	2.4	2.5	2.6
5	2.9	3.1	3.3	3.5	3.7	2.6	2.8	3.0	3.2	3.4	2.4	2.5	2.7	2.9	3.1
6	3.2	3.5	3.8	4.0	4.3	2.8	3.1	3.4	3.6	3.9	2.5	2.7	3.0	3.3	3.5
7	3.4	3.8	4.2	4.6	4.9	3.0	3.3	3.7	4.1	4.4	2.6	2.9	3.3	3.6	3.9
8	3.6	4.1	4.6	5.1	5.5	3.1	3.5	4.0	4.5	4.9	2.7	3.1	3.5	3.9	4.4
9	3.7	4.4	5.0	5.6	6.0	3.2	3.7	4.3	4.9	5.4	2.8	3.2	3.7	4.2	4.8
10	3.9	4.6	5.4	6.1	6.6	3.3	3.9	4.6	5.3	5.9	2.9	3.3	3.9	4.5	5.2

Notes: P(D) = frequency of disease in the total population; P(E) = prevalence of exposure in the total population; OR = odds ratio.

problem in 'cumulative incidence'-type studies in which controls are no longer at risk of becoming cases in the future, either because they have passed key milestones (e.g. birth or a specific age) or because they are protected from risk (e.g. by vaccination). Hogue *et al.* (1983) have shown that even with frequent outcomes the relative risk can be estimated from the odds ratio obtained from case-control studies if auxiliary information is available on the overall probability of disease *P(D)* or the proportion of exposure in the population *P(E)* (see Appendix 5.2).

Table 5.6 shows the adjusted *OR* in relation to the prevalence of exposure *P(E)* or probability of disease *P(D)*. The *OR* overestimates *R* with $OR > 1$, and the bias is most marked when *P(D)* is high, *P(E)* is low, and the level of risk is high. Similarly, the *OR* underestimates *R* when $\text{OR} < 1$ (not shown). Nevertheless, even without the rare-disease assumption, the *OR* can be adjusted using generally available ancillary data. This is a further argument for integrating case-control studies and population-based surveys, since the latter can provide the population estimates of disease or exposure frequency.

3.4.7. Attributable risk The odds ratio or relative risk are measures of association between exposure and outcome but do not reflect the public health relevance of an exposure, which is best measured by the attributable risk or excess disease attributable to exposure. In a cohort study, the attributable risk (δ) is simply the difference between disease rates in the exposed q_e and non-exposed q_o.

$$\delta = q_e - q_o. \tag{13}$$

In a case-control study attributable risk cannot be directly measured because disease rates among exposed and non-exposed are unknown. However, the proportion of disease due to exposure, known as the attributable risk per cent or aetiologic fraction and denoted by λ, can be estimated from:

$$\lambda = \frac{P(E/\bar{D})\,(OR-1)}{\{P(E/\bar{D})\,(OR-1)=1\}}, \tag{14}$$

where $P(E/\bar{D})$ is the proportion exposed among the non-diseased subjects. $P(E/\bar{D})$ can be estimated from the controls if these are representative of the general population or from survey and other ancillary data. Again, this provides a rationale for integrating case-control and survey studies. Furthermore, in an institution-based case-control study, if controls are selected from patients with diseases other than the condition under study, one cannot estimate $P(E/\bar{D})$ directly unless there are ancillary population data. This provides a rationale for linking institutional studies with population surveys.

If data on disease incidence or mortality are available, as would be the situation with a survey and linked case-control study, then the disease rates among the exposed, $P(D/E)$, and the attributable risk can be estimated as follows:

$$P(D/E)=\frac{P(E/D)\,P(D)}{P(E/D)\,P(D)+P(E/\bar{D})\,P(\bar{D})}, \tag{15}$$

where $P(E/D)$ and $P(E/\bar{D})$ are the proportions of cases and controls exposed, respectively, in the case-control study, $P(D)$ is the population-based disease or death rate, and $P(\bar{D})=1-P(D)$.

The attributable risk, δ is:

$$\delta=(OR-1)\times\frac{P(D)}{OR\{P(D/E)\}+\{1-P(D/E)\}} \tag{16}$$

(see Schlesselman, 1982: 40–5 for further details).

4. Summary and Conclusions

This brief review has presented arguments for the integration of epidemiologic case-control studies into demographic research on mortality. The main advantage of case-control studies is that the smaller sample sizes required allow collection of more detailed and complete information on the determinants of mortality, and powerful analytic methods are available for complex analyses. In addition, there are theoretical and practical advantages to integrating population-based surveys or vital-registration studies with the case-control approach. For example, the demographic data can be enriched by the detail of the case-control method, whereas selection bias in case-control studies can be avoided, and the population-based data allow estimation of rates or attributable risk required to evaluate the public health relevance of findings. Finally, although case-control studies are vulnerable to bias resulting from misclassification, inaccuracy of recall information, or confounding, these problems are common to most observational studies, particularly those dependent on retrospective data.

In summary, the case-control approach offers some unique advantages, and suffers disadvantages analogous to other non-experimental research methods. Although scepticism of the case-control method persists, the widespread use of this approach in epidemiologic research and the confirmation of case-control findings by subsequent cohort or experimental studies supports the validity of this method (Schlesselman 1982).

References

Black, R.E. (1984), 'Diarrhoeal Disease and Child Morbidity and Mortality', in Mosley and Chen (1984).

Breslow, N., and Day, N.E. (1980), *Statistical Methods in Cancer Research,* i: *The Analysis of Case-Control Studies*, IARC Scientific Publications, no. 32 (Lyons: International Agency for Cancer Research).

Briscoe, J., *et al.* (1985), *Measuring the Impact of Water Supply and Sanitation Facilities on Diarrhoea Morbidity: Prospects for Case-Control Methods*, WHO/CWS/85.3 (New York: WHO).

Copeland, K.T., *et al.* (1977), 'Bias due to Misclassification in the Estimation of Relative Risk', *American Journal of Epidemiology*, 105(5), 488–95.

Demographic and Health Surveys (1986), *Scientific Advisory Committee Meeting*, San Francisco, 31 Mar.–1 Apr.

D'Souza, S. (1984), 'Small-Area Intensive Studies for Understanding Mortality and Morbidity Processes: Two Models from Bangladesh: The Matlab Project and the Companiganj Health Project', in UN (1984).

Feinleib, M., and Detels, R. (1985), 'Cohort Studies', in W. Holland, *et al.* (eds.), *Oxford Textbook of Public Health*, iii (Oxford: Oxford University Press).

Gray, R.H. (1984), 'Case-Control Studies of High Risk Pregnancy', in *WHO Workbook on how to Plan and Carry Out Research on the Risk Approach in Maternal and Child Health including Family Planning*, WHO/FHE/MCH/IRA.84.1 (Geneva: WHO).

Hogue, J.R., *et al.* (1983), 'Estimators of Relative Risk for Case-Control Studies', *American Journal of Epidemiology*, 118(3), 396–407.

Kleinbaum, D.G., *et al.* (1981), 'Selection Bias in Epidemiologic Studies', *American Journal of Epidemiology*, 113(4), 452–63.

Last, J.M. (1983), *A Dictionary of Epidemiology* (New York: Oxford University Press).

Lilienfeld, A.M. (1976), *Foundations of Epidemiology* (New York: Oxford University Press).

McMahon, B., and Pugh, T.F. (1970), *Epidemiology, Principles and Methods* (Boston: Little, Brown and Company).

Mosley, W.H., and Chen, L.C. (eds.) (1984), *Child Survival: Strategies for Research*, Supplement to *Population and Development Review*, 10 (New York: The Population Council).

Orenstein, W.A., *et al.* (1984), *Field Evaluation of Vaccine Efficacy*, EPI/GEN/84/10.Rev.1 (New York: WHO).

Sacket, D.L. (1979), 'Bias in Analytic Research', *Journal of Chronic Diseases*, 32, 51–63.

Sartwell, P.E. (1979), Comment on 'Methodologic Problems and Standards in Case-Control Research', *Journal of Chronic Diseases,* 32, 42–4.

Schlesselman, J.J. (1982), *Case-Control Studies: Design, Conduct, Analysis* (New York: Oxford University Press).

Tecke, B., and Shorter, F. (1984), 'Determinants of Child Mortality: A Study of Squatter Settlements in Jordan', in Mosley and Chen (1984).

WHO (World Health Organization) (1977), *WHO Manual of Mortality Analysis* (Geneva: WHO).

UN (United Nations) (1983), *Manual* x: *Indirect Techniques for Demographic Estimation*, ST/ESA/SER.A/81 (New York: UN).

—— (1984), *Data Bases for Mortality Measurement*, ST/ESA/SER.A/84 (New York: UN).

Appendix 5.1

The approximate test of significance of the null hypothesis H_o, $OR = 1$ is:

$$\chi^2 = \frac{(|ad - bc| - \frac{1}{2}\, n)^2 n}{n_1 n_2 m_1 m_2}, \tag{A1}$$

and approximate 95% confidence intervals of the OR for large samples are:

$$\text{var}(1n\; OR) \cong (1/a + 1/b + 1/c + 1/d). \tag{A2}$$

Approximate upper and lower confidence limits (OR_u and OR_L) are:

$$OR_u \cong OR\ \exp[+Z\alpha\surd\{\text{var}(1n\; OR)\}], \tag{A3}$$
$$OR_L \cong OR\ \exp[-Z\alpha\surd\{\text{var}(1n\; OR)\}].$$

Appendix 5.2

The relationship between OR and R can be expressed in a series of equations as follows:

$$R = OR \cdot \frac{P(E/\overline{D})}{P(\overline{E}/\overline{D})} \quad \frac{P(\overline{E}/D)\, P(D) + P(\overline{E}/\overline{D})\, P(\overline{D})}{P(E/D)\, P(D) + P(E/\overline{D})\, P(\overline{D})}, \tag{A4}$$

and

$$R = OR \cdot \frac{P(E/\overline{D})}{P(E/\overline{D})} \quad \frac{1 - P(E)}{P(E)}, \tag{A5}$$

where P denotes the probabilities of disease D or non-disease $\overline{D}$, and exposure E or non-exposure $\overline{E}$. Thus, if the overall probability of disease in a population $P(D)$, or the probabilities of exposure $P(E)$ are known, and the proportion exposed or non-exposed among diseased and non-diseased individuals can be estimated from the case or control exposure data, the OR can be adjusted. $P(E)$ and $P(D)$ can be estimated from population surveys (Hogue *et al.* 1983).

6 Anthropological Approaches for the Study of Mortality and Morbidity

CHUKWUDUM UCHE

Department of Sociology and Anthropology, University of Benin, Benin City, Nigeria

This chapter reviews recent work by medical anthropologists and medical sociologists in the area of mortality and morbidity. The chapter is not exhaustive but attempts to maximize coverage of studies by non-demographers with the aim of highlighting significant methodologies and findings that may be of interest to demographers. If social scientists expect to have some influence in policy-making the analysis of mortality and morbidity in developing countries should utilize multi-disciplinary approaches, something easy to recommend but difficult to execute, and attempt to integrate perspectives, methodologies, and findings. The nature of the contexts in Third World countries with respect to nosologies, taxonomies of causative agents of illness and/or death, and health-related behaviours, including the quest for therapy in an environment of plural medical systems existing symbiotically, demands a broad-minded approach. Demographers should be aware of the contributions of other disciplines if we are to work together with other scholars at international, national, or local levels to reduce human suffering and increase human welfare. The goal of this chapter is to assist in the effort to increase our awareness of the achievements of our fellow social scientists.

This chapter is divided into seven sections. Section 1 is a critical review of anthropological methodology in this field. Section 2 discusses the classification of diseases, while Section 3 is on the types of therapy available. Section 4 deals with the determinants of so-called 'help-seeking' behaviour. Section 5 considers the environment of high morbidity and mortality, and Section 6 is concerned with multi-disciplinary aspects of morbidity and mortality. Section 7 concludes the chapter.

1. Methodology

In this section we shall concentrate on the methodologies used by medical anthropologists and sociologists in morbidity and mortality studies in Africa. A distinctive nature of these studies is the rejection of large-scale surveys that have become rather common for the study of fertility and family planning studies in the continent. When surveys are attempted at all, they are small-scale and are usually combined with participant observations and information resulting from intensive field-work. Examples of this approach include studies in the Ivory Coast, Ghana, Swaziland, and Liberia. Lasker's (1981) work in the Ivory Coast combined documentary and archival material, detailed interviews of health officials, population surveys, and field notes. Fosu's (1981) study of Ghana and Green and Makhubu's (1984) study of Swaziland used participant observation, informal interviews, random sampling, and structured questionnaires. In her Liberian study, Jackson (1985) ranged over a broad field: accidental sampling of clinic patients, interviews, participant and non-participant observations, ethnographic data, clinic examinations, and blood samples. However, some researchers rule out questionnaires completely. An example is a study in Mali by Hielscher and Sommerfeld (1985). In their study of concepts of illness and utilization of health facilities, the authors adopted informal interviews of individuals and groups in combination with participant observation.

The above paragraph shows the rather eclectic approach of medical anthropologists and sociologists in the matter of methods of investigation. We now inquire into how they select their subjects for study, as this has implications for the generalizability of their findings. Medical anthropologists and, to some extent, medical sociologists do not draw their subjects with representativeness in mind. Availability samples based on clinic attendance represent a major source of data (Swartz 1979; Jegede 1981; Jackson 1985). The convenience of the researcher also influences the choice of respondents, as, for example, in the study of social and environmental factors associated with diarrhoea in Bakau, Gambia (Pickering 1985) or the study of disease classification in rural Ghana (Fosu 1981). Fosu describes his choice of Berekuso village in Ghana as 'largely personal and convenient' (Fosu 1981). The atypical distribution or level of the phenomenon to be explained at times dictates the choice. This factor guided the choices of the two Nigerian villages in Uche's (1985) study of the environment of infant and child mortality.

Scholars working in this field can be criticized for producing and not adequately dealing with selection biases. However, it should be said on their behalf that they either deal with atypical cases, which rarely, if ever, are captured by simple random sampling, or they investigate in greater detail the occurrence of extreme cases revealed by a general demographic survey (Uche 1985). The relevance of the anthropological method to mortality study lies partly in the ability of this approach to unearth the operation of factors that have implications for the reduction of morbidity or mortality. These factors might be missed by a general demographic survey. Pickering's (1985) study of diarrhoea demonstrated that no single factor, for example, piped water, is sufficient to guarantee good health. She continued: 'It is the general raising of living standards, education and access to health care which taken together are effective in improving health' (Pickering 1985). Uche (1985) reached a similar conclusion for Nigerian villagers. It has also been shown that rejection of the free services of a health care scheme by a community was due to the conflict of ideas about birth and death existing between the community and the modern health scheme (Finerman 1984).

The importance of the anthropological method for morbidity and mortality studies is further enhanced by its ability to collect detailed and accurate data on aspects of morbidity or mortality. Caldwell *et al.* (1983) showed in their study of rural areas of South India that the survey method did not yield usable data on morbidity or timing of death or the community's conceptual framework concerning disease and its treatment. In their study, therefore, the authors combined the micro-approach of anthropology and surveys. Much of the strength of the anthropological method is due to the observational possibilities that prolonged residence in a community affords to a research team. If they become known and accepted by the local people, the researchers will be able to observe undisturbed all relevant community and individual activities and to interview people informally over many sessions in an endeavour to unravel intricate aspects of social organization.

In conclusion, a combination of the micro-approach of anthropology with the small-scale demographic survey seems suitable and appropriate for morbidity and mortality studies. The approach is imperative if we intend to go beyond the measurement of mortality levels and capture aspects of a community's conception of disease and the patterns of utilization of available plural health facilities and morbidity.

2. Classification of Diseases

Sub-Saharan Africans classify diseases according to the presumed causal agent. Thus diseases are classified as natural, supernatural, and mixed. The natural diseases are those that are attributable to germs, viruses, or human agents, while the supernatural ones originate at the level of gods, spirits, etc. Uche (1985) asked 929 ever-pregnant women to identify the causes of infant and childhood illness from the following list: germs, poisons from enemies, actions of witches, dead relatives, and spirits, curses from one's elders, eating tabooed food, and other causes. Respondents could pick as many of the items as they chose. Two major groups emerged. The 'germ theorists', that is respondents who picked only germs or viruses in the air, water, or food, numbered 150 or 16% of the 929 mothers. The other group numbered 727 or 78% of the sample. The 'core traditionalists', who subscribe exclusively to traditional explanations, and those who advocate neither traditional nor Western agents, numbered 52 or 6% of the respondents (Uche 1985). In the study of the Ghanaian village, Fosu (1981) observed the same taxonomy. To these villagers malaria–fever, dysentery, piles, tapeworm, hookworm, Guinea worm, nausea, toothache, diabetes, chickenpox, smallpox, eye diseases, and rashes all have a natural origin. On the other hand, measles, diarrhoea, headache, stomach-ache, and waist pains have a natural and supernatural basis. Finally, insanity, convulsions, epilepsy, pneumonia, asthma, leprosy, and paralysis have supernatural origins. A study of malaria in Liberian children and mothers showed that indigenous perceptions and symptoms of malaria correlated very significantly in both children and mothers with clinical and parasitological measures (Jackson 1985). In areas of endemic diseases, for example, folk symptomatology is as good as clinical evidence. In Uche's (1985) study, high- and low-grade fevers were the model symptoms observed by mothers during the latest illness of a surviving child. In general, the mothers' descriptions of the symptoms confirm the role of infections and of respiratory and digestive diseases in mortality and morbidity and tally with Nigerian hospital data (Hunpono-Wusu 1976).

In Latin America the trichotomy of different causal agents exists in conjunction with hot–cold classification. In the Brazilian study by Scheper-Hughes (1984) three causes of childhood (0–4 years) deaths were noted: natural, supernatural, and personal, with loci of responsibility as nature, God/saints, and human beings respectively. Gastro-enteritis, infectious and communicable diseases, and teething problems have their origins in germs and microbes. Infant deaths are often attributed to the action of God or saints and such deaths are seen as divine punishment directed against the parent(s). The last category refers to deaths traceable to poor child care, fear, shock, envy, malnutrition, dehydration, and selective neglect of frail children by their mothers. Foster (cited in Colson and De Armellada 1983) dichotomizes aetiologies into personalistic and naturalistic systems. However, his personalistic category includes human and non-human (ghost, ancestor, or evil spirit) or supernatural (deity or other powerful beings). Scheper-Hughes's (1984) classification is an elaboration of the much earlier taxonomy developed by Foster.

Aetiology in Latin America is discussed also within a humoral tradition of hot–cold classification. Diseases are classified as either hot or cold and so are remedies and foods. Hot–cold classifications have no relationship to thermal measurements, and generally agreed principles for categorization do not exist. Health is an equilibrium between hot and cold elements whose proportions 'are governed by intrinsic (inherited) body state, age, external environmental factors, activities, and the hot–cold qualities of ingestibles' (Messer 1981).

In Messer's Mexican study, we find that spearmint tea, oregano, and chillies are classified generally as 'warm', 'hot', or 'very hot'. On the other hand, citrus is categorized as 'very cold', 'cold', or 'cool'. In the middle of the continuum are unprocessed wheat, rice, and maize, which are 'temperate'. Green vegetables are 'cool' and beer is 'cold'. Biles and bloody dysentery are 'hot'. So are witchcraft and anger. Cold illnesses include green dysentery and over-indulgence. Children are cool and adults are warm, temperate, or cold, while old people are hot or cold depending on their past and present occupations (Messer 1981). In this system, illness is due to imbalance in the humours (blood, phlegm, yellow bile, and black bile) and treatment brings the system back to equilibrium. A 'hot' body is treated by administration of cool food and medicines. In short, health is restored by giving food and herbs of an opposite quality.

In Asia there is also a hot–cold classification. The elements (earth, water, wind, and fire and ether) that make up the body have to be kept in balance for good health. As before, health is restored by the appropriate quality of food and medicine. 'To accomplish a cure, a healer must select medicines appropriate to the patient's age, sex and activities and to the time of the year and climate' (Weisberg 1984). In Bangladesh, as in most Third World countries, each illness has its particular cause. Malnutrition, tetanus, abortion, eclampsia, and hysteria are attributed to bad air or an evil spirit, while diarrhoea and dysentery are caused by both bad air and poor diet. Malaria, dysentery, and tuberculosis are ascribed to germs by the educated (Ashraf *et al.* 1982).

The above review demonstrates that most Third World citizens do not adhere to the germ theory of diseases. For these people, diseases have natural, supernatural, and personal origins and therapy is sought with these dimensions in mind. Each disease has a source and the latter determines where to go for a cure. We turn to the quest for therapy in the next section.

3. Types of Therapy

This section will not ordinarily be of interest to professional demographers. However, the classification of causes of illness presented above demands some discussion of the various therapies available to people, since the efficacy of these remedies has an impact on the levels and trends of morbidity and mortality. Also, at the policy level, an appreciation of people's conception of origins of illnesses and their utilization of allopathic and traditional medicines could lead to better programming for better health for all. We now turn to an examination of the different therapies available to people in sub-Saharan Africa.

Given that illness is attributed to various agents, it is not surprising that a plural medical system exists. In this set-up a modern or allopathic system exists alongside the traditional system. In African societies every illness has a causal agent, natural, supernatural, personalistic, or a combination, and consequently, according to an Ivorian, every illness has a cure if you can find the right healer (Lasker 1981). We now turn to a consideration of medical practitioners, allopathic and non-allopathic, to whom Africans turn for succour during illness.

3.1. Herbalists

These are specialists who have knowledge of herbs with curative powers. Some devote their practice to a few diseases while others are general practitioners. Also, some travel around while most reside in their villages permanently. Finally, most herbalists are part-time practitioners while continuing their agricultural pursuits.

3.2. Diviners

Diviners are able to foretell the future, interpret the past, and prevent certain events from happening. They identify for their clients the cause of illnesses, misfortunes, or other problems, whether at home or at work, in the villages or in the urban areas where they may be residing or working. Having offered the diagnosis, they may also offer treatment by placating the gods or overcoming the spell of the witch or sorcerer.

3.3. Spiritualists

Following independence from colonial rule, many rural and urban areas of Africa have experienced the growth of syncretic sects and independent churches. Their sermons and other religious activities deal with issues of immediate relevance to their members that the more established churches do not address. Their songs and dances, revival meetings and retreats, well-publicized healing or miracle sessions, and the testimonies of people who were cured of serious or chronic ailments attract worshippers to them. In this period of rapid social change and economic depression, these religious bodies with their prophets and holy water provide succour and comradeship for citizens who are increasingly marginalized by forces of socio-economic change.

3.4. Islamic Medicine

Moslem scholars (marabouts) sell amulets or charms to protect one against dangers and ward off evil forces. They use spiritual powers to heal many complaints, whether their origins are presumed natural, supernatural, or personalistic. They are usually local people but their radius of operation may be wide. Murphy and Baba (1981) report about family members who trekked 15 kilometres through bush paths to be treated by a well-known marabout.

3.5. Bone-setters and Traditional Midwives

These are traditional medical practitioners whose specialized knowledge is orthopaedics or gynaecology. The first group concentrates on bone-setting and in this area they appear to have achieved a very high level of skill recognized even by modern medical doctors. It is not unusual for accident victims who face amputation prospects at government or other hospitals to abandon them for clinics run by bone-setters. Bone-setters have a very successful practice because of their scarcity value and their clientele cuts across socio-economic groups.

Traditional midwives, otherwise called traditional birth attendants, constitute another skilled group of traditional medical practitioners. They serve women of child-bearing age and children, and their work is doubly important since their clients cannot afford the services of modern hospitals. Unlike bone-setters, traditional midwives are being gradually integrated into the health care delivery system of governments.

3.6. Allopathic Medical Services

Western medical services are provided in government hospitals, clinics, and dispensaries. These are usually supplemented by facilities owned by religious organizations, communities, and private individuals. These facilities for scientific medicine are scarce all over sub-Saharan Africa and this can be illustrated by doctor–patient statistics. The doctor–patient ratio in Nigeria in 1979 was 1:12 550. This figure looks encouraging until we consider the distributions of doctors and population and then discover that while the doctor–patient ratio for urban areas was 1:2 789, for rural areas it was 1:100 461 in 1979 (Uche 1985). The doctor–patient ratio is only one aspect of accessibility. Others are distance, time spent, and costs of services including transportation and communication between providers and consumers of service (Lasker 1981). In a recent study of rural areas of northern Nigeria, it was found that utilization falls off exponentially with distance though this is moderated by the nature of the illness (life threatening or not), quality of service, sex, and age (Stock 1983).

3.7. Self-Medication, Drug Hawkers, and Patent-Medicine Stores

Self-medication is an indigenous practice in Africa. Adults usually know a few herbs which can be chewed, ground, and mixed with water, and drunk or spread over the ailing part of the body. Some traditional herbs and other pharmacopoeia are easily purchased at most markets for use at home.

The introduction of Western medicines had led to the emergence of patent-medicine stores and drug hawkers. These supply drugs for self-medication and even engage in diagnosis and treatment of ailments. In fact, there are far more patent-medicine stores than licensed pharmacies in Nigeria, and this appears to be true of most countries of sub-Saharan Africa.

A few observations are in order. Given the various sources of therapies available, it should be noted that people use more than one therapy for treating their condition. These sources are used either simultaneously or sequentially. However, the trend is towards scientific medicine. In our own study, mothers were asked about the type of advice they gave or received and the action taken in dealing with children's illnesses. Of the 929 women, close to a third report having received or given advice and having treated their children by relying exclusively on modern medicine and better feeding. This is despite the fact that only 16% of them believe that germs exclusively cause infant and childhood illnesses (Uche 1985). These data are consistent with Foster's observation that 'wherever we have recent data, the verdict appears nearly unanimous . . . scientific medicine has become the first choice for most illness' (Foster 1984).

Finally, the traditional healers are not replacing themselves by training up new recruits. Reviewing data from many Third World countries including Nigeria, Foster (1984) concludes that 'Everywhere that we have evidence over a generation or longer, the picture is the same: fewer traditional curers in training and in practice today than formerly.'

Having identified the various types of therapy available, it is desirable to find out who their patrons are and what are their most important characteristics. We consider these issues in the next section.

4. Determinants of Help-Seeking Behaviour

Medical anthropologists and sociologists do not normally study the determinants of mortality and so they usually rely on syntheses of demographic research results (Mechanic 1978). It is in the area of morbidity and especially behaviours related to illness and utilization of medical-care facilities that they have accumulated research results that may be of importance to demographers.

Kroeger (1983) has presented a synthetic review of the determinants of the use of medical care. These determinants have been grouped into three: predisposing, enabling, and health services system factors. Predisposing factors include socio-demographic characteristics, attitudes, and health-related decisions. Enabling factors include accessibility of regular source of care, health insurance, and income security. Health services system factors deal with the structure of the health care system (Kroeger 1983). We will examine each of these three determinants and their constituent elements in turn.

4.1. Predisposing Factors

4.1.1. Age and sex The use of traditional medical care cuts across age and sex categories. In certain societies the elderly and children utilize traditional healers more than other age groups.

4.1.2. Interaction with the social network Few quantitative data exist on the role of social networks during periods of illness. Kroeger (1983) reports that research by Abasiekong in rural south-eastern Nigeria showed that more than four-fifths of the heads of households affirmed that the family decided on the hospitalization of the sick and more than three-fifths of them wished to continue this system. Uche (1985) showed that parents, grand-parents, and siblings constituted the first source of advice to the mothers studied when their children fell ill. This group is closely followed by friends and to a much reduced extent by colleagues and medical personnel. In the Islamic society of northern Nigeria it is obligatory for a wife to have permission from her husband to attend hospitals (Murphy and Baba 1981).

4.1.3. Education Education has a mixed effect on utilization of modern medical services. In Taiwan, Korea, Thailand, Tunisia, Iran, and rural Mexico formal education was negatively correlated with use of traditional medicine. A study by Simpson carried out in Ibadan, the second-largest urban agglomeration in Nigeria, showed that there was no relationship between education and use of dispensaries and hospitals (Simpson 1980). In the Ivory Coast, 'education seems to have the effect of increasing the choices: the reliance on both kinds of medicine increased with education' (Lasker 1981).

4.1.4. Types of disease In Africa, the alleged causal agent determines the source of remedy consulted. Research has documented this in rural Ghana, Nigeria, Kenya, Chad, and northern Cameroon. In Ghana infections and digestive diseases receive modern remedies while muscle and skeletal ailments are treated by traditional methods. In rural Nigeria infectious diseases receive modern treatments while psychological conditions, for example sleeplessness and worry, receive traditional remedies (Kroeger 1983).

4.2. Enabling Factors

Distance, doctor–patient communication, quality of care, and costs are major factors affecting the quest for a particular type of therapy.

4.2.1. Distance In a study of rural accessibility to government-owned general hospitals in Bendel state, south-western Nigeria, Okafor (1984) found that households in the most deprived areas were on average 25 kilometres from the nearest hospital. Admittedly, this study did not consider the location of other modern care facilities, for example dispensaries, clinics, or hospitals owned by private or religious organizations. But, it is only in general hospitals that medical care was free in consonance with official policy. Also, in a study in Hadejia, Kano state, northern Nigeria, Stock (1983) found that utilization rates decreased exponentially with distance. However, other factors affect the impact of distance on utilization. One such factor is the nature of the illness. When the illness is non-threatening and other types of treatment are considered as viable alternatives, rates of utilization decrease sharply with distance. Where Western medicine is seen as decisive for recovery, distance does not affect utilization. Also, the more sophisticated the facility the more irrelevant is distance in determining utilization. Finally, males tend to travel further for treatment than females, and adults more than children. It has been reported that in rural Tanzania, Ethiopia, and the Ivory Coast the increased accessibility of modern health care facilities was instrumental in the flight from traditional services (Kroeger 1983).

4.2.2. Quality of care Quality of care is difficult to measure and, where people have only difficult access to alternative modern health care, their experiences take place mostly within the traditional system. What type of comparison within the modern services can residents of Hadejia be making with a doctor–patient ratio of 1:220 000 (Stock 1983)? However, since quality of service should correlate highly with level of care provided in a facility, we should expect a university teaching hospital to be more attractive than a general hospital. And this fits with the report by Stock (1983) that 'higher-order facilities are more successful than dispensaries in attracting patients from longer distances'. A similar finding has also been reported for another area of northern Nigeria (Murphy and Baba 1981). Kroeger (1983) notes that in India, Java, and Nigeria the poor quality of rural health services discouraged people from using them.

4.2.3. Costs High costs have been held responsible for the use of traditional as opposed to modern medical care. However, this is not usually the case. In Swaziland it has been observed that traditional treatment costs for a single condition could be as high as $US120–30 while consultation fees ranged between $US10 and $US20 (Green and Makhubu 1984). The advantages of visiting traditional healers lie elsewhere: they offer flexible payment over an extended period of time, the distance travelled by patients for most common ailments is relatively short, and the relations between patient and healer take place within the context of shared norms and beliefs without requiring, as does the relation between a doctor and a patient, reliance on an alien cultural product.

4.3. Health Services System Factors

According to Kroeger (1983) the 'concurrent or serial use of different healing systems seems to be a typical feature of transitional societies of the developing world'. Simpson (1980) has shown that when seriously ill the Yorubas of Ibadan, Nigeria, predominantly use self-medication followed by modern medicine. Half of those under 30 years of age use modern medicines while a third of them turn to self-medication. Those respondents who are at least 50 years old overwhelmingly use home remedies. Self-medication is used by 72% while only 22% use modern medicines. Uche's (1985) study shows that 62% of the mothers use a combination of modern and traditional remedies when their children are ill. Only 36% rely exclusively on modern medicines and improved feeding practices. Finally, Jagede (1981) has confirmed earlier findings indicating that irrespective of educational attainment sick Nigerians tend to use traditional and modern medicines either simultaneously or serially. This pattern of use is not just due to variation in accessibility but to the fact that Africans have a multi-causal concept of illnesses that is compatible with a multi-treatment system for dealing with diseases.

5. The Environment of High Morbidity and Mortality

In a 1978–9 study Uche (1985) reported on morbidity and mortality conditions in two Nigerian villages. In Ebe, a village in Anambra state, the infant mortality rate reached 275 per 1000 and the crude death rate 41 per 1000 while in the other village, Lokpaukwu, Imo State, the figures were 381 per 1000 and 35 per 1000 respectively. Part of the research agenda in the study was to uncover the type of environment that sustained such high levels of morbidity and mortality.

The general environmental conditions of the two settlements were similar. Lokpaukwu had had a rural health clinic for three years while the general hospital was located 20 kilometres away at Okigwe, the local government headquarters. Ebe's rural health clinic was about ready at the time of the study while general and specialized modern medical facilities were available 24 kilometres away at Enugu, Anambra state's capital. Drinking-water for Lokpaukwu was from a highly polluted stream while Ebe had had pipe-borne water since 1966. However, in both villages, rain-water was available for half the year and was used even at Ebe when the pipes were dry, a not infrequent situation. Modern toilet facilities were absent. The bush or river was used for the disposal of human waste in 79% of the households while pit latrines were used by 21% of the households. Housing conditions in the two villages presented a brighter picture than those suggested by medical care, drinking-water, and waste disposal. Rural households were not crowded. The mean number of rooms per rural household was 4.3 (and the median was 4) and these were used on the average by about six people. The main building in each household was for the head of household and its material nature as well as the possessions contained in it were a rough indicator of the economic status of the household. Of the main buildings 69% were roofed with corrugated-iron sheets while 30% had roofs of thatch or raffia. The remaining 1% of the main buildings had asbestos tiles for roofing materials. Thatched roofs helped to moderate the temperature of the houses during the dry and hot weather lasting from October to March. However, during the rainy season between April and September, thatched roofs absorb and retain water and enhance the dampness of buildings and by so doing promote cold-related illnesses. The vast majority of the main buildings roofed with iron sheets or asbestos would normally be sealed with asbestos boards. The public health hazard occasioned by use of asbestos materials had not yet become a matter of public concern or policy. Also, 61% of the main buildings had mud walls while 39% had walls made with cement blocks or baked bricks. It should also be noted that mud walls, like thatched roofs, promote dampness of dwellings and contribute to cold-related illnesses during the rainy season. Finally, the main buildings were fairly well ventilated with each having on average six windows of varying sizes. However, 17% of the main buildings had no windows at all.

In comparative perspective, the data for the two villages showed them to be at a disadvantage *vis-à-vis* the national average. While 46% of Nigeria's rural areas in 1980 had pit or pail toilets, only 21% of the households in the study village had similar toilets. Also, the households of the two villages were denser with six persons per household compared with a national average of five persons in 1980 (Uche 1985). This suggests that the physical environment of the study villages was worse than the national mean and therefore may lead to the higher levels of morbidity and mortality observed before.

6. Multi-disciplinary Orientations

There is enough room for collaborative work by anthropologists and demographers in the area of morbidity and mortality. The anthropological method, especially ethnography, is very apt in providing the 'contextualization' required in explaining morbidity and mortality, and in examining 'macro-level and micro-level interactions' (De Walt and Van Willigen 1984). That a movement in this direction is already taking place is shown by the work of Caldwell *et al.* (1983) which combined sample surveys and case-studies involving residence in the study area. This work, as noted earlier, demonstrates the strength of the ethnographic method in collecting morbidity and mortality data.

The International Review Group (1980) identified various priority areas for research in the developing countries. Ford and Arcury (1984) make a case for the involvement of anthropologists. In the area of mortality they identify the social contexts of mortality as largely unresearched. One such research could be the investigation of why education is so powerful in the reduction of household mortality. Is it because the educated possess special knowledge, say, nutrition information, that others do not have? Or is it because education alters the balance of power in the family to the advantage of mothers, who then share in decision-making in health-related matters and may even have the upper hand (Caldwell 1979; Caldwell and McDonald

1981)? It is not just food and health care that are shared in a family. Even in a relatively well-to-do household, resources (sleeping rooms, drinking-water, use of the only flush toilet, use of the car, etc.) may be unevenly shared not only between individuals of both sexes but also between different generations. This unequal distribution certainly has an impact on morbidity and mortality and helps to explain differentials observed among households. Yet, only a careful case-study approach can provide information with levels of detail that are sufficient to test these hypotheses. Thus, it is not surprising that the claim about the usefulness of anthropological method in mortality research is being received with less scepticism than before.

7. Conclusion

This chapter has examined the work of medical anthropologists and medical sociologists in the area of mortality, morbidity, and health contexts. The methods medical sociologists use in their work on mortality and morbidity in advanced countries are well known and their strengths and weaknesses are generally discussed (Mechanic 1978). Some of the most important among these methods are the clinic approach, double-blind techniques, controlled trials, population censuses, vital registration, and household interviews. Their shortcomings are exacerbated when one tries to use them in developing countries. Intimate familiarity with these techniques among demographers leads to overlooking the methods used by medical anthropologists and sociologists in developing countries; these are based on ethnographic methods, participant observation, extensive note-taking, and unstructured interviews. The most important advantage of the anthropological method is the ability to shed light on the social context within which the individual functions. In such a milieu, life events such as birth, illness, and death have a meaning and are rationalized and dealt with. These aspects of social life and behaviour are lost in large-scale demographic surveys. The use of small sample sizes in anthropological studies increases the probability of greater control by the researcher during field-work and so enhances the chances of obtaining high-quality data. The weakness of the anthropological method lies in the tendency of its practitioners to study extreme, atypical cases.

The health context of developing countries is one of a plural medical system in line with their multi-causal concept of diseases. Use of the various therapies, whether allopathic or traditional, is pragmatic and syncretic and mediated by accessibility and socio-cultural factors. However, the trend towards modern medicine appears to be unmistakable.

Demographers and medical anthropologists have a lot to learn from one another. Mutual benefits are likely to be maximized if each specialist acquires some of the basic skills of the other.

References

Ashraf, A., Chowdhury, S., and Streefland, P. (1982), 'Health, Disease and Health-Care in Rural Bangladesh', *Social Science and Medicine*, 16, 2041–54.

Caldwell, J.C. (1979), 'Education as a Factor in Mortality Decline: An Examination of Nigerian Data', *Population Studies*, 33(3), 395–413.

—— and McDonald, P. (1981), 'Influence of Maternal Education on Infant and Child Mortality: Levels and Causes in International Union for the Scientific Study of Population', *International Population Conference*, Manila, 1981 (Liège, Belgium: IUSSP), ii. 79–95.

—— Reddy, P.H., and Caldwell, P. (1983), 'The Social Component of Mortality Decline: An Investigation in South India Employing Alternative Methodologies', *Population Studies*, 37(2), 185–205.

Colson, A.B., and De Armellada, C. (1983), 'An Amerindian Derivation for Latin American Creole Illnesses and their Treatment', *Social Science and Medicine*, 17(17), 1229–48.

De Walt, K.M., and Van Willigen, J. (1984), 'Research Priorities for Medical Anthropologists in the 1980s', *Social Science and Medicine*, 18(10), 845–6.

Finerman, R.D. (1984), 'A Matter of Life and Death: Health Care Change in an Andean Community', *Social Science and Medicine*, 18(4), 329–34.

Ford, T.R., and Arcury, T.A. (1984), 'Population and Health in the Developing World: Research Perspectives for Medical Anthropologists', *Social Science and Medicine*, 18(10), 855–9.

Foster, G.M. (1984), 'Anthropological Research Perspectives on Health Problems in Developing Countries', *Social Science and Medicine*, 18(10), 847–54.

Fosu, G.B. (1981), 'Disease Classification in Rural Ghana: Framework and Implications for Health Behaviour', *Social Science and Medicine*, 15B, 417–82.

Green, E.C., and Makhubu, L. (1984), 'Traditional Healers in Swaziland: Toward Improved Co-operation between the Traditional and Modern Health Sectors', *Social Science and Medicine*, 18(12), 1071–9.

Hielscher, S., and Sommerfeld, J. (1985), 'Concepts of Illness and the Utilization of Health Care Services in a Rural Malian Village', *Social Science and Medicine*, 21(4), 469–81.

Hunpono-Wusu, O.O. (1976), 'Disorders which Shorten Life among Nigerians: A Study of Mortality Patterns in the Age Group 15–44 in Kaduna, Nigeria', *Tropical Geographical Medicine*, 28, 343–8.

International Review Group (1980), *Population Policy: Research Priorities in the Developing World* (London: Frances Pinter).

Jackson, L.C. (1985), 'Malaria in Liberian Children and Mothers: Biocultural Perceptions of Illness versus Clinical Evidence of Disease', *Social Science and Medicine*, 20(12), 1281–7.

Jegede, R.O. (1981), 'A Study of the Role of Socio-cultural Factors in the Treatment of Mental Illness in Nigeria', *Social Science and Medicine*, 15A, 49–54.

Kroeger, A. (1983), 'Anthropological and Socio-medical Health Care Research in Developing Countries', *Social Science and Medicine*, 17(3), 147–61.

Lasker, J.N. (1981), 'Choosing among Therapies: Illness Behaviour in the Ivory Coast', *Social Science and Medicine*, 15A, 157–68.

Mechanic, D. (1978), *Medical Sociology*, 2nd edn. (New York: Free Press).

Messer, E. (1981), 'Hot–Cold Classification: Theoretical and Practical Implications of a Mexican Study', *Social Science and Medicine*, 15B, 133–45.

Murphy, M., and Baba, T.M. (1981), 'Rural Dwellers and Health Care in Northern Nigeria', *Social Science and Medicine*, 15A, 265–71.

Okafor, F.C. (1984), 'Accessibility to General Hospitals in Rural Bendel State, Nigeria', *Social Science and Medicine*, 18(8), 661–6.

Pickering, H. (1985), 'Social and Environmental Factors Associated with Diarrhoea and Growth in Young Children: Child Health in Urban Africa', *Social Science and Medicine*, 21(2), 121–7.

Scheper-Hughes, N. (1984), 'Infant Mortality and Infant Care: Cultural and Economic Constraints on Nurturing in Northeast Brazil', *Social Science and Medicine*, 19, 535–46.

Simpson, G.E. (1980), *Yoruba Religion and Medicine in Ibadan* (Ibadan: Ibadan University Press).

Stock, R. (1983), 'Distance and the Utilization of Health Facilities in Rural Nigeria', *Social Science and Medicine*, 17(9), 563–70.

Swartz, M.L. (1979), 'Community and Healing among the Zaramo in Tanzania', *Social Science and Medicine*, 13B, 169–73.

Uche, C. (1985), *The Environment of Infant and Child Mortality: A Case Study of Nigerian Villages*, Population Studies Center Working Paper Series, no. WP-85-06 (Providence, RI, USA: Brown University).

Weisberg, D.H. (1984), 'The Practice of "Dr." Paep: Continuity and Change in Indigenous Healing in Northern Thailand', *Social Science and Medicine*, 18(2), 117–28.

7 Approaches to Assessment of the Cause Structure of Mortality: A Case-Study from Bangladesh

SUSAN ZIMICKI

Population Studies Center, University of Pennsylvania
Philadelphia, Pennsylvania, USA

Demographers have long been interested in establishing levels and age patterns of mortality in various countries, and where data on causal patterns are available have found them useful in explaining some of their observations. However, as Preston has pointed out, this type of causal analysis is 'biased toward states with relatively stable and well-developed government machinery, toward populations in which the exceptional peaks in mortality associated with epidemics are largely absent, and towards European experience' (Preston 1976).

To a certain extent one indicator of the health of a population is the existence of records to document that level of health. In most developing countries good vital registration systems are still lacking. Health information systems are rare and, even if they exist, may not serve the purpose of identifying all the important causes of mortality because they are based mainly on data from hospital and dispensary records, which are subject to selection bias (Leeuwenberg *et al.* 1984; Northrup 1986). What is needed is a reasonably accurate and potentially cross-nationally comparable source of data about the importance of various diseases as causes of death in the community.

'Lay reporting' (WHO 1978) has been proposed as one way to obtain this information. The WHO guidelines for lay reporting suggest classification of deaths into single-cause categories using information obtained in an unstructured interview in conjunction with a list of symptom associations for each cause of interest. A list of symptom associations is provided for some causes (e.g. for measles, 'fever with red rash disappearing after several days'; for meningitis, 'neck rigidity, vomiting, skin rash'), but researchers are encouraged to develop their own lists. The system of lay reporting used in the Matlab study area since 1975 is essentially the same as that proposed by WHO. Section 1 of this chapter describes this simple system, reviews some insights about recent changes in child mortality which have been obtained using it, and discusses its major weaknesses—biases arising from interviewer and classifier inconsistency and confusion of locally defined illness with medically defined disease, and lack of provision for multi-causality. Section 2 presents an experiment with a more structured lay-reporting approach, which was developed to overcome some of the weaknesses of the simple system. This new approach redirects attention from the illness(es) preceding death to the descriptive elements (symptoms and signs) used to make classifications of both illness and disease. The unstructured interview is replaced by a structured symptom-list questionnaire; and classifications are made on the basis of case definitions using decision rules. The results obtained for neonatal deaths, an age group for which the results obtained from the simple system are not very enlightening, indicate the usefulness of this new approach.

1. Simple 'Lay Reporting' in Matlab since 1975

The Matlab Demographic Surveillance System (DSS) provides a good example of a simple reporting system for cause of death. Under this system relatives' reports of the cause of death are collected and then classified into 27 categories.

1.1. Method

The Matlab DSS has been operating since 1966, with periodic censuses and ongoing registration of vital events (D'Souza 1984). Cause-of-death reporting has been part of the vital registration system throughout that time and since 1975 has been the responsibility of 12 field assistants and two office staff of the DSS. The field assistants interview family members about each death and complete the death report forms; the office workers use the information on the forms to classify the death.

The death report form is very simple. It consists of a section of demographic information, boxes for the field-worker to tick for 12 of the 27 categories of disease, and several lines for him to write 'symptoms leading up to death' (in English). At the home the field-workers often write the family's opinion about the cause of death in Bengali on the back of the form, and translate it later. These translated descriptions are usually very brief: 'Since his birth could not suck breast milk, and his colour became discoloured. People said evil spirit' (of neonate, age 2 days); 'Died of dysentery with fever and had been suffering from dropsy for about three or four days' (5-year-old); 'Old age is the major cause of his death. Could not take food properly due to poverty' (76-year-old).

Two coders in the data management branch review the form, and using all the information provided (the Bengali descriptions, the English descriptions, and the boxes the worker has checked) decide on the final classification of cause of death. The list of possible causes consists of a number of kinds of classification: specific diseases; circumstances of death; major

Table 7.1. *Classifications used in Matlab cause-of-death reporting*

Specific diseases	Major symptom
Smallpox	Fever
Measles	Diarrhoea
Tetanus	Dysentery
Cholera	Rheumatism
	Dropsy
Organ system	Jaundice
Respiratory	
Gastro-intestinal	Circumstances of death
Heart	Childbirth
Liver	Drowning
Venereal	Accident
Ear, nose and throat	Murder
Skin	Suicide
	Old age
Other	Unknown

symptom; organ system affected; as well as 'other', used if a definite other cause can be ascribed that does not fit into one of the categories (cancer or diabetes, for example) and 'unknown' (Table 7.1). There is some inconsistency of classification, due to the lack of provision for multiple causes of death and the lack of explicit rules for coding decisions: what to code, if, for example, a child's mother reports that he had dysentery and dropsy, a rather common combination?

For both eliciting information about circumstances of death and classification of causes, the system relies greatly on the common sense of the workers. With the lack of standardization of methods of collecting and classifying information, the potential for interviewer and classifier bias is obvious (Zimicki *et al.* 1985).

However, the problem of bias of individuals is aggravated by confusion between illness and disease, that is, between the lay and biomedical definitions of ill health (Helman 1985). Any ascertainment of cause of death based on reporting (as opposed to medical tests and autopsy) involves the translation of illness reports into disease categories. When the illness categories are strongly affected by medical definitions of disease, or when physicians collect the information and make the classification decision, as is the case in most countries with medical certification of death, this translation is not particularly problematic. Physicians have undergone a long period of training in eliciting reports and have a standardized view of causal pathways. With lay reporters, however, explanatory models derived from illness classifications can contaminate the process, particularly in countries where there are large differences in the explanatory models for illness and disease. This is complicated by the fact that there is rarely a one-to-one correspondence between illness and disease. An example of what can

Table 7.2. *Death rates attributed to tetanus, by age and year, and total neonatal death rate* (per 1000 live births)

Rate	Year								
	1975	1976	1977	1978	1979	1980	1981	1982	1983
Tetanus									
Day 0–3	7	8	8	16	13	17	14	17	14
Day 4+	20	15	18	30	23	30	24	24	18
TOTAL	27	23	26	46	36	47	38	41	32
Total neonatal	80	64	70	73	72	66	66	64	64

happen is the case of the classification 'tetanus' in Matlab.

Overall neonatal death rates declined from an average of 70 per 1000 live births in 1975–7 to an average of 64 in 1982–3, but in the same period the death rate ascribed to tetanus increased from an average of 25 per 1000 live births to an average of 37 (Table 7.2). Much of the increase was in the group of infants dying between birth and the fourth day of life—a time generally considered too early for tetanus (Newell *et al.* 1966). Also, over time a progressively larger portion of 'tetanus' deaths have occurred to males (Zimicki *et al.* 1985). In the absence of evidence of changes in birth practices, and with the rise in reported rates coinciding with the start of immunization against tetanus, it seems likely that most of this rise can be attributed to changes in reporting.

The domain of the disease (tetanus) overlaps that of the locally defined illness *alga*. *Alga* is evidenced by the infant convulsing or turning blue, in combination with an inability or refusal to feed (Rahman *et al.* 1982*a*). Deaths due to tetanus clearly belong in the category of *alga*, along with some deaths due to prematurity, congenital defects, and hyaline membrane disease, all of which can cause very early neonatal death and a predominance of male deaths. What creates the problem is the illogical inference that all or most *alga* deaths should be categorized as tetanus. With the institution of the current classification system in 1975 the interviewers were given some training about the medical definition of tetanus. With time this training has been gradually lost, and the definition of the cause-of-death category 'tetanus' has come to resemble more and more the illness *alga*. In addition, the increase in interest in tetanus on the part of researchers in 1978, when immunization of pregnant women was started, might have been translated into a higher 'index of suspicion' on the part of the interviewers.

1.2. Results: Insight into Changes in Child Mortality 1978–1983

In 1978 the ICDDR,B started a maternal and child health and family planning (MCH/FP) project in half of the Matlab project area, designated the 'treatment area' in contrast to the 'comparison area' (Bhatia *et al.* 1980). Interventions in the treatment area included twice-monthly visits of female village workers to each family and the institution of five MCH/FP clinics. Initially the female village workers provided contraceptives and treated simple complications of contraception, referring women to MCH/FP clinics for IUD insertion and treatment of any serious complications; they offered health and nutritional advice; and they distributed iron and folic-acid tablets to women during the last three months of pregnancy. With time their role expanded. Starting in January 1978 they offered tetanus immunizations to pregnant women (Rahman *et al.* 1982*a*) and began distributing oral rehydration solution (ORS) materials to home 'experts' whom they had trained in each extended household (Zimicki *et al.* 1984). Starting in March 1982 they initiated a programme of measles immunizations for all susceptible children in half of the treatment area (Rowley and Satterthwaite 1983). At the MCH/FP clinics, in addition to treatment of serious complications of contraception, women can obtain treatment for seriously ill children. It must be noted that women from the comparison area can obtain these services at the clinic in Matlab, but that accessibility of all clinics for all women is somewhat limited by the custom of purdah. All people in the study area have had access to the diarrhoea treatment centre in Matlab since it was initiated in the early 1960s; ambulance speedboats are provided for both treatment and comparison area populations. In addition, starting in 1982 DSS workers began providing packets of ORS in the comparison area.

The age-specific mortality rates for infants and children for the period 1978 to 1983 confirm the existence since 1979 of significant areal differences in mortality for neonates and 1-to 4-year-olds but no stable differences for post-neonates (Table 7.3). Neonatal mortality declined in both areas, but to a greater extent in the treatment area. Treatment and comparison area child mortality were the same in 1978, but diverged sharply in 1979, when treatment area mortality declined and comparison area mortality increased.

Table 7.3. *Trends in age-specific death rates in Matlab, 1975–1983* (per 1000)

Age group	Area	1975	1976	1977	1978	1979	1980	1981	1982	1983
Neonate	Total	79.5	64.4	70.3	72.8	71.9	65.9	66.3	63.5	64.0
	Treatment				67.2	69.4	58.7	64.0	58.1	56.4
	Comparison				77.8	73.6	71.7	67.5	68.1	70.3
Post-neonate	Total	112.2	38.5	43.3	47.0	44.4	37.9	43.6	49.0	42.0
	Treatment				45.9	44.7	32.9	38.6	47.5	41.8
	Comparison				48.0	44.1	42.1	47.8	50.2	42.2
1–4 years	Total	34.8	29.5	19.7	22.3	21.6	22.1	21.9	23.4	29.1
	Treatment				22.4	17.2	18.7	18.6	18.8	21.3
	Comparison				22.1	26.0	25.4	24.9	27.4	35.3

There was some convergence of rates during 1980–1, but in 1982 and especially in 1983 mortality increased to a far greater extent in the comparison than in the treatment area.

Using information about the cause structure of mortality we can gain better insight into the reasons for the observed changes in child mortality.[1] Even with the problems of lack of standardization and confusion between illness and disease, there are regular patterns of classification of local descriptions into the DSS categories.

For simplicity, only the major cause classifications for children are considered: dysentery, fever, respiratory disease, measles, diarrhoea, dropsy, and drowning. A review of the death report forms, particularly for 1979 and 1980, and extensive interviews with villagers and DSS staff about illness and disease provided information about correspondences between the DSS cause-of-death categories and possible medical classifications (Zimicki *et al.* 1985). Diarrhoea and drowning are the most sensitive and specific categories;[2] dysentery is fairly specific, including some cases of malabsorption as well as clinical dysentery, but less sensitive, as some deaths involving dysentery are reported as due to measles or dropsy. The measles category contains some deaths occurring after other diseases in which rash is a striking symptom, while some measles-related deaths have been classified as due to dysentery, respiratory disease, fever, or dropsy. Most of the child deaths attributed to dropsy involve oedema subsequent to malnutrition from malabsorption, anorexia (often secondary to dysentery or measles or both), or starvation. Deaths are classified in the fever or respiratory-disease categories if the appropriate symptom was prominent, but both these categories contain large portions of pneumonia deaths (Zimicki *et al.* 1985). With the exception of drowning, deaths in all the categories could be averted by health interventions; drowning is included in this analysis because it is an important cause of death for children and because it provides a measure of control of the quality of the data, since it should vary little between areas or, floods and droughts aside, with time.

To allow for any difference in age structure arising from the high levels of contraceptive use in the treatment area (Phillips *et al.* 1982) and for yearly fluctuations in the number of births, an age-standardized (standard = population of the combined area in 1978) death rate was obtained for each of the selected causes, the residual group, and for all causes combined. Examination of the rates for fever and respiratory disease revealed that there was a marked tendency for one to be high when the other was low. Because of this pattern and because a large portion of deaths reported as

[1] Unfortunately, for neonates the cause-of-death classifications are not very useful, since virtually the only causes coded are 'tetanus' and 'respiratory disease' and, as discussed above, there are serious questions about the classification 'tetanus'.

[2] A category is sensitive when all cases that truly belong in that category are classified into it (a high proportion of true positives); it is specific when all cases that truly do not belong to it are excluded (a low proportion of false positives).

'fever' are respiratory in origin, the two categories were combined for the analysis of areal differences.

1.2.1. Fluctuations in child mortality The pattern formed by the years with the highest rates for each cause suggest epidemics of diarrhoea in 1982 and 1983, dysentery in 1978 and 1983, and measles in 1979 and 1981 (Table 7.4). Dropsy was highest in both areas in 1980 and 1981. The pattern for respiratory disease is less clear, but 1982 is one of the years with the highest rates in both areas.

The diarrhoea and dysentery patterns observed in the death rates are confirmed by the attendance records of the Matlab diarrhoea treatment centre; a cholera epidemic started at the end of 1982 and continued through the spring of 1983, with a secondary peak in late 1983 (Khan *et al.* 1984); and dysentery admissions were high in 1978 (Hossain *et al.* 1981). The higher levels of dropsy in 1980 and 1981 may be connected with a shortage of food and the consequent malnutrition due to flooding in 1980. The apparent peaks in measles deaths are somewhat puzzling, since they do not coincide with peaks in measles incidence collected by a DSS survey carried out from

Table 7.4. *Age-standardized* cause-specific death rates for the age group 1–4, by year and area* (per 1000)

Cause	Year					
	1978	1979	1980	1981	1982	1983
Treatment area						
Diarrhoea	1.5	0.9	0.8	1.4	2.3	3.5
Dropsy	1.5	1.8	2.0	1.7	1.3	1.5
Drowning	2.3	2.2	2.2	3.0	1.3	3.2
Dysentery	6.1	3.6	3.5	3.7	4.5	5.8
Fever	3.0	1.7	2.8	1.7	1.7	1.8
Measles	1.8	2.3	1.6	2.2	1.0	0.6
Respiratory	1.5	2.1	1.3	1.5	3.1	1.6
All other	5.4	2.8	4.8	4.8	4.3	4.7
ALL	23.0	17.4	19.0	20.0	19.5	22.7
Comparison area						
Diarrhoea	1.5	1.3	2.1	1.6	2.9	4.2
Dropsy	1.7	2.5	2.9	2.8	1.3	2.0
Drowning	1.7	2.3	2.5	2.0	2.1	2.3
Dysentery	7.6	6.9	6.4	4.0	6.2	10.9
Fever	2.3	2.2	2.9	4.4	2.7	4.3
Measles	2.4	5.0	3.0	3.2	3.1	2.7
Respiratory	0.7	1.3	1.6	3.4	3.1	1.6
All other	5.1	4.3	4.3	4.8	6.5	8.1
ALL	23.0	25.8	25.7	26.2	27.9	36.0

* Standard = population of the combined area in 1978; standardization to allow for differences in age structure between the two areas.

1979 to 1982 (D'Souza 1984). However, measles incidence reporting is problematic; many illnesses with rashes are reported to be measles and the basic epidemiology of the disease in Matlab is still unclear: in a more careful survey carried out by female village workers in 1982, one subarea of the treatment area had over 2.5 times the reported incidence of the other three subareas (Rowley and Satterthwaite 1983). Reporting of measles deaths is likely to be more accurate than reporting of cases, since most of the other rash diseases have much lower case-fatality. The drop in measles in the treatment area in 1983 is understandable, since measles vaccination was started in half the treatment area in 1982. As expected, drowning is the most stable of all the rates, both across time and between the treatment and comparison areas.

On this evidence the rise in death rates in both areas in 1982 and 1983 can be attributed in part to the simultaneous occurrence of diarrhoea and dysentery epidemics.

1.2.2. Differences between treatment and comparison areas To examine the differences between the areas more easily, the cause-specific rates are grouped into three time periods: 1978, when mortality rates were the same in both areas; 1979–81, when rates were clearly different in the two areas but fairly stable; and 1982–3, when mortality increased in both areas (Table 7.5). Indications of the reasons for the difference in mortality between the treatment and comparison areas can be obtained by comparing the rank order of causes in each period and the direction and absolute magnitude of changes in cause-specific rates from one period to the next.

The cause-specific death rates for 1978 are quite similar in the two areas. Dysentery is the most important single cause, followed by the joint category fever–respiratory disease, then measles, dropsy, and diarrhoea. This rank order is the same for both areas. It remains essentially the same in 1979–81,

Table 7.5. *Age-standardized* cause-specific death rates in the age group 1–4, by period and area* (per 1000)

Cause	Treatment area			Comparison area		
	1978	1979–81	1982–83	1978	1979–81	1982–83
Diarrhoea	1.5	1.0	2.9	1.5	1.7	3.6
Dropsy	1.5	1.8	1.4	1.7	2.7	1.7
Drowning	2.3	2.5	2.3	1.7	2.3	2.2
Dysentery	6.1	3.6	5.2	7.6	5.8	8.6
Fever and respiratory	4.5	3.7	4.2	3.0	5.3	5.9
Measles	1.8	2.0	0.8	2.4	3.7	2.9
All other	5.4	4.1	4.5	5.1	4.5	7.3
ALL	23.1	18.8	21.1	23.0	25.9	32.0

* Standard = population of the combined area in 1978; standardization to allow for differences in age structure between the two areas.

when the only difference is that in the treatment area fever–respiratory disease and dysentery are equal in level because the level of mortality due to dysentery has declined. Between 1979–81 and 1982–3 the rank order of causes changes in both areas: dysentery remains the single most important cause, followed by fever–respiratory disease, but diarrhoea has become the third most important cause of death. In the comparison area measles is fourth, followed by dropsy, while in the treatment area measles is last.

Between 1978 and 1979–81 the rate of death due to dysentery fell in both areas, and measles and dropsy increased, but the decline is greater in the treatment area and the rise greater in the comparison area. Fever–respiratory disease and diarrhoea mortality increased in the comparison area, but decreased in the treatment area. The changes in 1982–3 death rates relative to 1979–81 levels are in the same direction in both areas: dysentery, fever–respiratory disease, and diarrhoea have increased, and measles and dropsy have decreased, with the magnitudes of the changes for dysentery, fever–resiratory disease, and measles favouring the treatment area. The increase in diarrhoea mortality is the same in both areas. The residual category in the comparison area is much higher than in 1979–81 and much higher than in the treatment area, but this does not appear to be due to any specific category (data not shown).

1.3. Discussion

The first possibility to consider is that the apparent similarity of the areas in 1978 is not representative of the true condition, but is the result of faulty data collection[3] or a temporary fluctuation, and that areal differences in mortality might have existed before the MCH/FP project. Unfortunately, because of the difficulty of matching all the event files needed to reconstruct the population in what became the comparison area, area-specific rates for the period before 1978 are unavailable. However, the areas are somewhat different socio-economically (Rahman *et al.* 1982*b*). But even if a differential existed before the start of the MCH/FP project, we can still explore the reasons for the fall in mortality in the treatment area between 1978 and 1979–81 and the small increase in mortality relative to that in the comparison area between 1979–81 and 1982–3.

Three possible explanations are (1) the risk of mortality declined in the treatment area relative to the comparison area because of areal differences in the incidence of disease starting in 1979; (2) specific medical interventions reduced case-fatality rates in the MCH/FP treatment area; and (3) an overall improvement in health, either exogenous or arising from

[3] At the end of 1978 the entire cohort of comparison area base-level DSS workers was replaced, at a time coinciding with a reduction in the number of villages covered in the registration system from 233 to 149. During the last quarter of the year the comparison area data show an extraordinarily low sex ratio at birth (Becker and Chowdhury 1981).

behavioural change stimulated by the MCH/FP project, rendered the treatment area population less susceptible.

There is no evidence of substantially different epidemic patterns in the two areas. Four of the cause categories considered—dysentery, fever-respiratory disease, measles, and diarrhoea—would be expected to reflect the presence of epidemics of specific diseases. Dysentery deaths declined in both areas between 1978 and 1979–81 and increased in both between 1979–81 and 1982–3; measles deaths increased in both between 1978 and 1979–81; and diarrhoea deaths increased in both between 1979–81 and 1982–3. The difference between the areas for diarrhoea in 1979–81 is very small and the difference for measles in 1982–3 is the effect of specific medical intervention.

The medical interventions available in the treatment area which might have affected child mortality include availability of antibiotics in the MCH/FP clinics starting in 1978, provision of ORS starting in 1979, and measles immunization of eligible children in half the treatment area starting in 1982. The effect of measles immunization can explain the much lower measles death rate in the treatment area in 1982–3. The evidence for the effect of the other interventions is less persuasive. The provision of ORS in the treatment area starting in 1979 coincides with a slight decline in diarrhoea mortality occurring at the same time as a slight increase in the comparison area. However, diarrhoea increased by the same amount in both areas with the cholera epidemic in 1982–3 even though ORS packets became available in the comparison area in 1982. This could be due to less effective distribution of the packets and less use of ORS in the comparison area, or underlying differences in the areas, or it could also indicate that the small areal difference in deaths ascribed to diarrhoea in 1979–81 was not due to ORS.[4]

The availability of antibiotics in MCH/FP clinics could be a factor in the contrasting trends for fever-respiratory disease in the two areas between 1978 and 1979–81, as well as for the greater treatment area decline in dysentery in that interval and the smaller treatment area rise in fever-respiratory disease between 1979–81 and 1982–3. However, although detailed clinic records (patients treated, condition treated, and treatment given) are kept, they have not yet been summarized, so it is unclear what proportion of the population might have been affected by this intervention.

The argument for an overall improvement in health as an explanation for the areal differences in mortality rests on consideration of the rank order of causes of death. Changes in mortality due to changes in disease incidence or case-fatality are manifested by changes in the relative ranks of causes. For example, the cholera epidemic resulted in diarrhoea moving from fifth place

[4] It is important to remember that in the context of the excellent diarrhoea treatment centres accessible to both treatment and comparison area populations it was not anticipated that ORS would have a significant mortality impact; it did reduce treatment centre attendance from the treatment area by about 30% in 1979–80 (Zimicki *et al.* 1984).

in both areas to third place in 1982–3 and the measles immunization starting in 1982 resulted in measles moving to last place in the treatment area while retaining its previous relative rank in the comparison area. In contrast, the 1978 rank order of causes, which was the same in both areas, did not change between 1978 and 1979–81 although mortality declined substantially in the treatment area. This suggests that a general, rather than specific mechanism was responsible for the mortality decline. One possibility is the nutritional and health advice given by the female village workers. This might be postulated to affect health status through behavioural changes similar to those hypothesized to account for the well-documented effect of mother's educational level (D'Souza and Bhuiya 1982).

The cause-of-death information obtained from the simple lay-reporting system can provide no definitive answer about the reason for the differences in child mortality in the treatment and comparison areas. However, this information is useful in that it rules out some explanations that might otherwise be quite appealing simply on the basis of timing. It suggests that the observed mortality difference between the areas is not the effect of ORS, and in fact not a simple effect of any single intervention, although antibiotics available in the MCH/FP clinics might be an important factor. Another possible explanation which should be investigated is that of a general health effect.

The argument for antibiotics could be sharpened if it were possible to know how much of the category fever–respiratory disease was attributable to pneumonia (which responds well to antibiotics), how much to whooping cough (which does not), and how much to other conditions. The simple lay-reporting system does not allow these finer distinctions. An even more serious problem, however, is the lack of provision in the system for multiple causes. Because it allows only a single cause to be coded for each death, the proportion of all deaths that might be affected by any particular epidemic or intervention is unknown. These weaknesses in a valuable system can be remedied by adopting a more structured approach.

2. An Experiment with a New Approach

With the simple approach currently used in Matlab part of the classification process is implicitly performed by the interviewers, who, when given an answer to the question 'What was the cause of death?', can accept that answer or choose to question further. The first concern of the proposed more structured approach is to distinguish clearly between the data collection and classification processes. In 1982 a new method of data collection was tried. The purposes were to improve consistency of reporting and to provide sufficient information to physicians or other experts to use in making decisions about assignments of deaths to cause categories. The long-term goal is to devise an alternative system of classification that can be used by lay classifiers or computers. This system would be more consistent and would more accurately reflect the multi-causal process of mortality. We have recently been experimenting with algorithms that could be used by lay classifiers or computers.

2.1. Methods

2.1.1. Data collection It was decided that the best way to improve consistency of data collection while avoiding problems that might be created by confusion between illness and disease would be to collect information about the descriptive elements (symptoms and signs, here for simplicity called 'symptoms') used by those classifying illness and those classifying disease.

To avoid giving interviewers extensive training about symptoms associated with various diseases, and to overcome the problem of interviewers only asking about certain things or stopping after the family had mentioned one illness, the interview was structured as a symptom list. It differs from symptom list questionnaires used by other investigators (see Chowdhury and Kahn (1980) and Puffer and Serrano (1973) for examples) in three major respects: it is written completely in Bengali; the questionnaire is highly structured; and an effort was made to collect information on the timing of symptoms.

Bengali is a language with many regional and socio-economic variants. Some of the words used by villagers in Matlab for symptoms are unfamiliar to people living in Dhaka, and for some symptoms—in particular convulsion—a number of different words are used in Matlab, and no single word is understood by every villager. Even when interviewers are from the area, social class and education differences lead to the use of different words, and two interviewers may not translate from English in the same way. Thus it was decided to write the questionnaire in Bengali. A great deal of time was spent talking to villagers about illnesses, noting what words are used to describe symptoms, and what symptoms were mentioned frequently. Using this information two questionnaires were drafted, a general form and one for neonates.

The first versions of the questionnaires tested were essentially symptom check-lists; the interviewers asked directly only about the presence and timing of a symptom, but were instructed to record any other volunteered information. This approach resulted in a lot of writing for the interviewers, who did not know what the significant points were and recorded everything that was said. It also led to frustration on the part of the physician classifiers, who occasionally had a lot of information about a symptom, but not the exact point that they considered critical. In lieu of training the interviewers to know what was significant, the questionnaire was revised to include auxiliary questions for symptoms, which would be asked if the symptom was reported as having been present. Certain signs that had been mentioned frequently during the interviews with villagers and that appeared to be diagnostically useful were incorporated into the symptom list as auxiliary questions.

Information about timing of symptoms is particularly important for determining the order of onset of diseases in the case of deaths in which multiple causes have played a role. And, in turn, determining the order of onset can be critical for classification decisions. For example, a rash that appears after several days of fever and lasts several days may be measles; a rash that appears on the same day as a fever and lasts only a day is probably not. However, collecting information can be complicated, particularly if it is to be coded for computer entry; for each symptom time of onset and duration are of interest, and there are a number of ways of describing each. Thus, symptom onset can be described in reference to an outside event (a holiday, for example), relative to the onset of other symptoms, or relative to death. Duration can be described in absolute terms or as ending at a time relative to death. For simplicity, the question about onset was phrased 'How long before death did this start?'; the question about duration was 'How long did this last?', and provision was made in the coding plan for responses in terms of absolute duration or time relative to death. For neonates even more types of response to symptom-timing questions are possible, since answers can be given in terms of time relative to birth or death. Questions were framed to leave either possibility open.

The final version of the general interview was structured so that after verification of demographic information the first question was 'What was the cause of death?' The question was asked in such a way as to elicit a short reply. Following this are questions about the circumstances of death—if it was sudden, due to an accident, or, if to a married woman of child-bearing age, if she was pregnant, recently pregnant, or had missed any menstrual periods in the time just before death. Then the interviewer read a list of 16 symptoms (fever, pain, rash, cough, trouble breathing, vomiting, diarrhoea or dysentery, other defecation problems, trouble with urination, bleeding, oedema, swollen abdomen, jaundice, paralysis, convulsions, weight loss) asking for each if it was present before death. If the respondent said 'yes', auxiliary questions were asked about the timing and characteristics of the symptom, and about other related signs and symptoms (Figure 7.1). For children less than 3 years old a question was asked about whether or not they were breast-fed or had been recently weaned, and for all deceased persons the respondent was asked to give an opinion about whether the person had been fat, normal, or thin before getting ill. The name of the respondent and relationship to the dead person was recorded, and interviewers were asked to record their assessment of the reliability of respondents.

The basic structure of the final version of the neonate form was the same; the first question after the demographic information had been verified was 'What was the cause of death?' This was followed by questions regarding the length of the pregnancy (as a control all women were asked the length of a normal pregnancy) and whether the mother had had problems or been ill

Symptom	Present?		How long before death did it start ?	How long did it last?		
	No	Yes		Until death	How long before death did it stop ?	
Fever						Was the fever ☐ high or ☐ low ? Did it occur ☐ all the time ☐ some of the time. When ? ________
Pain						In what part of the body was the pain ? ________ Was it ☐ severe ☐ not so severe ?
Rash (such as chickenpox or measles)						Was the rash ☐ large ☐ small ☐ both ? Was there water inside ? ☐ No ☐ Yes Did the rash itch ? ☐ No ☐ Yes Where on the body was the rash ? ________
Cough						Was this whooping cough ? ☐ No ☐ Yes Did the person vomit after coughing ? ☐ No ☐ Yes Was there blood in what was coughed out ? ☐ No ☐ Yes
Trouble breathing						Did the person's chest hurt ? ☐ No ☐ Yes Was it easier to breathe sitting up ? ☐ No ☐ Yes Was there a sound with breathing ? ☐ No ☐ Yes Was the skin pulled between the ribs ? ☐ No ☐ Yes
Vomiting						Was there blood in the vomit ? ☐ No ☐ Yes Was the vomit black ? ☐ No ☐ Yes
Diarrhoea or dysentery						Was this ☐ diarrhoea ☐ dysentery ☐ both diarrhoea and dysentery ? Was there blood in the stool ? ☐ No ☐ Yes Was there mucous in the stool ? ☐ No ☐ Yes How many times a day defecation ? ________ Was defecation painful ? ☐ No ☐ Yes Did the eyes become sunken ? ☐ No ☐ Yes

Fig. 7.1. A portion of a page translated from the experimental general form

during it. The next section contained some questions about previous births and early childhood deaths of those children, about the delivery and birth defects noticed, the size of the child, and its age at death.

These questions and questions about four symptoms—trouble breathing, abnormal colour, convulsions, and paralysis—were asked for all neonatal deaths; questions about the other nine symptoms (tetany, vomiting, fever, cough, rash, oedema, abnormal sleepiness, loss of consciousness, lump or swelling anywhere in the body), breast-feeding history and normality of eye movements, urination, defecation, and cord drying were asked only for children who survived more than a day.

Even in high-mortality populations, death is a fairly rare event. To save the interviewers' travel time and reduce costs the general sample included all reported deaths occurring between October 1981 and July 1982 in the largest and median villages (by 1981 mid-year population) in each of the 14 subareas of Matlab. For neonates, for whom we wished an enhanced sample, the same time frame was used, but deaths from other villages in each subarea were also included. Because of the notification system for death used in Matlab at that time, the speed of reporting varied across villages, so the study closed with some otherwise eligible deaths having been excluded from the sampling frame. Thus no population-based rates can be calculated directly from the results. However, for the general form the age–sex distribution of the dead persons for whom completed interviews were obtained corresponds almost exactly to the age–sex distribution of all deaths for 1982.

2.1.2. Classification The classification system being experimented with assumes multi-causality and is based on case definitions. The aim is to have for each cause an assessment of the probability that it was a factor in the death. Thus if there is strong evidence against its being a factor in the death, the cause is *ruled out*; if strong evidence exists for its presence it is considered a *highly probable cause*; if some evidence exists but it is not definitive the cause is considered a *possible* factor. There is also a default category for cases when there is insufficient evidence either to suggest the existence of a cause or to rule it out.

As an example some causes of neonatal deaths will be used, since this age group has been particularly problematic for the simple Matlab lay-reporting system. Four causes will be considered: tetanus, prematurity, pregnancy complications, and delivery problems. The decision rules used are given in Table 7.6. The probability categories for each cause and, within them, the rules are arranged in priority order. Thus for tetanus a death satisfying the first criterion to be ruled out (the child breast-fed and continued feeding until death) is classified into that group and not tested against further criteria. If it did not satisfy this first rule, it was tested against the second, and so on. Thus a death was tested against successive rules until it either satisfied one of them or, if it satisfied none of the rules, was assigned to the default category of insufficient evidence for that cause.

Table 7.6. *Criteria for assessing four possible factors in neonatal death, and proportion of neonatal deaths fulfilling criteria, by area*

Cause	Probability	Decision rules	Deaths fulfilling criteria (%) (Totals = 100%)	
			Treatment (no. = 34)	Comparison (no. = 88)
Tetanus	Ruled out	Breast-fed and continued until death	41.2	36.4
		No major symptoms or minor symptoms	—	—
		Age < 3 days	17.6	17.0
	Highly likely	Age > 2 days; breast-fed and stopped; major symptom	2.9	11.4
	Possible	Age > 2 days; breast-fed and stopped; minor symptom	14.7	6.8
		Age > 2 days; never breast-fed; major symptom	2.9	10.2
		Age > 2 days; never breast-fed; minor symptom	11.8	11.4
	Default	Meets none of the above criteria	9.9	7.8
Prematurity	Ruled out	Declared length of pregnancy = 10 lunar months	44.1	34.1
	Highly likely	Declared length of pregnancy < 9 lunar months	32.5	20.5
	Possible	Declared length of pregnancy = 9 lunar months	23.5	45.5
	Default	Length of pregnancy unknown	—	—
Pregnancy complications	Ruled out	No reported severe illness during pregnancy or symptoms of pre-eclampsia	47.1	48.9
	Highly likely	Reported severe illness during pregnancy and symptoms of pre-eclampsia	14.7	13.6
	Possible	Symptoms of pre-eclampsia	17.6	11.4
		Reported severe illness during pregnancy	20.6	23.9
	Default	Fulfils none of the above criteria	—	2.3

Table 7.6. *cont.*

Cause	Probability	Decision rules	Deaths fulfilling criteria (%) (Totals = 100%)	
			Treatment (no. = 34)	Comparison (no. = 88)
Delivery complications	Ruled out	Normal presentation; labour reported as 3–24 hours' duration; declaration of no complications	14.7	36.4
	Highly likely	Other than normal presentation	20.6	6.8
		Labour reported as > 48 hours' duration	8.8	18.2
		Labour reported as < 1 hour's duration	5.9	5.7
		Declared delivery complication	23.5	21.6
	Possible	Labour reported as 25–48 hours' duration	5.9	3.4
		Labour reported as 1–2 hours' duration	20.6	8.0
	Default	Fulfils none of the above criteria	—	—

The decision rules for tetanus are based on a four-element case definition: age, breast-feeding history, 'major' symptoms (convulsions, tetany), and 'minor' symptoms (abnormal colour, abnormal eye movements, paralysis). The critical age is taken as the fourth day after birth; a child younger than this is highly unlikely to have tetanus. For breast-feeding history there are three possibilities: the infant never breast-fed; the infant started to breast-feed and breast-fed until death; the infant started to breast-feed and stopped before death. Of these three the one classically used in case definitions of tetanus is the infant breast-feeding normally and stopping before death (Gray 1985). However, in Bangladesh, and perhaps in other areas where the colostrum is considered bad for the child, breast-feeding is often not started until after the third day of life, and sometimes much later. Thus a child who develops tetanus may never have been breast-fed; some rules for the 'possible' category were constructed to allow for this. The third and fourth elements of the case definition concern the 'minor' and 'major' symptoms. The major symptoms are more specifically linked to tetanus and provide a greater surety of its presence than the minor symptoms.

The decision rules for prematurity are based only on the declared length of pregnancy. The relative size of the child had been asked during the interview, but the responses do not seem to be discriminatory.

For the category of pregnancy complications there are two elements: severe illness and symptoms of pre-eclampsia (swelling of hands, face, severe headache over a period of time, convulsions, or unconsciousness during the pregnancy). Because of the variation in apparent seriousness of illness reported as severe and in the duration and intensity of pre-eclamptic symptoms reported, neither element alone is taken as definitive.

The assessment of the existence of delivery complications is based on three elements: presentation of the child, duration of labour, and reported delivery complications. The presentation criterion is straightforward. The question on labour specifies the duration of bearing-down pains, but the range of responses received suggests that the question was not well understood. In addition, digit heaping of the responses suggests that the reporting of time is not very accurate. Thus the limits chosen for abnormally long or precipitate labour are somewhat arbitrary. The declaration of delivery complications includes particularly long or painful labour, unconsciousness of the mother, excessive loss of blood, breech delivery, twin births, and vaginal tearing.

To allow for multi-causality, deaths are considered separately for each cause; thus the death of a child whose mother had symptoms of pre-eclampsia during pregnancy and had convulsions during labour would have both pregnancy complications and delivery complications rated as highly likely. In like manner the death of a child who was born of a nine-lunar-month pregnancy, who never breast-fed, had convulsions, and died at age 4 days would have both prematurity and a history of delivery complications, and thus get highly likely scores for both those causes. No attempt has been made to determine relative importance of causes.

2.2 Results

Examples of the kinds of history that can be obtained with the new questionnaires are given in Figure 7.2. Interview time averaged about 20 minutes for both types of form. The median number of symptoms reported for the general form was five; this did not vary significantly by age of the person who died or with time between death and interview (up to ten months). In general, recall about circumstances of death was not problematic. In fact, very often respondents reported episodes involving diarrhoea, dysentery, or rash that occurred several months before the death. The questions about timing produced good order-of-magnitude results, but for symptoms starting more than one month before death information about relative timing of onset was lost because it was not asked explicitly, and responses were recorded in terms of months. This lack of information occasionally made the classification process more difficult.

Application of the decision rules for tetanus, prematurity, delivery complications, and pregnancy complications to information obtained about 122 neonatal deaths result in the classifications shown on the right-hand side of

Female died 24–12–81 3 years old

History

The respondent declared that dysentery was the cause of death.
Fever started 23 days before death and lasted until death. It was high and continuous. *Cough* started 22 days before death and lasted until death; the respondent declared that the child did not have whooping cough; the child did not vomit after coughing, nor was there blood in the sputum. *Rash* started 15 days before death and lasted eight days. It was small in size, covered the body, did not itch and contained no water. *Dysentery* started 19 days before death and lasted until death. There was blood and mucus in the stools, and the child went to stool more than ten times a day. *Weight loss* started 22 days before death and lasted until death. It was rapid; the child did not want to eat.

Classification:

Ruled out: accident, acute respiratory disease, diarrhoea
Highly likely: measles, dysentery, malnutrition
Possible: —

Classification under simple system: dysentery

Male died 31–12–81 1 day old

History:

The respondent declared that *alga* was the cause of death.
Pregnancy. The infant was born after a ten-month pregnancy during which the mother had no severe illness and experienced no symptoms of pre-eclampsia. The mother has had three other live births; none of these infants died during the first 40 days of life. *Delivery*. The infant's head presented first. The mother declared that there were delivery complications; labour lasted 72 hours. The infant had no visible birth defects. *Symptoms*. The infant had trouble breathing from birth until death, and was an abnormal colour (dark) from birth until death. The infant was never breast-fed, had no convulsions or paralysis.

Classification:

Ruled out: tetanus, prematurity, pregnancy complications
Highly likely: delivery complication, respiratory
Possible: —

Classification under simple system: tetanus

Fig. 7.2. Histories obtained using the new questionnaire

Table 7.6. Delivery complications were a highly likely factor in a high proportion of deaths in both areas, followed in magnitude by prematurity, pregnancy complications, and tetanus. Tetanus was ruled out for 41% of the treatment area deaths and for 36% of the comparison area deaths, and was considered to be a highly likely factor in only about 3% of the treatment area but about 11% of the comparison area deaths. Prematurity also shows a difference by area, with the treatment area having more deaths in which it was highly likely to be a cause (eight-lunar-month pregnancies), and fewer in which it was a possible cause, but about 10% more deaths in which it was ruled out as a possible factor. As defined, pregnancy complications show virtually no differences between areas. Delivery complications are a highly likely cause in more than 50% of the neonatal deaths in both areas; of the types of delivery complication, precipitate labour and declared delivery complication show no difference by area, while there is a clear excess of abnormal presentation in the treatment area.

Only 32% of all neonatal deaths had none of the four causes rated as a highly likely factor; for almost all of these at least one cause was rated possible. Two factors rated as highly likely in 24% of the deaths, and three factors in 3%. The combination of delivery complication and prematurity accounted for a little over half of all the multiple categories.

2.3. Discussion

The results obtained using the questionnaires suggest that this is a useful way to collect information about illness before death, at least in cultures where there is no reluctance to speak about the circumstances of death. The first major advantage of this more structured method over the simple method is that it standardizes the interview, ensuring that questions are asked about all important possibilities, and that the interviewers' or respondents' explanatory models based on illness classifications do not bias the information obtained for disease classification. A second advantage is that much more information is obtained and recorded, so that classifications can be made with more assurance and multiple classifications can be made. The basic information is permanently available, so that classification rules can be reformulated and deaths reclassified as more knowledge is obtained about clinical manifestations and cultural perceptions of disease or as interest in cause categories changes.

It is reassuring to note that intervals of up to ten months between death and interview do not affect the amount or the quality of information obtained. Since distress is caused by asking about the death too soon after it, in prospective studies where the period between loss and follow-up is not a problem it is probably better to wait several months.

The problem of loss of information about relative timing of symptoms could be resolved by asking about timing after all the symptom questions have prompted the respondent to remember everything that happened. An additional possibility would be to provide the interviewer with a time grid to fill out. However, this would add to the training time for interviewers and might in some cases give the impression of more precision than is truly the case. Another possible approach would be to ask the respondent which symptom appeared first, ask the auxiliary questions about it, then ask which symptom came next, etc.—an augmented serial account of events and symptoms leading up to death. However, if paper forms are used this approach involves frequent page-turning on the part of the interviewer, with the possibility of some symptoms being skipped. This approach would also require that the interviewer ask about the unvolunteered symptoms at the end, to be sure that they really had not been present, and if one had been forgotten, then the relative order would have to be revised. This approach may become feasible as mini-computers become available for survey work.

The major barrier to use of this system in other settings is formulation of the questionnaire. In addition to issues of language, choice of symptoms and signs to ask about may require some anthropological study. For example, work by Jackson in Liberia suggests that anorexia and joint pains are culturally recognized symptoms of malaria that have a high correlation with clinically severe disease, while fever, vomiting, and convulsions all over-predict medical severity (Jackson 1985). This information would be important for constructing a questionnaire for Liberia that would allow assessment of the importance of malaria as a cause of death.

Validation of this system is important, but technically difficult, since necessary supporting evidence for any classification can rarely be obtained from the health system. The information obtained using the structured questionnaire is similar to that obtained in in-depth interviews by physicians, but there remains a question of whether two interviewers of the same respondent are sufficiently independent measures. Another possibility that has not yet been tried would be to use the questionnaire to ask retrospectively about non-fatal episodes of serious disease that could be clinically validated.

The results of applying some classification rules to the information obtained using the neonate questionnaires add a great deal to our understanding of neonatal mortality in Matlab. They suggest that the observed difference between the areal neonatal death rates is largely the result of the tetanus immunization programme. Prematurity and delivery complications are clearly very important factors in both areas. There are several possible explanations for the pattern of areal differences observed for prematurity. One is that because the menstrual history of all married women in the treatment area is recorded monthly by the female village workers, the length of treatment area pregnancies may be known with more accuracy. In the comparison area, where menstrual histories are not recorded, some of the eight-lunar-month pregnancies may have been mistakenly reported as nine-month pregnancies. The deaths of those neonates would then be classified as having prematurity as a possible rather than as a highly likely cause. An alternative explanation, which would also explain why prematurity is ruled out in a higher proportion of treatment than comparison area deaths, is that one effect of the health advice and iron and folic-acid tablets distributed in the treatment area has been to reduce prematurity overall, but that some infants who would have otherwise died *in utero* are surviving to be live-born, and then dying. The validity of this hypothesis can only be known if the approximate length of gestation is known for all infants, not just for those who die. Since 1985 the DSS birth report form has included a question about length of pregnancy, and this information is expected to be available soon.

3. Conclusion

The simple lay-reporting approach currently used in Matlab provides information that can be useful in assessing the trends and differentials in mortality, but its usefulness varies with the causes that are affected and the similarity of the illness and disease classifications for those causes. Thus the system works well for assessing the year-to-year and areal differences due to dysentery or diarrhoea, for which the illness and disease definitions coincide, but is virtually useless for assessing differences due to neonatal tetanus, because the illness *alga* has a very different domain than the disease tetanus. However, even for diseases for which it works well the simple approach may give misleading results when used to predict the effect of interventions because it allows only a single cause to be coded for each death. Thus the proportion of all deaths associated with any individual disease remains unknown. As a result, for example, the impact of a measles vaccination programme might be greater than predicted from results of the simple system because some 'dysentery' and 'respiratory-disease' deaths in which measles was an unclassified underlying cause are averted. Similarly, the effect could be less than expected if, for example, some deaths classified as due to 'measles' occurred in children who had other unclassified serious problems such as whooping cough and malnutrition that were not affected by the vaccination.

The more structured approach redirects attention from the illness(es) preceding death to the descriptive elements (symptoms and signs) that are used to make classifications of illness and disease. While information obtained using the questionnaire can be used to assign single causes for each death, the cause structure of mortality may be better described by a system which assigns a probability that each cause was a factor in a particular death, and cumulates this information over a population.

References

Bhatia, S., Mosley, W.H., Faruque, A.S.G., and Chakraborty, J. (1980), 'The Matlab Family Planning–Health Services Project', *Studies in Family Planning*, 11(6), 202–12.

Chowdhury, S.A., and Khan A.H. (1980), *Causes of Death, Companiganj, Bangladesh 1975–78* (Dacca: Dacca Christian Commission for Development in Bangladesh).

D'Souza, S. (1981), 'Population Laboratories for Studying Disease Processes and

Mortality: The Demographic Surveillance System, Matlab, Comilla, Bangladesh', Paper presented at the *Seminar on Methodology and Data Collection in Mortality Studies, IUSSP*, Dakar, 7–10 July.

—— and Bhuiya, A. (1982), 'Socio-economic Differentials in a Rural Area of Bangladesh', *Population and Development Review*, 8, 753–69.

Gray, R.H. (1985), 'The Integration of Demographic and Epidemiological Approaches to Studies of Health in Developing Countries', Paper presented at the *IUSSP General Conference*, 5–12 June, Florence (Liège: IUSSP, Ordina Éditions).

Helman, C.G. (1985), 'Communication in Primary Care: The Role of Patient and Practitioner Explanatory Models', *Social Science and Medicine*, 20(9), 923–31.

Hossain, K.M.B., Glass, R.I., Huq, M.I., Yunus, M., and Hossain, M.M. (1981), 'Surveillance of Shigellosis in Matlab: A Five-Year Review', *International Conference on Shigellosis*, Cox's Bazar, Bangladesh, 15–20 June.

Jackson, L.C. (1985), 'Malaria in Liberian Children and Mothers: Biocultural Perceptions of Illness vs. Clinical Evidence of Disease', *Social Science and Medicine*, 20(12), 1281–7.

Khan, M.U., Samadi, A.R., Huq, M.I., Yunus, M., and Eusof, A. (1984), 'Simultaneous Classical and El Tor Cholera in Bangladesh', *Journal of Diarrhoeal Disease Research*, 2(1), 13–18.

Leeuwenberg, J., Muller, A.S., Voorhaeve, A.M., Gemert, W., and Kok, P.W. (1984), 'The Epidemiology of Measles', in J.K. Van Ginneken, and A.S. Muller, *Maternal and Child Health in Rural Kenya* (London: Croom Helm).

Newell, K.W., Lehmann, A.D., Le Blanc, D.R., and Osoria, N.G. (1966), 'The Use of Toxoid for the Prevention of Tetanus Neonatorum: Final Report of a Double-Blind Controlled Field Trial', *Bulletin of the World Health Organization*, 35(6), 863–71.

Northrup, R.S. (1986), 'Decision Making for Health Care in Developing Countries', in UN, *Consequences of Mortality Trends and Differentials* (New York: UN).

Phillips, J.F., Stinson, W.S., Bhatia, S., Rahman, M., and Chakraborty, J, (1982), 'The Demographic Impact of the Family Planning Health Services Project in Matlab, Bangladesh', *Studies in Family Planning*, 13, 131–40.

Preston, S. (1976), *Mortality Patterns in National Populations* (New York: Academic Press).

Puffer, R.R., and Serrano, C.V. (1973), *Patterns of Mortality in Childhood*, appendix 1 (Washington, DC: Pan American Health Organization).

Rahman, M., *et al.* (1982*a*), 'Use of Tetanus Toxoid in the Prevention of Neonatal Tetanus: Reduction of Neonatal Mortality by Immunization of Non-pregnant and Pregnant Women in Rural Bangladesh', *Bulletin of the World Health Organization*, 60(2), 261–7.

—— D'Souza, S., and Karim, R. (1982*b*), 'Village Characteristics and Contraceptive Practices', Report presented at the *Seventh Annual Contributors' Conference, Bangladesh Fertility Research Program*.

Rowley, D., and Satterthwaite, A.S. (1983), *Review of Measles in Matlab, 1982–1983*, mimeo (Dhaka).

WHO (World Health Organization) (1978), *Lay Reporting of Health Information* (Geneva: WHO).

Zimicki, S., Yunus, M., Chakraborty, J., and D'Souza, S. (1984), *A Field Trial of Home-Prepared Oral Rehydration Solution in Rural Bangladesh*, Center for Population Studies Discussion Paper no. 84(4) (Cambridge, Mass.: Harvard University Press).

—— Nahar, L., Sardar, A.M., and D'Souza, S. (1985), 'Cause of Death Reporting in Matlab', *Demographic Surveillance System, Matlab*, xiii, ICDDR,B Scientific Report no. 63.

8 Assessing Probable Causes of Death Using a Standardized Questionnaire: A Study in Rural Senegal

MICHEL GARENNE and OLIVIER FONTAINE

Institut Français de Recherche Scientifique pour le Développement en Coopération (ORSTOM), Dakar, Senegal

---

Mortality remains strikingly high in many developing countries. According to figures for 1984 prepared by the US Bureau of Census, some 73 countries around the world appear to have a life expectancy below 60 years. Mortality is especially high in the rural areas of these countries, where the coverage of public health services is very low. Little information is available on levels of mortality and on causes of death in most remote places of Africa, Asia, and Latin America. In fact, vital registration is virtually non-existent in many places and official statistics are often unreliable. This lack of information has important implications for policy since a better knowledge of causes of death can help to direct treatments towards young children and mothers who most need them. A better knowledge of causes of death can also help evaluate the impact of programmes directed towards reducing mortality. Since no other source of information is currently available in remote places, ways for improving the knowledge of the health status of people of low-income countries seems to be an important research issue.

There have been several attempts to improve the quality and quantity of information on causes of death in places where no reliable system of registration is available. Biraud (1956) proposed the use of symptoms to help the recording and assessment of possible diagnoses of death. He suggested training non-medical personnel to record at least:

- sex and age at death
- circumstances of the death (accidents, violence, sickness)
- main symptoms, their localization and duration
- epidemic diseases prevalent.

Biraud's aim was to obtain 'community diagnoses' possibly gathered by civil servants. With a prior knowledge of the country, and of its customs and pathology, he estimated that important information on causes of death could be gathered this way. Biraud's questionnaire was simple but his approach was to focus on adults rather than on children, among whom more than 50% of deaths take place. Thirty years later one has to recognize that little has been achieved in setting up efficient systems of recording deaths or improving the assessment of causes of death in developing countries.

WHO has documented ways of improving lay reporting on causes of death and morbidity (WHO 1978). This document provides a comprehensive list of 123 diseases and 20 external causes, plus health services. Directed towards health personnel it appears, unfortunately, to be poorly adapted for research in the field. The diseases are assumed to be specified from a list of symptom associations susceptible of recognition by primary health services personnel. Some important diseases, like measles, are analysed with only one symptom, which seems too little to achieve high reliability. There are many diseases listed that occur only infrequently. Furthermore, the minimal mortality list groups together measles and chicken-pox, which can be easily separated, and it does not have categories for neonatal deaths, in particular neonatal tetanus, a very frequent cause of death. Lastly, they do not provide systematic ways of interviewing people on causes of recent deaths.

The need for more standardized procedures is particularly acute for researchers involved in projects on maternal and child health in rural places of developing countries. Other projects in developing countries have relied either on the report of the family, as in Bangladesh (D'Souza 1981), or on the constant involvement of a medical team in the field, as in Guatemala (Mata 1978) or in the Gambia (McGregor 1976), or on a combination of both as in Kenya (Van Ginneken and Muller 1984). The authors of this article are in fact two of the investigators of a research project on the relationship between nutritional status and mortality of children aged 0–5 years. The project is based on a comprehensive record of demographic events in a population of about 23 000 people in rural Senegal (Niakhar) (Garenne 1984). The assessment of causes of death in this population is of great importance for analysing the risk factors associated with nutritional status in relation to pathology.

The questionnaires devised for this study benefited from the experience of other researchers working in similar conditions. Two questionnaires for general and neonatal causes elaborated at ICDDR,B Bangladesh by Susan Zimicki as well as the RAMOS questionnaire for maternal mortality used by FHI were of great help at the beginning of this study. Four draft questionnaires were tested on more than 200 cases among people living in the rural areas of Sine-Saloum in Senegal prior to the adoption of the final version of the questionnaire that is discussed below.

1. The Approach

According to WHO guidelines, causes of death are defined as 'all those diseases, morbid conditions or injuries which either resulted in or contributed to death and the circumstances of the accident or violence which produced any such injuries' (WHO 1975: 763). This definition includes all the complex processes experienced by the patient prior to death. Of course it is out of question to analyse all these processes through retrospective interviews and only careful autopsies conducted by well-trained personnel can allow one to assess scientifically all the causes of death. The ICD classification of causes of death includes some 931 categories (three-digit codes, not counting E and V codes) and for each death there can be three to four 'diseases or morbid conditions or injuries'. This multiplies the possible coding or assignment of exact causes of death beyond the boundaries of feasibility of any lay classificatory system.

However, in high-mortality populations a very small number of diseases account for a very large number of deaths, and, furthermore, most of them are simple infectious diseases, usually caused by one or few pathogenic agents. Hence, an important amount of information on mortality could be gathered if it were possible to identify these most common causes. And in fact in high-mortality populations and especially for childhood deaths, the most common causes of death are infectious, with typical clinical signs that people are able to notice and remember. For instance, the main killers of children in Senegal are diarrhoea and pneumonia–bronchopneumonia–bronchitis whose symptoms are fairly easy to describe. As these two diseases already account for half of the deaths of children between 1 month and 5 years of age, the possibility of identifying them appears extremely valuable.

The approach to the construction of a questionnaire for assessing causes of death from lay reporting is first to list a series of causes that are likely to occur frequently in the population. This will be done for various age groups and both sexes. Then, a series of symptoms associated with each of the common causes is elaborated. Symptoms that will be kept have to be sufficiently specific to pin-point a probable cause as well as to exclude the other possible causes already listed. The classification of causes of death includes a category of 'undetermined' causes that are assumed to be different from the specified causes. This aspect is of particular importance for the validity of the classification. As will be seen below, the questionnaire produces more information on certain diseases than expected, sometimes from additional evidence provided by health personnel. These diseases are either rare diseases or more frequent ones that were assumed to be difficult to identify from retrospective interviews. They should not be included in the final results since their statistical value is questionable. This is the case, for instance, with tuberculosis, which can sometimes be identified from the questionnaire but which is likely to be missed in a retrospective investigation.

Of course, the results that are obtained with this type of questionnaire depend on the ability and willingness of people to talk about recent deaths in their families. In various ethnic groups in Africa for instance this investigation is extremely difficult due to taboos and general reluctance to speak about deaths. Furthermore, some ethnic groups classify symptoms in ways that only remotely relate to Western classifications and it is sometimes difficult to have people make a distinction between similar symptoms that they consider equivalent. Our experience in Sine-Saloum, however, as well as the experience of others in various countries indicates that a large body of information can be gathered using this procedure.

2. The Questionnaire

The main guideline to the elaboration of the questionnaire developed in Senegal is to make use of all the information available that appears to be relevant for assessing the probable cause of death. The main items of information are the following:

- the age at death, particularly for neonates;
- the duration of illness, needed for standardization of the coding and also relevant for many causes such as malaria or maternal deaths;
- the symptoms: these are the basis for the diagnosis;
- the history of the disease: this item is particularly rich in information, although rarely sufficient for diagnosis. It summarizes the main symptoms that have caught the family's attention and provides the sequence of events between the first symptom and death. In addition, the open question that probes into the history of the disease often gives some evidence of contamination for communicable diseases.
- the treatments: this is important for assessing causes of death such as malaria. For instance, cases of acute fever among children who took quinine or chloroquine could not be attributed to malaria;
- any evidence of contamination: this can be a determining factor for assessing causes of deaths from communicable diseases such as measles, whooping cough, and cholera. In some cases symptoms are not always as unequivocal as required for diagnosis, but clear evidence of contamination in addition to a few symptoms can make the assessment of a probable cause of death less problematic.

Since symptoms selected for diagnosis vary widely according to the type of disease under investigation, four types of questionnaire have been designed:

- for neonates, i.e. children who died within the first four weeks of life;
- for children, i.e. children who died between 28 days and 14 years;
- for maternal deaths, i.e. women who died within the first six weeks after delivery;
- for other adult deaths, i.e. men or women above 15 years of age.

For convenience maternal-deaths questionnaires were included in the adult questionnaire for women and questionnaires for men were printed separately. The four types of questionnaire have the same structure:

- a first page, with all the information to be coded, including identification, age, and sex of respondent, the date of death, place of death, a summary of the treatments, and the cause the family thought to be responsible for the death. A separate part is for final assessment of probable causes of death;

- the history of the diseases leading to death and the treatments;
- a list of main symptoms or conditions. For each symptom follows a series of questions probing into its nature. These include the duration of illness, the dates of onset and termination, and the treatments for the symptom;
- other symptoms or conditions that may add useful information, as well as chronic diseases and other regular treatments;
- a check-list for establishing the diagnosis.

2.1 Questionnaires for Neonates

Questionnaires for neonates were designed to assess five main causes: neonatal tetanus, pneumonia of the new-born, birth trauma, congenital defects, and a composite group of causes that were impossible to separate: prematurity and low birth weight. The diagnosis of neonatal tetanus was made according to the following criteria:

- death from three to 20 days after birth;
- evidence of tetanus symptoms: refusal of the breast, tetanic crisis, rictus, infection of the umbilical cord;
- no evidence of vaccine against tetanus.

A good indication was also the family's recognition of a magical cause, referred to as *kumalass* in Sereer. The fact that tetanus is often attributed to magic created some difficulties in eliciting response to the questions. In several instances the investigators had to return to the family for checking purposes to attempt to obtain a better description of the symptoms.

Pneumonia of the new-born was diagnosed from:

- fever until death;
- symptoms of pneumonia within at least one day prior to death, such as: rapid breathing, difficult breathing, palpitating nostrils.

Birth trauma and congenital defects were assessed from the mother's report of the history of delivery and the child's aspect after birth.

Prematurity and low birth weight were assessed from the mother's report either of marked prematurity (low number of weeks of pregnancy) or that the child was markedly small, or of cases of twin births or triplets. Deaths among these children usually occurred within three days of birth.

2.2. Questionnaires for Children

These were designed to assess the role of a series of infectious diseases thought to be important; diarrhoea, pneumonia–bronchopneumonia–

bronchitis, measles, whooping cough, meningitis, tetanus, malaria, epilepsy. Additional questions were directed towards malnutrition.

The categories of diarrhoea and dysentery are probably the easiest to determine. The main difficulty in attributing a cause to diarrhoeal diseases is the elimination of the other possible causes that can be associated with diarrhoea, such as malaria or measles. Criteria that were retained for classifying diarrhoea as the main cause were:

- declaration of diarrhoea;
- evidence of abundant stools until death;
- no evidence of other disease mentioned in the check-list;
- signs of dehydration such as thirst, sunken eyes, or sunken fontanelle.

Pneumonia, bronchopneumonia, and bronchitis diagnoses were based on the evidence of fever until death and indications of the following symptoms: rapid breathing, breathing like a little dog, difficult breathing, palpitating nostrils, insuction, acute cough. The symptom reported should have lasted at least 24 hours prior to death to avoid confusion with signs of agony.

Measles deaths were easy to determine since people recognize accurately the disease. Criteria were:

- declaration of measles by the parents;
- death within six weeks of the beginning of the fever and rash;
- evidence of an epidemic in the village, or evidence of contamination outside the village;
- symptoms in the following sequence: fever, running nose, red eyes, rash starting on the face, rash inside the mouth (Koplik spots), rash all over the body, peeling skin;
- no evidence of excluding symptoms, such as water in the pimples.

Whooping cough is also easily recognized by the population. In addition to the declaration by the family, the following criteria were used:

- death during the period of cough (100 days after the start);
- evidence of an epidemic in the village, or evidence of contamination outside the village;
- long-lasting cough together with symptoms such as whoop, vomiting, red eyes.

Malaria is one of the most difficult diseases to evaluate from this type of questionnaire. Criteria that were used were:

- high fever with sweat or chills;
- death within three days of the beginning of fever;
- no evidence of adequate prevention or treatment at least 12 hours prior to death, such as chloroquine or quinine.

The fact that the death occurred to a healthy child and that it surprised the parents or that the death occurred during the rainy season were considered additional evidence, but were not taken as decisive criteria.

Meningitis appeared to be more difficult to assess than other causes, probably because people tend to ascribe these signs to magic causes. Criteria for the diagnosis were:

- fever until death;
- evidence of symptoms of meningitis: head bent backwards, arms and legs bent, swollen fontanelle. Typical convulsions, headaches, and photophobia were used as additional evidence.

Declaration of paramedical personnel was also considered valuable. No case of meningitis has been diagnosed outside the epidemics of 1983 and 1984.

Epilepsy: like measles, epilepsy is well identified by people. Criteria to attribute the death to this cause were:

- declaration by the family;
- symptoms of epilepsy, usually the report of a typical crisis by the family.

Evidence of treatment against epilepsy was often a help for diagnosis, since most people had been treated prior to death.

The diagnosis of malnutrition was possible because during the project nutritional status was regularly assessed for all children below the age of 5, and they were examined by a physician at the time of the nutritional assessment. However, the questionnaire often provided evidence about weight losses prior to death. Death was ascribed to malnutrition if there was no evidence of acute diseases such as measles or whooping cough in the period required for attributing the death to another cause (see Section 5).

2.3 Maternal Deaths

Maternal deaths were divided into two categories: antepartum, i.e. during pregnancy, and postpartum, i.e. within six weeks of delivery. Questions concerning maternal deaths were directed towards identifying eclampsia obstructed labour, severe bleeding, infections, and pulmonary embolism.

Eclampsia and pre-eclampsia were identified from signs of hypertension (swollen legs, arms, body, face, hearing loss) and from a typical crisis. In most cases these women had been examined by a physician during pregnancy and had received treatment for hypertension. Obstructed labour and severe bleeding were reported by the family. Deaths usually occurred during or within three days of delivery. Puerperal infections were identified from evidence of fever and usually belly-aches until death. Pulmonary embolism was identified in cases of sudden death from no other cause with accompanying signs of violent chest-aches. In several instances it was possible to get more accurate detail from the health centre where the women had been brought. All maternal deaths were assessed with the collaboration of a specialist working in a nearby town hospital (Kaolak).

2.4. Other Adult Deaths

Adult deaths other than maternal deaths were the most difficult to identify from the questionnaires. The only diseases that were possible to assess with a reasonable degree of confidence were cholera, tuberculosis, leprosy, epilepsy, and accidents.

Cholera is characterized by severe diarrhoea or vomiting, usually occurring simultaneously. Symptoms required for diagnosis were: abundant watery diarrhoea (like rice-water) or vomiting, without fever and obvious signs of dehydration, death occurring within three days, and good evidence of contamination or of an epidemic (all cholera deaths took place in 1985 and are not reported in this study).

Tuberculosis, leprosy, and epilepsy have typical symptoms. In addition, they last for a long period and they strike mostly young adults, who visit the physician and receive adequate treatment. Evidence of this treatment such as prescriptions or empty boxes of drugs are often the best indicators of the disease for adults.

In some cases it was possible to ascribe death to cardiovascular diseases. But, as will be explained later, the distribution of causes of adult deaths is not as reliable as those for infants and children.

3. Importance of the Enumerator and of the Respondent

In remote parts of developing countries a constant obstacle to surveys and inquiries is communication with people who have a different value system. The choice and training of enumerators is fundamental to the quality of the answers. Enumerators should be of the same ethnic group as the people who are surveyed, if possible of the same village or the same region, since words that designate diseases or symptoms may vary from place to place even within the same ethnic group. Enumerators also need to be able to translate the answers into a more widely used language, usually a European language. Enumerators working at the Senegalese Project have at least four years of secondary schooling and one of them has seven years. All have received intensive training to carry out this type of survey. Special care is placed on the translation of answers given in Sereer. Enumerators are also asked to write down Sereer names of diseases as often as possible for checking purposes. The training also involves the clear identification of morbid symptoms as well as preparation in the approach to those who are surveyed. The relationship between the enumerator and the interviewee is determinant for the quality of the answers. In-depth knowledge of the customs and taboos of the people under investigation is also required for good reporting.

The choice of the respondent is as important as the choice of the enumerator. In most cases when the death of a child is to be investigated, the best respondent will be the mother if she is alive and present. When the mother is reluctant to answer, or absent, or unable to answer for a variety of reasons, the father may give reliable answers. In cases of fostering, i.e. when the child is not living with his parents, good answers can also be obtained from the person who takes care of the child, usually the grandmother or the mother's sister. Reports from mothers are usually very detailed and accurate, especially when the disease is recognized by the population. Mothers notice with precision the symptoms that occur in a child as well as changes in the child's health status that are often not noticeable to a foreign observer. Accurate and detailed reports of mothers are a valuable source of information when barriers of taboos and shyness are overcome.

Deaths of adults are far more difficult to investigate, since nobody pays as much attention to other adults as a mother to her child. Among adults maternal deaths appeared to be the easiest to investigate since at delivery women are taken care of by other women in the family and by traditional or modern midwives. Other women who have taken part in the delivery are usually a reliable source of information for maternal deaths. Deaths of young adults due to other causes often result from long-lasting sicknesses that have prompted individuals to visit a physician or a nurse and to receive treatment. For diseases other than those easily recognized by the population (epilepsy, leprosy, or cholera), empty boxes of drugs and prescriptions are probably the best source of information. The most serious problems in the investigation of deaths occur among the elderly. Among the Sereer, old people who feel sick often hide their disease, do not go to dispensaries, refuse to take drugs, and declare they want to die within the village like their ancestors.

Table 8.1 shows how important is the age effect on the quality of reporting on causes of death. Among neonates information is obtained in 97% of cases and a diagnosis is made in 73.2% of reports. Among children aged 1 month to 5 years, 96.8% of questionnaires are properly filled in and diagnosis is possible from the questionnaire in 78.5% of cases. Among

Table 8.1. *Quality of answers, by age at death and type of questionnaire, Niakhar, 1983–1984*

	Information supplied				No information supplied		Total
	Probable cause specified		Cause undetermined	Total			
	(no.)	(%)	(no.)	(%)	(no.)	(%)	(no.)
Neonates	71	73.2	26	97.0	3	3.0	100
Children 0–5	328	78.5	90	96.8	14	3.2	432
Children 5–14	19	47.5	21	97.6	1	2.4	41
Maternal deaths	9	64.3	5	100.0	0	0.0	14
Adults 15–49	11	28.9	27	74.5	13	25.5	51
Adults 50 +	22	21.6	80	60.0	68	40.0	170
TOTAL	460	64.9	249	87.7	99	12.3	808

children of 5 to 14 years reports were properly completed for 97.6% of cases and a diagnosis was possible in 47.5% only. For maternal mortality, questionnaires were always filled in and a diagnosis was possible in 64.3% of cases. Among adults aged 15 to 49 years, questionnaires were answered in 74.5% of cases, but diagnosis was made in only 28.9%. Among men and women older than 50 years of age, 40% of the questionnaires were returned with no answer, and among those that were filled in a probable cause of death was assigned in only 21.6% of cases.

The choice of the respondent also has an impact on the quality of the answers. Respondents have been divided into only two groups, close relatives and others, since there was no evidence of any difference between close relatives. Close relatives are defined as mother or father for children, grandmother for foster-children, and as spouse, parents, or children for adults. The quality of answers is always better when close relatives have

Table 8.2. *Quality of answers, by respondent, age at death, and type of questionnaire, Niakhar, 1983–1984*

	Information supplied				No information supplied		Total
	Probable cause specified		Cause undetermined	Total			
	(no.)	(%)	(no.)	(%)	(no.)	(%)	(no.)
Neonates							
Mother, father, grandmother	55	68.8	25	100.0	0	0.0	80
Other relatives	16	94.1	1	85.0	3	15.0	20
Children 0–5							
Mother, father, grandmother	280	90.2	74	97.8	8	2.2	362
Other relatives	48	72.7	16	91.4	6	8.6	70
Children 5–14							
Mother, father, grandmother	15	51.7	14	100.0	0	0.0	29
Other relatives	4	36.4	7	91.7	1	8.3	12
Maternal deaths							
Mother, spouse	5	71.4	2	100.0	0	0.0	7
Other relatives	4	57.1	3	100.0	0	0.0	7
Adults 15–49							
Spouse, parents	7	41.1	10	77.3	5	22.7	22
Other relatives	4	19.0	17	72.4	8	27.6	29
Adults 50+							
Spouse, children	8	18.6	35	63.2	25	36.8	68
Other relatives	14	23.7	45	57.8	43	42.2	102
TOTAL							
Close relatives	370	69.8	160	93.3	38	6.7	568
Other relatives	90	50.3	89	74.6	61	25.4	240

answered the questionnaire, whatever the age at death (see Table 8.2). Results are statistically significant when all age groups are considered together. Close relatives rarely refuse to give details of the death (6.7%, against 25.4% for other relatives) and on average the questions are better answered, since a diagnosis can be made in 69.8% of cases when close relatives answer rather than in only 50.3% when other relatives answer.

4. Delay in Answering

The delay in answering the questionnaires also has an impact on the quality of the answers. People interviewed too early after a death may be more reluctant to talk about it. On the other hand, if the visit occurs too long after the event, people may have already forgotten some of the details of the sequence of events that are necessary for assessing the cause of death. The routine demographic survey in Niakhar is done once a year only, and on average deaths have been investigated 5.2 months after the event. Table 8.3 indicates that the quality of the answers does vary significantly with the delay after death. The best period to ask the questions seems to be between three and nine months after death. Questionnaires on deaths investigated too early are twice as likely to be returned unanswered (17.9%) and among those answered the cause is assessed less frequently (57.9%) than those investigated between three and nine months (between 62.8% and 70.1%). Deaths of children investigated more than nine months after were also poorly reported: the questionnaire was not properly filled in in 17% of cases, and a cause could be attributed in only 62.8% of cases. Differences between those interviewed three to nine months after the death and the others are statistically significant ($P>0.05$).

Table 8.3. *Quality of answers, by delay between death and interview, Niakhar, 1983–1984*

	Information supplied				No information supplied		Total
	Probable cause specified		Cause undetermined	Total			
	(no.)	(%)	(no.)	(%)	(no.)	(%)	(no.)
< 3 months	110	57.9	80	82.1	40	17.9	230
3–5 months	185	70.1	79	90.8	27	9.2	291
6–8 months	116	65.5	61	91.7	16	8.3	193
> 9 months	49	62.8	29	83.0	16	17.0	94
TOTAL	460	64.9	249	87.7	99	12.3	808

5. Coding

The questionnaires used in this study are considered by the authors to be a minimum for assessing causes of deaths with a reasonable degree of confidence. However, they lead to more information than a single probable cause of death. They allow us to understand the sequence of the diseases that resulted in the death of a child. For instance, cases of measles followed by acute diarrhoea or pneumonia are usually well reported. This raises an important problem for coding answers. Is the recorded death to be attributed to measles or to diarrhoea? Does one need to code all the diseases leading to death as recommended by WHO?

In the pilot study, we preferred to follow exactly the sequence of morbid episodes that resulted in death. For instance, a sequence of measles followed by diarrhoea could be coded as having measles as the first disease and diarrhoea as the second but without assigning priority to either of the two. However, as we show below, the information on the exact sequence of diseases obtained from this type of questionnaire is rather poor and, except in common cases involving sequences such as measles and diarrhoea, there was virtually no information on conditions other than those related to the main cause. We later adopted a hierarchical classification of 'probable main cause', 'probable immediate cause', and 'probable associated cause'. The definition of 'probable main cause' is not equivalent to that of the underlying cause defined by WHO; however, both classifications appear as equivalent in many instances, since most cases are simply attributed to one cause. This hierarchical classification proved to be far more useful for tabulations and comparisons with other sources, since only the main cause is usually published in mortality data.

Selection rules for considering the main cause, the immediate cause, and the associated cause were adopted. These rules are applied only in a small number of diseases, in particular measles, whooping cough, diarrhoea, and pneumonia. The main cause was coded measles when the death occurred during the six weeks after the onset of main conditions. The main cause was coded whooping cough when the death occurred during the 100 days of cough following the onset of the disease. In cases of a double infection of measles and whooping cough, the first disease to occur was coded as the main cause and the second as the immediate cause. In cases with diarrhoea or pneumonia as complicating factors the complication was coded as an immediate cause if it lasted until death or as an associated cause if it did not. In cases of double infection with diarrhoea and pneumonia, the first disease to occur was coded the main cause and the other was coded as the immediate cause. Malnutrition was coded as the main cause of death if no main disease such as measles, whooping cough, or other diseases of children was considered as the cause of death. Malnutrition could be considered as the main cause if diarrhoea or pneumonia was the immediate cause. If another disease such as measles was considered the main cause, malnutrition was considered an associated cause. Measles was considered an associated cause if the death occurred between six weeks and six months after the onset of the rash. Whooping cough was coded as an associated cause if the death occurred between 100 days and six months after the onset of the cough.

Tetanus is coded as such for umbilical tetanus, tetanus after vaccination, circumcision, or after a minor cut. In cases of more serious injury, the accident is coded as the main cause and tetanus as an immediate cause. Epilepsy is coded as the main cause even if death is caused by an accident resulting from epilepsy; in this case the accident is an immediate cause. Mental disorders are always considered as an associated cause, never as a main cause.

6. Results

Depending upon the age at death and the type of questionnaire a number of causes were assessed with a reasonable degree of confidence.

Among neonates, 26 deaths were attributed to neonatal tetanus, 32 to prematurity or hypotropia, ten to pneumonia of the new-born, and two to congenital defects. No analysis of immediate cause was made. Associated causes were mostly due to the death of the mother, which often implies the death of the child. In rural places orphans have little chance of survival. They can receive assistance from charitable organizations, which exist and are efficient in the country, but in many cases the child has to survive on cows' milk for the first four weeks of life after which he might be breast-fed by his grandmother or his mother's sister.

Among deaths of children less than 5 years old, 132 were attributed to diarrhoea and 68 to pneumonia–bronchopneumonia–bronchitis, the two leading causes of deaths. Thirty-nine were attributed to measles, 19 to whooping cough, 38 to malaria, and 13 to malnutrition. Among rarer causes of death there was one case of food poisoning, one case of meningitis, one case of tetanus, two cases of septicaemia, three cases of chickenpox, two cases of hepatitis, one case of anaemia, two cases of epilepsy, one case of late effect of meningitis, one case of hydrocephalus, and one case of intestinal obstruction. In addition there were two accidents.

Among deaths of children of 5–14 years there were four cases of diarrhoea, three cases of meningitis, one case of measles, six cases of malaria, three cases of pneumonia, and two accidents.

Six out of the 14 maternal deaths recorded were possible to attribute to a cause directly from the report in the questionnaire. Furthermore, in seven cases there was a visit by a physician since maternal deaths often occur in a hospital. Among antepartum deaths one was an extrauterine pregnancy with haemorrhage, one was due to pre-eclampsia, and one had no report from the family since it occurred outside the village and the case was not found at the hospital. Among the 11 postpartum deaths, two were due to pulmonary embolism, two to postpartum haemorrhagia, one to placenta praevia, one to eclampsia, one to postpartum infection, and one to uterine failure.

Among deaths of young adults, two were attributed to tuberculosis, one to septicaemia, two to epilepsy, two to pneumonia, one to nephrotic syndrome, and three to accidents.

Among the elderly, only few probable causes of death were specified: one case of diarrhoea, one food poisoning, four tuberculosis, one leprosy, one septicaemia, one hemiplegia, three hypertensive heart diseases, one myocardial infarction, one intracerebral haemorrhage, four pneumonia, one asthma. In addition there were three accidents.

The analysis of immediate causes of death adds only little to our knowledge of the main causes (see Table 8.4). Diarrhoea was found to be followed by pneumonia in seven instances among 137 cases. Whooping cough was followed in five instances by diarrhoea, in two by pneumonia, and in one case by measles. Among 40 cases of measles, six were followed by diarrhoea, three by pneumonia, and one by whooping cough. Of 13 deaths attributable to malnutrition as a main cause three were followed by diarrhoea and two by pneumonia. Finally, of 77 cases of pneumonia-bronchitis three were followed by diarrhoea. All in all, among 494 probable causes of death assessed from the questionnaire, only in 50 (10%) has a probable immediate cause of death been added.

Associated causes also added little to the probable main cause (see Table 8.5). Diarrhoea was associated in 22 instances (one with whooping cough, nine with measles, one with malaria, and nine with pneumonia). Malnutrition was also found to be often associated with other causes of death: in 11 cases with diarrhoea, in one case with whooping cough, and in one case with pneumonia. In our project at least the main purpose of using associated

Table 8.4. *No. of deaths according to probable main cause and probable immediate cause, Niakhar, 1983–1984*

Main probable cause	Probable immediate cause				
	Diarrhoea	Pneumonia	Other specified	None	Total
Diarrhoea	0	7	1	129	137
Malaria	0	0	0	44	44
Measles	6	3	1	30	40
Meningitis	0	0	0	4	4
Whooping cough	5	2	1	11	19
Other infectious	2	0	0	17	19
Malnutrition	3	2	2	6	13
Pneumonia-bronchitis	3	0	0	74	77
Maternal	0	0	1	13	14
Neonatal	0	0	0	100	100
Other specified	0	0	1	16	17
Accident	0	0	3	7	10
Undetermined	6	0	1	211	218
No answer	0	0	0	96	96
TOTAL	25	14	11	758	808

Table 8.5. *No. of deaths, by probable main cause and probable associated cause, Niakhar, 1983–1984*

Probable main cause	Probable associated causes						
	Diarrhoea	Measles	Whooping cough	Malnutrition	Other specified	None	Total
Diarrhoea	0	7	2	11	5	112	137
Malaria	1	1	0	0	0	42	44
Measles	9	0	0	0	0	31	40
Meningitis	0	0	0	0	0	4	4
Whooping cough	1	0	0	1	0	17	19
Other infectious	0	1	0	1	1	16	19
Malnutrition	1	0	0	0	3	9	13
Pneumonia–bronchitis	9	1	0	1	1	65	77
Maternal	0	0	0	0	1	13	14
Neonatal	0	0	0	0	2	98	100
Other specified	0	1	0	1	1	14	17
Accident	0	0	0	0	1	9	10
Undetermined	1	0	1	0	5	211	218
No answer	0	0	0	0	4	92	96
TOTAL	22	11	3	15	24	733	808

causes is to allow evaluation of routinely recorded late effects of measles, whooping cough, and malnutrition on mortality. The fact that the child has had a history of chronic diarrhoea is more difficult to interpret since there is no record of this disease in the population as a whole. However, it should be noted that many cases of deaths associated with several diseases or symptoms have resulted in an 'undetermined' cause.

7. Discussion

This type of standardized questionnaire has an advantage over unstructured interviews: it does not require a well-qualified physician to conduct the interviews as is done in some research projects. The expertise of a physician is required only for reading the completed questionnaire and for checking in the field when needed.

This type of questionnaire also produces multiple types of evidence for assessing causes of death. In fact, the procedure relies on several articulate layers of evidence. The first is derived from the declaration of the family. The second step is the use of a systematic questionnaire aimed at assessing a probable cause. A third is the clinical examination by a physician prior to death. A fourth is the in-depth study of the case prior to death, with biochemical examinations. A fifth and final level is the autopsy, which is probably the only rigorous way of assessing causes of death.

In developing countries mortality surveys often record causes that people are able to identify, such as diarrhoea or measles, and causes that they may learn to identify at a health centre when the person has been brought there, such as tetanus or meningitis. Although often this record may be of interest, it is seldom reliable since categories used by laymen do not necessarily coincide with modern medical categories. Furthermore, since there is no systematic attempt to analyse deaths, many causes end up in the ill-defined category and analysis is made difficult even though the causes of death that are recorded are properly assessed.

However, since labelling of most probable causes of death relies on the family's recognition of the disease, it is important to show that the questionnaire has improved the quality of the answers. Table 8.6 indicates the correspondences between most common causes of neonatal deaths as they are declared by the family and probable causes assessed from the questionnaire. Some diseases are well reported but others poorly identified. Neonatal tetanus appears as very poorly reported by families mostly because people ascribe it to a magical cause. Prematurity–low birth weight–hypotropia are more consistently identified. In 27 out of 32 cases there is complete coincidence but nine cases have been attributed to other causes. Chest-ache also appears as linked with pneumonia, but only five out of ten cases are consistently identified.

For children less than 5 years old (see Table 8.7), diarrhoea is reported by the population (117 out of 132 cases are correctly identified). Probable diagnosis of measles and whooping cough are, by definition, closely linked to spontaneous reports of parents since the probable cause assessed from

Table 8.6. *No. of neonatal deaths, by probable main cause and cause declared by the family*

Probable main cause	Cause declared by family							
	Chest-ache	Fevers	Tetanus	Whooping cough	Premature, low weight	Other	Unknown	Total
Pneumonia	5	0	0	2	1	0	2	10
Tetanus	0	1	6	0	3	5	11	26
Congenital defects	0	0	0	0	0	1	1	2
Premature, low weight	0	0	0	0	27	0	5	32
Other specified	0	0	0	0	0	1	0	1
Undetermined	2	1	1	0	5	4	13	26
No answer	0	0	0	0	0	0	3	3
TOTAL	7	2	7	2	36	11	35	100

Table 8.7. *No. of deaths of children aged 1 month to 5 years, by probable main cause and cause declared by the family, Niakhar, 1983–1984*

Probable main cause	Cause declared by family											
	Cough chest-ache	Diarrhoea	Fever malaria	Measles	Meningitis	Whooping cough	Crisis	Weight loss	Other	Accident	Unknown	Total
Diarrhoea	1	117	2	0	0	0	0	2	5	0	5	132
Malaria	1	1	21	0	0	0	12	0	2	0	1	38
Measles	0	8	1	30	0	0	0	0	0	0	0	39
Meningitis	0	0	0	0	0	0	1	0	0	0	0	1
Whooping cough	1	2	1	1	0	13	1	0	0	0	0	19
Other infectious	0	3	0	0	0	0	3	0	2	0	1	9
Malnutrition	0	2	2	0	0	0	0	3	5	0	1	13
Pneumonia–bronchitis	29	5	21	0	1	0	4	0	6	0	2	68
Other specified	0	0	0	0	0	0	2	1	3	0	1	7
Accidents	0	0	0	0	0	0	0	0	0	2	0	2
Undetermined	9	9	23	0	2	0	10	2	26	0	9	91
No answer	0	5	1	0	0	0	0	0	0	0	8	14
TOTAL	41	152	72	31	3	13	33	8	49	2	28	432

the questionnaire relies upon parents' declaration of the disease. Discrepancies occur mostly from a confusion between main and immediate cause and, sometimes, from poor reporting of cough or crises for pertussis and of fever for measles. Cough, chest-aches, and cold, along with malaria–fever–hot body are the most confusing categories where the benefits of the questionnaire are the greatest. Malaria would have been misdiagnosed in 70% of cases and pneumonia–bronchitis in 30% of cases if only spontaneous reports of the family had been taken into account. Not surprisingly, accidents are consistently reported in both cases. Although the case of meningitis among children less than 5 years old would have been missed, cases of death due to meningitis among older children are properly reported by the family.

If correspondences can be established for diseases in children, little can be done for adults, due to the poor performance of the questionnaire. With respect to maternal death no comparisons were done since relatives do not report on these deaths.

It should also be noted that a problem of translation can often complicate a comparison between what people say and what is later analysed. Causes recorded by the enumerator were translated into French and not always accurately. For instance, cases of epilepsy were properly reported by the family but enumerators had written 'crises' on the questionnaire, as they would have done translating another word used to designate other types of convulsion due to hyperthermia, malaria, or other causes.

Epidemiological evidence added a lot to the accuracy of identification of probable causes of death. This was particularly true for measles and whooping cough. Results from Machakos indicate that tuberculosis can be identified as a cause of death for children less than 5 years old if and when longitudinal data on morbidity are recorded. This also seemed to be the case in an ongoing survey of morbidity in Senegal.

It can be argued that little reliability can be expected in reports from people with little or no modern education. This was not the case, however, in this investigation, which suggests that this difficulty may be frequently overestimated. Most of the reports obtained in this survey are consistent and results are similar to those found by other more in-depth surveys. When a cause could not be assessed from a properly completed questionnaire it was usually because a series of complex symptoms pointed either to a non-identifiable disease (several diagnoses possible) or to a series of diseases that were too complex to be properly assessed.

The reading and interpretation of the questionnaire involves subjectivity and judgement. Although there were strict guidelines, questionnaires analysed by two different persons rarely end up with the same diagnosis. This leads to an inflation of the category 'undetermined cause'.

It would be extremely valuable to compare results of such a study with autopsies. Naturally, there would be certain discrepancies, if only because clinical signs do not always signal the disease. However, Van Ginneken and

Muller (1984) found that clinical signs were at least as good as most sophisticated biochemical examinations for measles and whooping cough. On the other hand, whatever the discrepancies, the distribution of the deaths according to main probable cause is still a valuable tool. In studies of deaths over time, before and after intervention, between control and intervention areas, differences in the distribution of probable causes of death will provide important clues for evaluating the underlying causes associated with the observed differences.

8. Conclusion

The results of retrospective interviews recorded with a standardized questionnaire in rural West Africa suggest that very valuable information can be gathered with a simple methodology and adequately trained enumerators. Physicians are needed only to read the questionnaires and assess 'probable causes of death'. Results of surveys applied over a two-year period on a population of 23 000 people indicate that causes of deaths of neonates, of children below the age of 5 years, as well as of maternal deaths can be assessed with a reasonable degree of confidence. Causes of deaths of older children and of young adults are more difficult to assess, and reports on deaths of the elderly were too poor to be of any use.

For all ages combined, 87.7% of questionnaires were properly filled in and a cause could be identified in 64.9% of those cases. The report and assessment was better for neonate, young children, and maternal deaths. The quality of answers was better when close relatives answered the questionnaire and when the interview took place not too soon and not too long after death; the optimal delay seemed to be between three and nine months after the death.

Among the causes of death that were identified from the questionnaires the most common were: diarrhoea and dysentery, pneumonia–bronchopneumonia–bronchitis, malaria, measles, whooping cough, malnutrition, tetanus, and a composite category of prematurity–low birth weight–hypotropia for neonates. Among other infectious diseases meningitis, septicaemia, chicken-pox, hepatitis, tuberculosis, and leprosy were identified. Among other diseases (other than maternal deaths) epilepsy, hemiplegia, hypertension, myocardial infarction, intracerebral haemorrhage, asthma, and anaemia were identified. It should be noted that some of these causes have been assessed from physicians and found either in health centre registers or in prescription records.

A classification of causes into 'probable main cause', 'probable immediate cause', and 'probable associated cause' added little information to the general picture. Major associations were found between diarrhoea, pneumonia, measles, whooping cough, and malnutrition.

Although the reliability of this approach can always be questioned in the context of low levels of modern education the results we obtained seem to be in line with other investigations. The use of a standardized questionnaire does provide at least some evidence supporting an assessment of probable causes of death. Its limitations are a function of the list of causes of death under investigation.

References

Biraud, Y. (1956), 'Méthode pour l'enregistrement par des non-médecins des causes élémentaires de décès dans des zones sous-développées', WHO unpublished document /HS/60.

D'Souza, S. (1981), *A Population Laboratory for Studying Disease Processes and Mortality*, ICDDR,B Special Publication no. 13.

Garenne, M. (1984), 'The Concept of Follow-up Survey and Its Implications for Data Collection: Example of Using a Computerized Questionnaire for Improving the Recording of Early Deaths in Rural Senegal', Paper presented at the *IUSSP Seminar on Micro-approaches*, Canberra, 3–7 Sept.

McGregor, I. A, (1976), 'Health and Communicable Diseases in a Rural African Environment', *Oikos*, 27, 180–92.

Mata, L. (1978), *The Children of Santa Maria* (Cambridge, Mass: Harvard University Press).

Van Ginneken, J. K., and Muller, A. S. (1984), *Maternal and Child Health in Rural Kenya* (London: Croom Helm).

WHO (World Health Organization) (1978), *Lay Reporting of Health Information* (Geneva: WHO).

9 Reliability of Lay Reporting of Morbidity and Cause-of-Death Data: An Evaluation of Reported Cases and Deaths from Measles in Rural Senegal

PATIENCE W. STEPHENS

Office for Population Research, Princeton University, Princeton, New Jersey, USA

---

1. Introduction

The collection and classification of morbidity and cause-of-death data in less developed contexts (LDCs) raise major questions regarding accuracy and reliability. The unavailability of modern medical facilities for the bulk of the population in LDCs implies that many illnesses are treated either at home or by local traditional medical practitioners. Consequently no records are kept of illnesses or, in the event of death, of the cause of death. Even in urban areas which are served by modern medical facilities, morbidity and mortality data are likely to be available only for cases treated in hospitals, and are therefore severely biased by selection effects.

The dearth of morbidity and mortality data in LDCs has led, in recent years, to emphasis on and recommendations for the effective use of lay field personnel in the data-gathering effort (WHO 1978, 1980; Kupka 1981). Lay reporting involves the use of trained, non-medical interviewers in efforts to gather health-related data. If successful, such efforts have the potential to generate health data for many contexts for which they have never been previously available.

In order to justify the commitment of resources to the collection of health data by lay personnel it is necessary to demonstrate that this method of data collection can yield reliable data for use in health planning.

This paper evaluates the reliability of morbidity and mortality data for one disease—measles—collected by lay interviewers in a rural area of Senegal. The data are evaluated primarily using consistency checking, relying on what is known about the nature and epidemiology of measles. Based on the findings of this research, we offer suggestions for improving the lay reporting of health-related data.

2. The Survey and Data

The data which are evaluated in this paper were collected during multi-round surveys in the Sine-Saloum region of Senegal between 1962 and 1982. (This series of surveys was initiated by Pierre Cantrelle (Cantrelle 1969).) The analysis is performed on data for all (6 779) children who resided in Ngayokheme—a rural zone in the Sine-Saloum region—during the study period. Ngayokheme is divided geographically into eight villages, which are primarily administrative or tax units. In turn, villages are divided into a total of 30 hamlets. Each hamlet consists of several fenced-in compounds,

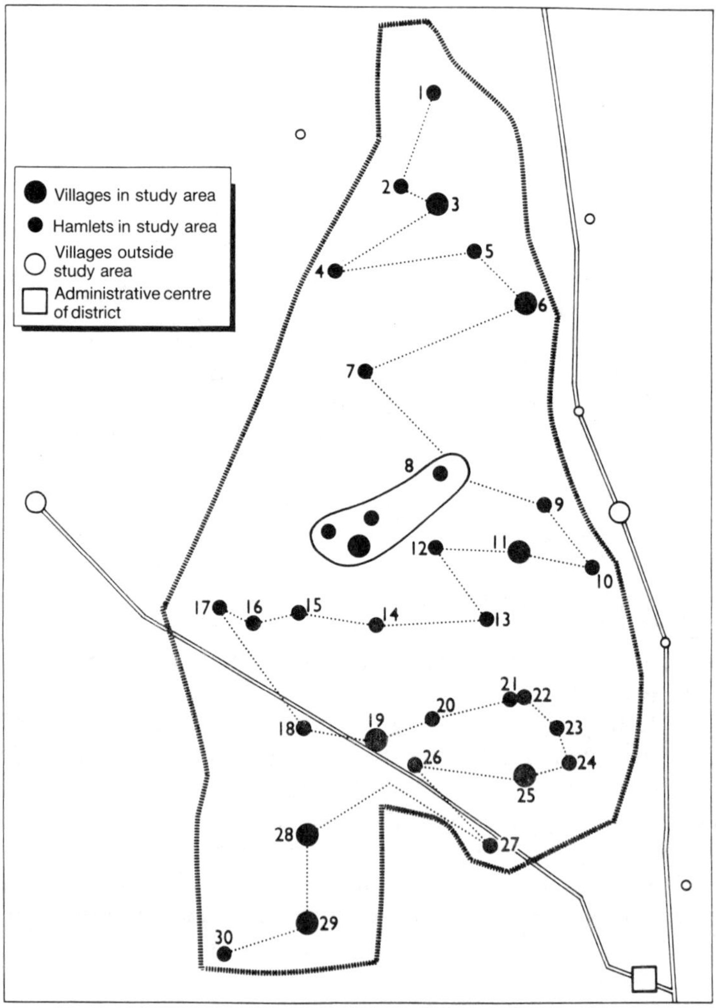

Fig. 9.1. The study area

which are very well defined. Well over 100 compounds were in existence in the study area at the end of the survey (1982). Figure 9.1 shows the location of villages and hamlets in the study area.

After an initial enumeration of the population in December 1962, interviewers visited each household annually and collected information on basic demographic events, and on cases of and deaths from measles since the last interview. Data were also obtained on whether a child had been vaccinated against measles since the last interview. Four measles vaccination campaigns took place in the study area in 1969, 1974, 1979, and 1981.

All data were gathered by trained, non-medical field interviewers. The diagnosis of measles morbidity and mortality was based on the symptoms which were reported to have been observed at the time of illness or death. Interviewers assigned cause-of-death and morbidity data based on a checklist of symptoms. The month and year of occurrence of each event of interest was recorded. Although respondents could not give accurate exact dates on which events occurred they often knew the number of lunar months elapsed since their occurrence.

3. The Nature and Epidemiology of Measles

Measles (morbilli) is an 'acute exanthemous disease . . . marked by fever and other constitutional disturbances, a catarrhal inflammation of the respiratory mucous membranes, and a generalized maculopapular eruption of a dusky red colour, followed by a branny desquamation' (*Stedman's Medical Dictionary* 1982: 834). The disease is also characterized—and is usually easily identified in its early stages—by the appearance of reddish spots (Koplik's spots) on the buccal mucous membranes (ibid.). Measles is an airborne disease. Its virus is transmitted through the air via the trachea or larynx (Fraser and Martin 1978; Reinish and Minear 1978; Dekkers 1981), or through the conjunctiva (Papp 1956; Florman and Agatson 1962). Once in place, the virus goes through a period of incubation of between ten and 12 days during which no clinical signs of infection are apparent. This is followed by two to four days of prodromal syndromes of catarrh and about two days of acute illness during which the exanthem erupts and effloresces. The fourth stage is characterized by desquamation of the exanthem and, in uncomplicated cases, convalescence (Sydenham 1939; Panum 1940; May 1958; Morley 1973; Reinish and Minear 1978).

Measles is thus a disease with clear and well-established symptoms, making it relatively easy to recognize. In addition to the clarity of its clinical manifestations, several other peculiarities of measles aid in its identification and facilitate the evaluation of measles-related data.

First, it is known that a second attack of measles in the same individual is rare, since the disease confers immunity through incidence (Panum 1940; Miller *et al.* 1974; Fraser and Martin 1978). Vaccination against measles is also an effective way of acquiring immunity if the vaccine is administered at the right age and is potent at the time of administration. This implies that repeated reports of multiple cases of measles in the same individual should render the data suspect. Similarly, information on vaccination or on measles incidence should be questioned if many cases of measles occur after vaccination.

A second known feature of measles is its age selectivity with few or no cases—but high fatality rates—below the age of 6 months (Hayden 1974; Barkin 1975; Sinha 1977; Kenya Ministry of Health and WHO 1977). Measles is otherwise a childhood disease with most susceptible children contracting it before their fifth birthday (Borgono and Greiber 1971; Guyer and McBean 1981).

Related to the fact that measles confers immunity through incidence is the tendency for epidemics to 'fade out' as the number of susceptible persons approaches zero (Frost 1979; Langmuir 1980). Once this incidence level is reached and no measles cases are observed in a population, the recurrence of another epidemic would require not only a creation of a new pool of susceptible persons but also the reintroduction of the virus from outside through migration or other external contacts. This feature of measles implies that if reasonably accurate data are available by specific localities of occurrence within a larger geographical area, it should be possible to observe some sequence and timing of measles cases as epidemics 'move in' and 'out' of each locality.

To what extent do the data on measles morbidity and mortality in Ngayokheme conform to the known behaviour of measles as discussed in this section?

4. Methods of Analysis

To evaluate the accuracy of the morbidity data gathered in Ngayokheme, two main approaches are adopted. First, a detailed examination of the sequence and timing of measles cases in the study area is used to ascertain whether the pattern is consistent with the epidemiology of the disease as discussed previously. This analysis is performed for each of the 30 hamlets for the first four years of the survey (1963–6). Hamlets, rather than compounds, were chosen as the smallest geographical unit of interest, because their more restricted number eliminates the need for an otherwise unwieldy tabulation of measles cases by month and year of occurrence. Where necessary, references are made to the behaviour of measles within compounds.

A second stage of evaluation involves the analysis of the data with respect to age, vaccination, and prior measles status to determine the conformity of the data to expectations discussed previously. This consistency checking is used to assess the validity of the measles mortality data and to further evaluate the information on measles cases. It is expected that, if the data are being accurately recorded, they will show vaccinated children and those who have previously contracted measles to have been spared subsequent attacks.

5. Results

Table 9.1 shows the distribution of all measles cases in the study area by year of occurrence. It is apparent that measles was occurring in epidemic proportions in Ngayokheme until 1966 when, apparently as a result of the first measles vaccination campaign, the annual number of measles cases was reduced to near-zero levels. Epidemic levels of incidence are again noted after 1970, although measles never again appears to have assumed the importance it did during the first four years of the study.

Table 9.1. *No. of measles cases by year of occurrence, Ngayokheme, 1963–1980*

Year	No. of cases	Year	No. of cases
1963	226	1972	70
1964	132	1973	120
1965	202	1974	73
1966	192	1975	44
1967	16	1976	43
1968	9	1977	108
1969	16	1978	189
1970	3	1979	101
1971	60	1980	42
		1963–80	1645

A more detailed breakdown of the cases of measles by month, year, and hamlet of occurrence is provided in Table 9.2. This table covers only the first four years (1963–6) of the study. These four years cover the period just prior to the first major decline in measles incidence. Other years are not individually examined to reduce the volume of the data presented. The table also shows the number of children who were susceptible to measles during the period in question. It should be noted that susceptible children are those who were known to have neither contracted measles nor been vaccinated against it. Children whose measles and vaccination status were reported as unknown were assumed to have been susceptible to measles. The figures shown in Table 9.2 pertain to successive adjacent hamlets as shown by the dotted line on the map in Figure 9.1.

It will be noted that measles appears to occur in the study area in clusters of adjacent hamlets. However, this pattern is not very evident when the figures for 1963 and 1964 are examined. This appears to be the result of large numbers of cases which are not assigned by month of occurrence. When measles cases with unknown months of occurrence are examined, however, a clustering of cases into adjacent hamlets is evident even for 1963

Table 9.2. *No. of measles cases, by hamlet, by month and year of occurrence, 1963–1966*

Year and month	Hamlet no.																													
	1	2	3	4	5	6	7	8	9	10	11	12	13	14	15	16	17	18	19	20	21	22	23	24	25	26	27	28	29	30
1963																														
January	—	—	—	—	—	—	—	—	—	—	—	—	—	1/2*	—	—	—	—	—	—	—	—	—	—	—	—	—	—	—	—
February	—	—	—	—	—	2/1	—	1/39	—	—	—	—	—	—	—	—	—	—	—	—	—	—	—	—	—	—	—	—	—	—
March	—	—	1/10	3/9	—	—	—	1/43	—	5/1	—	—	—	—	—	—	—	4/1	—	—	—	1/6	—	—	—	—	—	1/10	3/33	—
April	—	—	—	—	—	—	—	—	—	1/14	1/5	—	—	—	—	—	—	—	—	—	—	—	—	—	—	—	—	—	—	—
December	—	—	—	—	—	—	—	1/63	—	—	—	—	—	—	—	—	—	—	—	—	—	—	—	—	—	—	—	—	—	—
Unknown	—	—	6	1	—	—	2	50	—	7	32	—	9	—	1	1	—	—	13	2	—	—	5	3	—	—	—	1	17	3
All year	—	—	7	6	1	6	22	60	—	14	34	—	9	1	1	1	1	4	13	2	—	6	5	3	—	—	—	3	24	3
1964																														
January	—	—	1/4	—	—	—	—	—	1/1	—	1/8	—	—	—	—	—	—	—	—	—	—	—	—	—	—	—	—	—	—	—
February	—	—	—	—	—	—	—	—	—	—	—	—	—	—	—	—	—	—	—	—	—	—	—	—	—	—	—	—	—	—
March	—	—	—	8/22	—	—	—	—	—	—	—	—	—	—	—	—	—	7/12	—	—	1/1	—	—	—	—	—	—	—	—	—
April	—	—	—	2/22	—	—	—	23/88	—	—	—	—	—	—	6/22	—	—	—	1/17	—	—	—	—	—	—	—	—	—	—	—
December	—	—	—	—	—	—	—	—	—	—	—	—	—	—	—	—	—	—	—	—	—	—	—	—	—	—	—	—	—	—
Unknown	—	—	1	3	1	—	—	14	—	1	3	—	3	2	—	9	—	—	5	2	1	1	—	2	4	—	—	—	3	—
All year	—	—	3	15	1	—	—	51	1	1	4	—	6	2	7	9	—	8	6	2	2	1	2	2	5	1	—	—	3	—
1965																														
January	—	—	—	—	—	—	—	—	—	—	—	—	—	—	—	—	—	—	—	—	—	—	—	—	—	—	—	—	—	—
February	—	—	—	1/18	—	—	—	—	—	—	—	—	—	—	—	—	—	—	—	—	—	—	—	—	—	1/12	—	—	—	—
March	—	—	—	—	—	—	—	5/87	—	—	2/26	—	—	7/14	13/18	1/18	—	3/11	—	2/10	—	7/15	—	18/19	—	—	—	59/68	18/33	5/8
April	—	—	2/38	—	—	—	—	—	—	—	3/31	—	—	—	1/22	—	—	—	—	—	—	—	—	—	—	—	—	10/68	1/41	1/9
December	—	—	1/42	—	—	—	—	—	—	—	—	—	—	—	—	—	—	—	—	—	—	—	—	—	—	—	—	—	—	—
Unknown	1	—	1	2	—	—	—	2	—	—	1	—	—	1	1	5	—	—	—	—	—	—	1	1	—	—	—	4	1	—
All year	1	—	4	3	—	—	—	8	—	—	9	—	—	9	15	6	—	5	—	2	—	8	1	19	—	1	—	77	27	7
1966																														
January	—	—	—	—	—	—	—	—	—	—	—	—	—	—	—	—	—	—	—	—	—	—	—	—	—	—	—	—	—	1/5
February	—	—	—	—	—	—	—	5/54	6/3	13/8	—	—	—	—	—	—	—	—	6/8	13/8	—	5/9	8/10	—	—	7/1	—	—	11/18	—
March	—	—	2/5	—	—	—	—	—	1/52	1/3	2/18	2/4	—	—	—	1/10	2/8	2/9	3/8	7/9	—	4/10	9/3	—	—	3/1	1/3	6/39	36/33	3/7
April	—	1/16	—	1/5	—	—	—	—	—	—	1/21	1/3	—	—	—	—	—	4/12	2/11	—	—	1/12	1/15	1/15	—	—	—	—	—	—
December	—	—	—	—	—	—	—	—	—	—	—	*—	—	—	2/11	—	—	—	—	—	—	—	—	—	—	—	—	—	—	—
Unknown	—	—	—	—	—	1	—	—	—	—	—	—	1	—	—	—	—	—	—	1	—	—	—	—	—	—	—	—	1	3
All year	—	1	2	1	—	1	—	5	7	13	4	3	1	—	2	—	2	6	11	21	—	10	18	1	—	10	1	6	48	7

* No. of cases/no. susceptible.

and 1964. The clustering of measles cases strongly suggests that the cases of measles recorded by interviewers are, in fact, measles. If measles cases were recorded in the study area in a more scattered manner, it would indicate the presence of random errors with possible measles-like rashes being diagnosed as measles.

When the number of measles cases recorded are related to the number of susceptible children during the period, it is clear that in most cases the volume of measles per hamlet per month does not exceed the number of susceptible children. In a few of the cases where measles cases exceed the number susceptible at the beginning of the period, questions must be raised regarding the data on either measles and vaccination history or the diagnosis of measles. It is worth noting, however, that most of the instances where numbers of measles cases exceeded the number of susceptible children occurred in 1966, the year of the first vaccination campaign. Given that only the month and year of occurrence of events are given by respondents, it is not possible to establish precedence when vaccination and measles occur in the same month. A child may therefore be excluded from the list of susceptible children during a particular month but contract measles in that month.

It should be noted that, in many cases, even after an apparent measles epidemic the number of susceptible children is not exhausted. An examination of measles incidence across compounds, however, shows that close to the entire pool of susceptible children is exhausted once measles is introduced. Further, many compounds, even within the same hamlet, escape measles while others may experience fairly major outbreaks. Thus hamlets in which measles occurs in a period may still have a large pool of susceptible children who escape the disease. Presumably there is only limited contact between compounds and this prevents a major spread of measles across compounds.

The results pertaining to vaccination are, perhaps, the clearest indication that the reporting of measles in Ngayokheme was quite accurate. A simple

Table 9.3. *Measles incidence reported among vaccinated and unvaccinated cohorts of children, by age, Ngayokheme*

Age (years)	Vaccinated		Unvaccinated	
	No. of cases	Per 1000 population	No. of cases	Per 1000 population
0	1	212.8	44	23.6
0.5	2	58.5	68	42.5
1.0	2	30.9	64	48.3
1.5	1	11.4	79	72.0
2.0	2	20.6	55	59.7
2.5	3	26.3	67	87.4
3.0	1	8.0	44	68.0
3.5	1	7.5	36	64.1
4.0	1	7.2	45	91.5
4.5	2	13.9	34	78.2

tabulation of the cases of measles recorded among vaccinated and unvaccinated cohorts of children lends credence to the accuracy of the data. These results are shown in Table 9.3. It is clear that vaccination against measles protects against the disease. This suggests that both vaccination and measles data were quite accurately recorded in the survey. Although 90 children in the study area were reported to have contracted measles after vaccination, it is conceivable for a genuine measles case to be observed after vaccination if the measles virus was in the stage of incubation at the time the vaccine was administered. There is therefore little reason to question the reliability of the measles data on the basis of the apparent failure of measles vaccination.

One striking fact from the analysis is the strong seasonality of measles in the study area. From Table 9.2 it is apparent that measles in Ngayokheme seldom occurs outside the months of January to April—the middle of the dry season. This pattern is maintained for all the years of the survey and is also consistent with earlier findings of strong seasonality of measles incidence (Gans *et al.* 1961; Meyer *et al.* 1964). The fact that measles outbreaks are not reported as more sporadic throughout the year suggests that rashes that may be associated with other viral diseases which occur randomly through the year are not being diagnosed incorrectly as measles.

There are indications that measles deaths, like measles cases, were quite accurately diagnosed by lay reporters in Ngayokheme. As would be expected if the data are accurate, all measles deaths in the study area can be associated with a measles case in either the same or previous month. As with cases, measles deaths are also concentrated in the months of the dry season—January to May.

To assess the reliability of the measles mortality data, the measles death rates for age groups of children were computed. The results are shown in Table 9.4 and are presented separately for males and females. In general, both male and female measles death rates are high during the first years of life and lower at higher ages. There is, however, much fluctuation in the figures with a peak death rate from measles at 2.5 years for girls and 1.5 years for boys. Above the age of 5 years children seldom are reported to have died from measles. One surprising fact, however, is the fairly large number of children (45) noted to have contracted measles under the age of 6 months (with a case-fatality rate of 15.6%). As noted earlier, measles seldom occurs before this age. However, the high fatality rate associated with early measles in Ngayokheme suggests the possibility that most of the cases reported were genuine measles. It is, however, likely that the diagnosis of measles at ages under 6 months was not as accurate as at other ages.

6. Discussion

Given the findings presented in the last section, it may be inferred that although data collected through lay reporting in Ngayokheme may not be

Table 9.4. *Measles incidence and death rate, by age and sex, Ngayokheme*

Age (years)	Males				Females			
	Incidence rate		Death rate		Incidence rate		Death rate	
	No. of cases	Per 1000 susceptible	No. of cases	Per 1000 susceptible	No. of cases	Per 1000 susceptible	No. of cases	Per 1000 susceptible
0	19	20.73	1	1.09	26	27.50	6	6.32
0.5	25	31.95	7	8.79	43	52.51	11	13.19
1.0	34	52.99	4	6.01	30	43.94	7	9.85
1.5	46	86.58	12	22.95	33	53.38	8	10.00
2.0	28	63.77	3	8.23	27	55.94	6	13.29
2.5	32	87.42	7	19.01	35	87.29	9	22.16
3.0	21	68.82	3	10.89	23	67.27	2	7.55
3.5	11	41.57	1	3.02	25	84.08	2	8.36
4.0	17	71.90	0	0.0	28	109.56	2	9.33
4.5	13	61.08	1	3.56	21	94.56	3	13.63
5.0	18	95.73	0	0.0	21	108.12	1	3.74

perfectly accurate, they appear to show a pattern consistent with the known epidemiology of measles. The reporting of measles cases, in particular, may have been particularly accurate but the reporting of measles deaths slightly less so. The data certainly provide good estimates of the importance of measles in Ngayokheme.

The examination of the spread of measles across hamlets suggests that measles cases were not observed in a sporadic manner across the study area. Instead, during specific periods measles cases were concentrated in clusters of adjacent hamlets. It is not surprising that measles was not limited to one cluster of adjacent hamlets during each period. With migration and other forms of mobility, cases of measles could be introduced into hamlets which may be quite a distance from where the original case was identified.

It would be expected that not all types of morbidity and cause-of-death data collected through lay reporting are likely to be as accurate as the measles data appear to be. One fact about measles and the data-gathering effort in Ngayokheme contributes in a major way to the apparent high degree of accuracy of the data. As noted earlier, measles is an easily recognizable disease and is known and feared by mothers in the study area. Further, its highly seasonal pattern of occurrence facilitates its identification. Evidently the ability to recall the symptoms surrounding a child's illness and death depends, in part, on the alertness of the mother to these symptoms. In Ngayokheme, this alertness was facilitated by the fact that the disease in question was well known and recognized. This suggests that the quality of data gathered through lay field interviewers may be improved if respondents are also made aware of a need to observe symptoms surrounding illness and death.

Evidence that measles-related data can be accurately reported based simply on clinical signs has previously been shown by Gemert *et al.* (1977). Their demonstration, matching diagnoses based on clinical signs to those based on laboratory testing, showed that in most cases of clinically identified measles in the Machakos district of Kenya, the diagnosis was correct. In Machakos, as in Ngayokheme, measles was a well-known disease and showed a seasonal and geographic distribution similar to those found for Ngayokheme.

It should be emphasized that the morbidity and mortality data for Ngayokheme appear to be reliable even though the data were derived from interviews which were spaced by 12-month intervals. Given that events and dates which occurred in a year prior to an interview may be forgotten, this study suggests how accurate health data can be produced by a more continuous lay reporting system. If lay reporting of health data is adopted within a context of more regular surveys or a continuous registration system, more accurate information on the dates on which events occurred and a high level of completeness in the recording of events may be achieved.

7. Conclusions

Based on an assessment of the quality of measles morbidity and cause-of-death data collected in Ngayokheme by lay reporters, this paper concludes that the reporting of health data by lay field-workers has the potential to increase the availability of reliable data for less developed contexts. The analyses performed in the paper suggest that data on measles morbidity and mortality as recorded by the field-workers are certainly accurate indicators of the importance of measles in Ngayokheme. The reliability of the measles data probably derives from the fact that measles was very well recognized and feared by residents of Ngayokheme as a common childhood disease. Evidently, other diseases, especially those which do not show as many clinical signs as measles, are likely to be less accurately recorded by lay field-workers. The recording of even these diseases, however, could be reasonably accurate with a more continuous system of data gathering and education of respondents about the clinical signs surrounding illness and death.

References

Barkin, R.M. (1975), 'Measles Mortality: An Analysis of the Primary Cause of Death', *American Journal of Diseases of Children*, 129(3), 307–9.

Borgono, J.M., and Greiber, R. (1971), 'Measles: A Five-Year Experience with a Vaccination Program in Chile', in *International Conference on the Application of Vaccines against Viral, Rickettsial and Bacterial Diseases of Man*, Scientific Publication no. 229 (Washington, DC: Pan American Health Organization).

Cantrelle, P. (1969), *Étude démographique dans la région du Siné Saloum (Sénégal): État civil et observation démographique* (Paris: ORSTOM).

Dekkers, N.W.H.M. (1981), *The Cornea in Measles* (Dordrecht: Dr W. Junk Publishers).

Florman, A.L., and Agatson, H.J. (1962), 'Keratoconjunctivitis as a Diagnostic Aid in Measles', *Journal of the American Medical Association*, 179, 568–70.

Fraser, K.B., and Martin, S.J. (1978), *Measles Virus and Its Biology* (New York: Academic Press).

Frost, W.H. (1979), 'Some Conceptions of Epidemics in General', *American Journal of Epidemiology*, 103(2), 141–50.

Gans, B., Macnamara, F.N., Morley, D.C., Thompson, S.W., and Watt, A. (1961), 'Some Observations on the Epidemiology of Measles in West Africa', *West African Medical Journal*, 10, 253–7.

Gemert, W., Valkenburg, H.A., and Muller, A.S. (1977), 'Machakos Project Studies: Agents Affecting Health of Mother and Child in a Rural Area of Kenya, ii: The Diagnosis of Measles under Field Conditions', *Tropical and Geographical Medicine*, 29(3), 303–13.

Guyer, B., and McBean, A.M. (1981), 'The Epidemiology and Control of Measles in Yaoundé, Cameroun, 1968–75', *International Journal of Epidemiology*, 10(3), 263–9.

Hayden, R. J. (1974), 'The Epidemiology and Nature of Measles in Nairobi before the Impact of Measles Immunization', *East African Medical Journal*, 51(2), 199–205.

Kenya Ministry of Health and the World Health Organization (1979), 'Measles Immunity in the First Year after Birth and the Optimum Age for Vaccination in Kenyan Children', *Bulletin of the World Health Organization*, 55, 21–31.

Kupka, K. (1981), 'Lay Reporting of Health Information', *World Health Forum*, 2(2), 211–17.

Langmuir, A.D. (1980), 'Changing Concepts of Airborne Infection of Acute Contagious Diseases: A Reconsideration of Classic Epidemiologic Theories', *Annals of the New York Academy of Sciences*, 353, 35–44.

May, J.M. (1958), *The Ecology of Human Disease*, Studies in Medical Geography, no. 2 (New York: MD Publications).

Meyer, H.M. (jun.) *et al.* (1964), 'Response of Volta Children to Live Attenuated Measles Virus Vaccine', *Bulletin of the World Health Organization*, 30, 769–81.

Miller, F. J. W., Knox, E. G., Court, S. D. M., and Brandon, S. (1974), *The School Years in Newcastle Upon Tyne, 1952–1962* (London: Oxford University Press).

Morley, D.C. (1973), *Paediatric Priorities in the Developing World* (London: Butterworth).

Panum, P.L. (1940), *Observations Made during the Epidemic of Measles on the Faroe Islands in the Year 1846* (New York: American Publishing Association).

Papp, K. (1956), 'Experiences prouvent que la voie d'infection de la rougeole est la contamination de la muquese conjonctivale', *Revue d'Imer Ther Antimier*, 20, 27–36.

Reinish, E.H., and Minear, R.E. (1978), *Health of the Pre-school Child* (Chichester: Wiley).

Sinha, D.P. (1977), 'Measles and Malnutrition in a West Bengal Village', *Tropical and Geographical Medicine*, 29, 125–34.

Stedman's Medical Dictionary (1982), 24th edn. (Baltimore, Md.: William and Wilkins).

Sydenham, T. (1939), 'Selected Writings', in *Medical Classics*, 4(4), 304–20.

WHO (World Health Organization) (1978), *Lay Reporting of Health Information* (Geneva: WHO).

—— (1980), *Lay Reporting of Health Information for Primary Health Care*, Report of a meeting (SEARO), New Delhi, 27–31 Oct.

Part II

Models and Methods of Analysis Applied to Mortality

10 A Promising Source of Data for Measuring Childhood Mortality in Africa: Maternity or Pre-natal Consultation Registers: The Case of Abidjan, 1980

PHILIPPE FARGUES

Institut National d'Études Démographiques (INED), Paris, France

---

1. Introduction

Public health services in Africa rarely provide death statistics. Available publications contain only statements of their equipment and personnel, which are sometimes accompanied by a count of nurse and physician consultations and sparse morbidity data. In some countries, however, health services do collect data which, if properly processed, make mortality analysis possible. This is the case of the death registers filled in upon delivery of the burial certificate. These registers sometimes provide much more complete data than the vital records, and may even contain information on the cause of death, as is the case in Bamako. This chapter does not discuss the methods of processing such data, as they are the same as those used for assessing the coverage of vital registration and are amply expounded elsewhere. Instead, we consider a source which seems very promising: maternity registers and pre-natal or infant welfare consultation registers.

The nature of a pregnancy and delivery and the health of the new-born child are not unrelated to the history of the preceding births. Thus, frequently at the time of a pre-natal, maternity, or infant welfare consultation, the physician or nurse register systematically information on this earlier history, in particular the survivorship of the children previously born. When this information is recorded in a health book that passes on to the child's family, its use for demographic purposes requires a sample survey, which is a rather cumbersome data collection procedure. When, on the other hand, it is recorded in a register which stays within the health unit, it provides a continuous system of data collection, eliminating the complications and expenditures of a population survey.

Introduced by Brass (1969) 15 years ago, the idea of using the data collected regularly at the time of a delivery for measuring childhood mortality has only recently been developed for the first time (Macrae 1982). These new methods are presented in this chapter. We deal in turn with three types of data likely to be provided by these registers:

1. proportion of children dead by age of mother, all birth orders combined;
2. proportion of children dead among first and second preceding births (that is, last- and next-to-last-born);
3. distribution of dead children by age at death, all birth orders combined.

We focus only on information involving *preceding* births and, therefore, excluding the current birth in the case of maternity registers and the last-born child (necessarily surviving) in the case of infant consultation registers.

2. Proportion of Children Dead: All Birth Orders Combined

Apart from a few rare exceptions, all population censuses and surveys conducted in Africa record the proportions dead among children ever born. Derived from the now 'classic' Brass method, several simple techniques can be used for converting this proportion $d(a)$ among women aged a into an estimate of the probability of dying between birth and an age $x(a)$, ${}_{x(a)}q_0$ in the life table of the children born to these women. Using multiplying factors k_a, where

$${}_{x(a)}q_0 = k_a d(a)$$

we observe a fixed relation between a (ages 15–19, 20–24,. . ., 45–49) and $x(a)$ (age 1, 2,. . .,20). These methods, however, do not permit a straightforward conversion of the proportions of children dead computed from the data in a maternity register into probabilities of dying.

2.1 Characteristics Specific to a Maternity Sample

The women recorded in a maternity register must necessarily be at the end of pregnancy. Instead of being randomly distributed over the reproductive life cycle, as is the case with a random sample of women taken in the total population, these women are all observed at the same point in this cycle: confinement, which is the condition for entering a maternity hospital. For this reason, we call the sample *conditional*. Figure 10.1 gives a theoretical representation of the basic differences between the two sampling procedures. The conditional sample has three important characteristics:

2.1.1. Elimination of the first year of the retrospective period At the time of maternity, the last preceding child is at least 9 months old. In the conditional sample, all the preceding births have thus already been exposed to the risk of dying during the first nine months of life and even during the

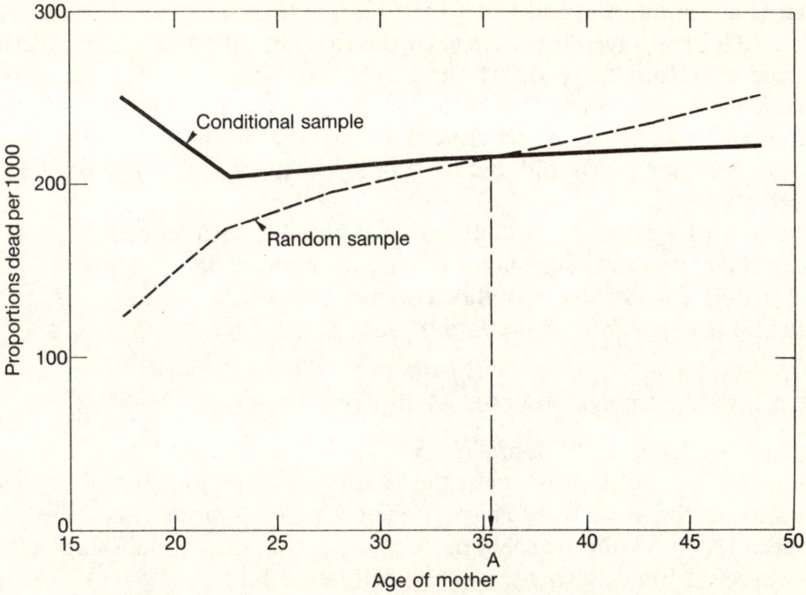

Fig. 10.1. Proportions of children dead, by age of mother, in a conditional and a random sample. Theoretical representation

entire first year if we consider postpartum sterility. The proportions dead are therefore higher, *ceteris paribus*, than in the random sample. Not represented in the conditional sample, the children under 1 year of age represent in the random sample a proportion which increases with the youth of the mothers. As mortality is highest during the first months of life, it follows that an increase in the mother's age has less effect on the increase in *d(a)* in the conditional sample. Consequently the slope of the proportions dead on age of mother should be gentler in this latter sample.

2.1.2. Elimination of women who are no longer fertile By interviewing only women who are at the end of pregnancy, one neglects the children of women of the same age who have stopped child-bearing. These children are older and, *ceteris paribus*, experience higher proportions dead. This second bias works in the opposite direction to the first. After a certain age *A*, probably around age 35, beyond which sterility gradually affects all women, the second effect must overtake the first and the lines representing the relation between age of mother and proportions dead in the conditional and unconditional samples must cross.

2.1.3. Predominance of the first birth effect among younger women The younger the women, the higher the proportion of first-born among the preceding births. In Africa, mortality among first births is appreciably higher than among the following birth orders. In the conditional sample, these first births have all experienced the first, maximal-mortality year of life, and therefore the proportions of children dead should normally be higher among women aged 15–19 than those aged 20–24. This is not the case for the random sample, where the children of the youngest women are, on average, less than a year old and have not yet experienced the entire infant mortality period.

The conditional sample is contaminated by other sources of bias as well; in particular, it is not representative since it involves only the women who give birth in a maternity hospital. However, these biases have no specific implications regarding the method of data processing.

2.2. A First Approach: Adapted Multiplying Factors

Proposed by Brass and Macrae (1984, 1985) the first method consists of recreating the conditions whereby the classic Brass methods can be applied, by reconstructing what would have been observed in a random sample if this had been taken within the same population as the conditional sample. The data observed for a given age a are as follows:

- N: size of the sample of women. Apart from the cases of multiple birth and stillbirth, this corresponds to the number of current births;
- NP: preceding births;
- D: deaths among the preceding births.

If instead of being selected at the time of a maternity, the women were randomly distributed in relation to current birth, the birth would already have occurred in $N/2$ of cases and not yet have occurred in the remaining $N/2$ cases. The total number of children would thus be equal to $NP+N/2$. To D, the number of deaths among NP, would be added D', the number of deaths among the last-born $N/2$. Having been exposed to the risk of death for a shorter period of time than their elders, these last-born children would have lower proportions dead:

$$\frac{D'}{N/2} < \frac{D}{NP}.$$

If we call d and d' the proportions dead among children of women in the conditional and random samples respectively, then

$$d' < d \text{ and } d' < D/NP.$$

As D' is not zero, we have

$$d' > D/(NP+N/2).$$

These inequalities give

$$d' = D/(NP+\alpha N) \text{ where } 0 < \alpha < 1/2. \qquad (1)$$

Brass and Macrae propose a unique value of $\alpha = 0.2$. To apply the classic method to the proportions thus corrected and calculate the k_a coefficients, we must also know the average parity of the women in the first three age groups in the random sample. For the women in the maternity sample, it would be:

$$P = (NP + N/2)/N \text{ or } P = NP/N + 0.5. \quad (2)$$

By definition, these women have all been exposed to the risk of conception. They therefore represent the sub-population of 'married' women. For populations where 'a significant proportion of women are not exposed to the risk before age 20' (Brass and Macrae 1984), the authors recommend multiplying P_2 and P_3 obtained through equation (2) by the proportion of married women at these ages. A source other than the maternity register is naturally required for this information.

Table 10.1 illustrates this technique as applied to data collected in Bamako's maternity hospitals (Hill 1986). It offers an enormous advantage: access, by means of a series of very simple calculations, to a well-established and unquestionably effective method. The advantage of being so easy to use, however, is somewhat offset by certain drawbacks, three of which are discussed below:

Table 10.1. *Application of the adapted multiplying factors method to maternity data, Bamako*

	Age of mother (a)				
	15–19	20–24	25–29	30–34	35–39
No. of women (N)	1353	1560	1636	965	536
Preceding births					
Total (NP)	444	2390	5244	4898	3805
Dead (D)	103	504	1110	1118	834
Proportion of children dead (per 1000)					
Crude (D/NP)	232	211	212	228	219
Reconstructed ($D/NP+0.2N$))	144	186	199	220	213
Parity 'married' women ($NP/N+0.5$)	0.828	2.032	3.705	5.576	7.599
Proportion of 'married' women	0.304	0.650	0.835	—	—
Re-estimated parity, all women*	0.252	1.322	3.094	—	—
Multiplying factor (Trussell West) (k_a)	0.954	1.041	1.019	1.034	1.055
Age (x_a)	1	2	3	5	10
Death probabilities (per 1000) ($_{x_a}q_0$)	137	194	203	227	225

* Ratio of 'parity "married" women' to 'proportions "married" women'.

Source: Hill 1986.

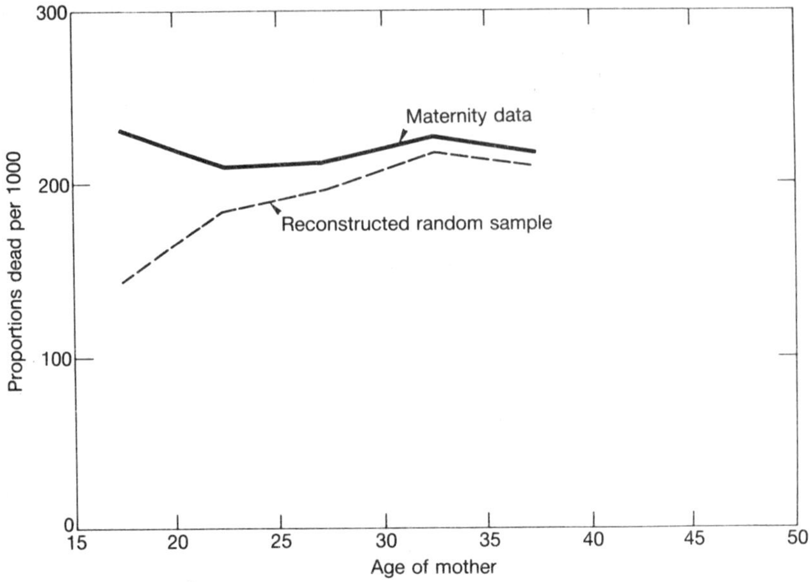

Fig. 10.2. Proportions of children dead, by age of mother, maternity data, Bamako, 1985. *Source*: Hill 1985

1. Although the 'reconstructed' random-sample data are treated as though they come from a real random sample, one of the sources of bias in the conditional sample remains: the selection of currently fertile women and the elimination of the cases of secondary sterility. This is shown by comparing Figures 10.1 and 10.2: the expected crossing does not occur, because equation (1) necessarily leads to lower proportions dead in the 'reconstructed' random sample than in the original conditional sample. Eventual excess mortality among the children of the most fertile women may, however, reduce this bias.

2. The use of the α parameter introduces a non-negligible error in the estimates of the first two probabilities, ${}_1q_0$ and ${}_2q_0$. If, instead of fixing it arbitrarily at 0.2, we opt for alternative values between 0.1 and 0.3, the proportion d' will also be located in an interval which, however, will be wider since D is small compared to N (Table 10.2, columns 2 and 3).

3. The estimation of the parties in the random sample from those in the conditional sample, which is necessary for applying the Brass method or the Trussell variant, depends on two unknowns:

(*a*) the β *parameter* for correcting the fertility-related selection effect in the conditional sample. Equation (2) is in fact only a specific case of the more general expression of the form

$$P = NP/N + \beta,$$

(Fargues 1985) where β is fixed at 0.5.

Table 10.2. *Additional error introduced by the adapted multiplying factors method, Bamako*

	Age of mother (a)				
	15–19	20–24	25–29	30–34	35–39
Proportions dead (per 1000) ($D/(NP+\alpha N)$)					
$\alpha = 0.3$	121	176	194	215	210
$\alpha = 0.1$	178	198	205	224	216
Parity of women 'exposed' ($NP/N+\beta$)					
$\beta = 1/3$	0.661	1.865	3.538	5.409	7.431
$\beta = 2/3$	0.995	2.199	3.872	5.743	7.765
Proportion of women 'exposed' (per 1000)					
min.[a]	304	650	835	—	—
max.[b]	536	825	917	—	—
Age (x_a)	1	2	3	5	10
Multiplying factors (Trussell West) (k_a)					
min.[c]	0.737	0.963	0.990	1.023	1.040
max.[d]	1.007	1.059	1.025	1.036	1.055
Death probabilities (per 1000) ($_{x_a}q_0$)					
min.[e]	89	169	192	220	218
max.[f]	179	210	210	232	228
Additional relative error (%)	±34	±11	±4	+3	±2

Note: The minima and maxima correspond to the following combinations:

[a] 'non-married' women non-exposed to the risk of conception;

[b] 'non-married' women exposed to the risk of conception (33% before age 20, 50% after age 20);

[c] = 2/3 and maximum proportion of women exposed;

[d] = 1/3 and minimum proportion of women exposed;

[e] = 0.2 and *Ka* minimum;

[f] = 0.1 and *Ka* maximum.

(*b*) the exact proportion of women exposed to the risk of conception: to use the proportion of married women as a proxy for exposure to conception as was the case for Bamako (Hill 1986) is unsatisfactory for most populations in Black Africa, where illegitimate births are frequent and categories between celibacy and marriage, ignored by population censuses, do actually exist. In fact, the proportion of women who are 'non-married' but exposed to the risk of conception may well be as high as a third of 'non-married' under age 20 and half of those over age 20.

The combination of errors of types 2 and 3 leads to a relatively wide range of estimated probabilities for women aged under 25: 34% for $_1q_0$ and 11% for $_2q_0$ (Table 10.2, final column). As biases of type 1 may be substantial beyond age 30 thus invalidating the estimate of $_5q_0$, it seems reasonable to recommend this method for estimating $_3q_0$.

Table 10.3. *Symbols for mean length of exposure and proportions dead according to birth order*

Birth order	Mean length of exposure	Proportions dead
n	I_n	$Q(kI_n)$*
$n-1$	$2I_n$	$Q(2I_n)$
1	nI_n	$Q(nI_n)$

* Where $0 < k < 1$.

2.3 A Generalization of the Method: The Calculation of Exact Ages

Faced with such problems in recreating the conditions necessary for applying the usual methods, we may consider another procedure. Rather than transforming the crude data into a hypothetical one, we could attempt to determine the relation between the two series specific to the population studied (Fargues 1985). For the time being, let us continue to consider the case of a maternity register. Let $F_{a,n}$ be women of age a and of parity n (excluding the current birth). Assuming that I_n is the length of the interval between successive births and that this does not vary with birth order, the first preceding birth (or last-born child) goes back I_n years on average, the second preceding birth $2I_n$ years, and so on until the birth of the first-born that goes back nI_n years. If, at the same time, we assume a life table which is invariable for all these children (defined by the probability $Q(x)$), the proportions dead by birth order are as shown in Table 10.3. A correcting factor k, around 0.8 (Brass and Macrae 1984), is used to take into account the fact that the first preceding birth occurred during an interval (between nine months and five years) which corresponds to ages when the instantaneous risk of dying decreases very rapidly with age.

The mean number of children dead per woman of parity n is thus:

$$D_n = Q(kI_n) + Q(2I_n) + \ldots + Q(nI_n). \tag{3}$$

Regardless of parity, women of age a experience $\sum_n F_{a,n}.D_n$ children dead out of a total of $\sum_n . F_{a,n}$ children born alive. The proportion of children dead among women of age a is thus equal to:

$$d(a) = \frac{\sum_n F_{a,n}.D_n}{\sum_n n.F_{a,n}}. \tag{4}$$

This proportion corresponds to a death probability $Q(x_a)$. To determine x_a we use a 'standard mortality' and apply equations (3) and (4) to the following data:

1. the $F_{a,n}$ distribution in the register studies;
2. the value of I_n in the reference population;
3. the series $Q^*(x)$ from a reference life table, such as Brass's 'African' model.

We thus obtain the proportion of children dead $d^*(a)$ that would be observed in the population studied if its mortality followed the pattern in the Brass model. The ages x_a can then be obtained by simply relocating $d^*(a)$ in the series $Q^*(x)$ (see Figure 10.3). We have demonstrated elsewhere (Fargues 1985) that the ages x_a obtained in this way are practically independent of the reference life table chosen and are strongly dependent on the length I_n of the mean interval between successive births, but only slightly dependent on the parity-related variations of this interval, given the mean I_n. The latter parameter is thus the only information from a source other than the register studied required for applying this method.

In contrast to the previous method, the modified one is not limited to data collected at the time of a maternity: it is easily extended to calculating proportions dead from data collected some time before or after the confinement.

Where data are collected at the time of a pre-natal consultation the women are observed at a mean distance G from the normal term of their pregnancy, and equation (3) becomes:

$$D_n = Q[k(I_n - G)] + Q(2I_n - G) + \ldots + Q(nI_n - G), \qquad (5)$$

with the coefficient k being now smaller than in (3).

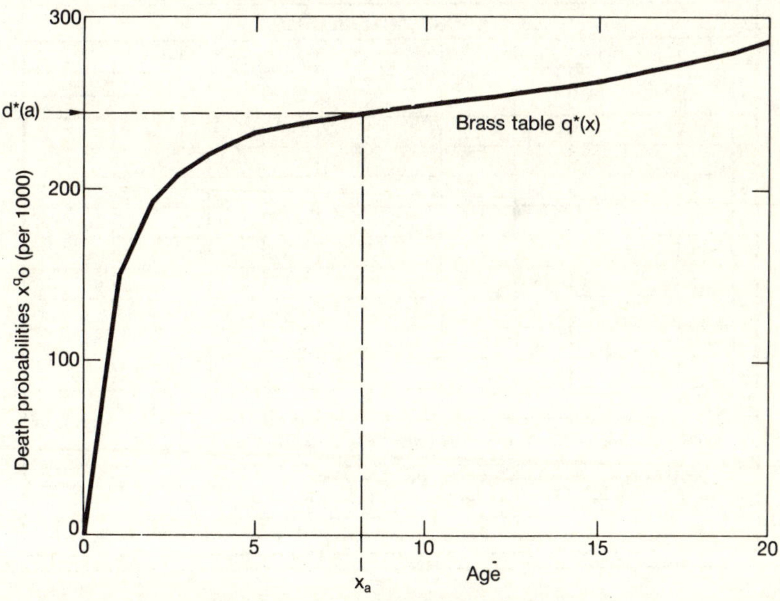

Fig. 10.3. x_a determined by using $d^*(a)$

When data on preceding births are collected at the time of the first infant health consultation for the new-born with mean age A, equation (3) becomes:

$$D_n = Q[k(I_n + A)] + Q(2I_n + A) + \ldots + Q(nI_n + A) \qquad (6)$$

and coefficient k is now larger than in (3).

This method has been applied to the register kept over the last ten years (with varying regularity) by the Pre-natal Care Centre of the National Institute for Public Health, Abidjan. To date only a small, experimental fraction of this register has been studied: 4110 records for 1980, representing less than half of the total consultations for that year. Although this is acceptable for illustrating the utility of such registers it is clearly insufficient for a detailed analysis of differential mortality.

A copy of the forms used in the Pre-natal Centre is displayed in Figure 10.4. As is often the case with documents not originally designed for statist-

REPUBLIQUE DE COTE D'IVOIRE

CONSULTATION PRÉNATALE PROTECTION MATERNELLE ET INFANTILE

MINISTERE DE LA SANTE PUBLIQUE
Département ______
Centre de ______

No. ______ Année ______
A B C D

NOM: ______ NÉE: ______ AGE: ______
ADRESSE: ______ ETHNIE: ______
PROFESSION: ______ EMPLOYEUR: ______
NOM DU MARI: ______ ADRESSE: ______
PROFESSION: ______ EMPLOYEUR: ______

GESTATIONS ANTÉRIEURES

No.	DATE	DUREE			Mode d'Accouch	Complications Hémorragie Infect. Eclamp. Trav. prolongé Autres.	ENFANT				
		A vortem	Prématur	A terme			né-vivant	Mort-né	Poids	DECEDE	
										AGE	CAUSE
1											
2											
3											
4											
5											
6											
7											
8											
9											
10											

Antécédents gynécologiques (DPR - Caractère du cycle -Durée) Histoire de Stérilité ______

Antécédents familiaux Pathologies familiales Hist. de gemellaire ______

Antécédents chirurgicaux ______

Fig. 10.4. The official form presented at the Ivory Coast Pre-natal Care Centres

ical purposes, these are subject to certain pitfalls which we will not discuss here. It suffices to say that they have the decided merit of having been elaborated prior to the formulation of techniques discussed here and that they are likely to contribute original data on mortality in the Ivory Coast.

Table 10.4 shows the crude data collected from the 4110 forms, along with the ages x_a calculated by applying equations (4) and (5) to the age and parity distribution of these women and to the probabilities from the Brass African standard life table. As the estimated date of confinement is indicated on the forms, the G parameter could be calculated precisely: at the time of the first pre-natal visit, the mean duration of pregnancy was 5.5 months. In the sub-population studied, the length of the mean interval ranged between 30 and 36 months and centred on 33 months, the Ivory Coast national average. The findings are as follows: the proportion dead among all children ever born for women aged 20–4, 102 per 1000, is an estimate of the probability of dying between birth and an age between 2.5 and 3 years, that is, ${}_{2.7}q_0$ on average. It is noteworthy that the ages x_a estimated from the conditional sample cover a smaller range than those for a random sample: 2.1 to 9.8 years in the former against 1 to 20 years in the latter.

Like the first method, the second entails an element of uncertainty. In the Brass and Macrae technique the probability of dying before a given birthday is located within a range. In this second method it is the age up to which a given probability applies which is estimated by a range. Similarly, in the first method the range narrows as the mother's age increases whereas in the second method the opposite occurs. To compare the two types of uncertainty, we recalculate by interpolation on the figures in Table 10.4,

Table 10.4. *Death probabilities before age* x *and corresponding ages* x *computed from pre-natal consultation data, Abidjan, 1980*

Age of mother a	No. of women N	Preceding births		Death probabilities (per 1000) D/NP $_{x_a}q_0$	Age x_a Mean birth interval (years)		
		Live-born NP	Dead D		30 months	33 months	36 months
15–19	1084	543	54	99	1.9	2.1	2.3
20–24	1204	1842	188	102	2.5	2.7	3.0
25–29	765	2251	259	115	3.6	4.0	4.5
30–34	339	1495	191	128	4.9	5.8	6.7
35–39	155	862	139	161	6.8	8.3	9.7
40–44	54	342	47	(137)	7.9	9.8	11.6
45–49	8	45	5	(111)	7.9	9.8	11.7
Unknown	501	1083	170	*	*	*	*
TOTAL	4110	8463	1053	*	*	*	*

Notes: * computation impossible; ( ) proportion dead not significantly different from that observed at 35–9 years with a level of 95%.

Table 10.5. *Death probabilities interpolated between birth and second, third, and fifth birthdays, pre-natal consultation data, Abidjan, 1980*

Probabilities (per 1000)	Mean birth interval			Relative error (%)
	30 months	33 months	36 months	
$_2q_0$	100	99	98	±1
$_3q_0$	108	105	102	±3
$_5q_0$	130	122	118	±5

Table 10.6. *Death probabilities between birth and age* x *by mother's ethnic group, pre-natal consultation data, Abidjan, 1980*

	Age of mother (a)				
	15–19	20–24	25–29	30–34	35–39
Preceding births					
Live-born (NP)					
Akan	145	510	617	385	309
Other Ivory Coast	275	841	1095	741	380
Alien	119	472	521	345	159
Dead (D)					
Akan	13	47	66	44	52
Other Ivory Coast	29	93	124	90	59
Alien	12	46	69	52	28
Death probabilities (per 1000) (D/NP) ($_{x_a}q_0$)					
Akan	90	92	107	114	168
Other Ivory Coast	105	111	113	121	155
Alien	101	97	132	151	176
Age (years) (x_a)					
Akan	2.06	2.86	3.96	5.81	10.54
Other Ivory Coast	2.09	2.74	4.11	5.81	6.85
Alien	2.14	2.65	3.87	6.12	8.32

Note: Ages x_a are computed using equation (5), with a mean birth interval constantly equal to 33 months.

'Death probabilities', the conventional probabilities up to age 2, 3, and 5 obtained by our method. We then observe (see Table 10.5) a narrower range for the estimates of $_2q_0$ and $_3q_0$ and a wider one for the estimate of $_5q_0$.

As is the case in data from a random sample, the proportions dead among preceding births to women who come for a pre-natal visit are of interest for assessing mortality differentials. In Table 10.6, for example, we observe that the Akans, an indigenous ethnic group living in southern Ivory Coast, have lower mortality than the alien ethnic groups. The social homogeneity observed for the women who come to the Centre for consultation (mostly working class) suggests the influence of regional factors. In the south, and

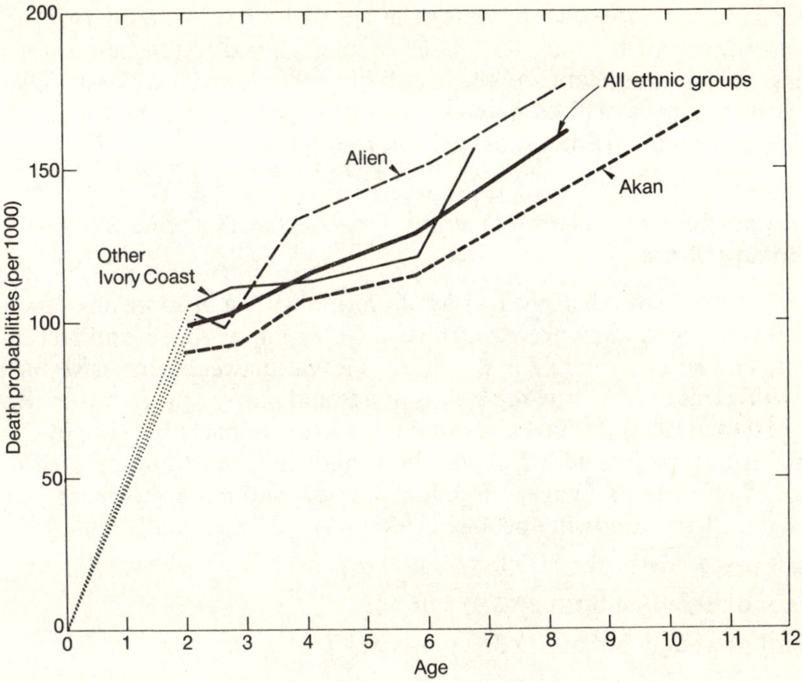

Fig. 10.5. Probabilities of dying ${}_xq_0$ before age x, by mother's ethnic group. Pre-natal consultation data, Abidjan, 1980

in Abidjan in particular, there is better coverage of both health and education services. For all the women who have immigrated from rural areas or other countries, who may represent a high proportion of all the women consulting, retrospective data of this kind reflect a situation which is midway between that of their place of origin and that of the place where the data are collected (Abidjan), since a fraction of all children were born before the immigration to Abidjan.

Fig. 10.5 shows that the estimates of x_as are quite sensitive to variations in the $F_{a,n}$ distribution. For instance, if we had used only the crude data for comparing the Akans with the other autochthonous ethnic groups (Table 10.6), we would have concluded that mortality between ages 0 and 10 (mothers aged 35–9) is higher among Akan children. Our method reveals that, due to the fertility differentials by ethnic group, we should compare mortality between ages 0 and 10.54 years for the Akans with mortality between ages 0 and 6.85 years for the other ethnic groups. The comparison shows that Akan mortality levels are lower than in the other groups. This illustrates how misleading can the crude proportions *d(a)* be as indicators of mortality and, simultaneously, confirms the utility of the method proposed here. We should caution, however, against uncritical applications of this method. In fact, note that adopting the same length of birth interval for all groups may lead to under-estimates of mortality differentials: since the timing of an infant death and the length of birth interval are closely related, the estimates of x_a will be over-estimated for groups with high mortality, resulting in under-estimates of their mortality levels.

3. Proportions of Children Dead by Birth Order: First and Second Preceding Births

Equation (3) shows that given constant mortality the proportions dead are $Q(0.8I)$ among the first preceding births, $Q(2I)$ among the second preceding births, and so on, where I is the mean interval between successive births. Assuming this to hold true regardless of age and parity of the mother, Brass and Macrae (1984) derived a technique which is remarkably simple. As in Africa it hovers around 2.5 years, the proportions dead among preceding births of mothers of all ages combined provide, without adjustments, good estimates of the following probabilities:

first preceding birth: $Q(0.8I)$ or ${}_2q_0$
second preceding birth: $Q(2I)$ or ${}_5q_0$
third preceding birth: $Q(3I)$ or ${}_{7.5}q_0$
etc.

Going back beyond the third preceding birth is probably too risky. For one thing, the higher n the longer the variance of the actual distribution of the date of the nth preceding birth around a mean date ($n.I$ years before the survey). In addition, women who are not removed when considering higher values of n are selected for higher fertility and the latter is related quite strongly to the mortality levels to which their children have been exposed.

Brass and Macrae recommend adding two questions to those usually asked in maternity hospitals: are your last and next-to-last live-born children still alive today? Although apparently very simple, these questions left an element of uncertainty when asked during the Bamako survey (Hill

Table 10.7. *Childhood mortality estimated from proportions dead among first and second preceding births, Bamako*

Mother	Births		Deaths		Proportions dead	
	First preceding birth	Second preceding birth	First preceding birth	Second preceding birth	First preceding birth	Second preceding birth
Literate	914	624	85	83	0.093	0.133
Illiterate	3861	3113	594	537	0.154	0.173
TOTAL	4775	3737	679	620	0.142	0.166

Source: Hill 1986.

Table 10.8. *$_2q_0$ and $_5q_0$ estimates obtained by two methods, Bamako*

Death probabilities (per 1000)	$_2q_0$	$_5q_0$
Adapted multiplying factors	0.194	0.227
Proportions dead by birth order	0.142	0.166

Source: Hill 1985.

1986). This survey provides two estimates for $_2q_0$ and $_5q_0$ (see Tables 10.7 and 10.8). The relative difference observed between these two estimates (about 0.3) is too high to be uncritically accepted. Is this due to the fact that the reference period for the second estimate is slightly more recent than for the first? The interval between these two periods does not, however, exceed a year. This cannot, therefore, account for the fact that the absolute differences between the first and second estimates are 52 per 1000 and 61 per 1000 for $_2q_0$ and $_5q_0$ respectively. Since it is unlikely that the first method over-estimates mortality, we suspect that the second introduces a downward bias. One possible explanation is that the data on last- and next-to-last-born children overrepresents children who have survived.

4. Distribution of Dead Children according to Age at Death and Age of Mother

When the outcome of each pregnancy and the trajectory of each live birth (survival, death, and age at death) are recorded, we obtain complete birth histories. Birth histories present no particular difficulty for the construction of life tables. However, recording of these histories is a long and painstaking operation which cannot simply be added to the questions usually asked at the time of maternity or pre-natal consultation, particularly in view of the conditions under which these interviews are normally conducted in Africa. For instance, the form filled in at the Ivory Coast Pre-natal Centre leaves a line for each previous pregnancy to enable a full reconstruction of the woman's birth history and, although the information thus obtained proves to be always somewhat incomplete, it is nevertheless sufficient for estimating a partial life table for the earlier periods of life.

We define the following quantities:

- a is the current age of mother;
- y is her age at the time of the preceding births, $13 \leq y \leq a-1$;
- x is the child's age at death, as reported by mother;
- $D_{a,x}$ is the number of children of mothers of current age a, who died at age x;
- P_a is the mean parity of women of current age a;
- F_a is the number of women of age a.

Since only children born x or more years before the consultation can have died at age x, we will have

$$x \leq a-y \text{ or } y \leq a-x$$

for children born to women of current age a when they were aged y. If mortality, represented by the functions $l(x)$, $q(x)$, and $d(x)$ in the life table, and fertility, represented by $f(x)$, which is assumed to be null before age 13, were invariable in the recent past, the following equalities should hold:

$$D_{a,x}=F_a\int_{13}^{a-x} f(y)\, S(x)\, q(x)\, dy,$$

$$D_{a,x}=F_a d(x)\int_{13}^{a-x} f(y)\, dy,$$

$$D_{a,x}=F_a\, d(x)\, P_{a-x}.$$

The deaths in the table could thus be deduced from the observed series as follows:

$$d(x)=\frac{D_{a,x}}{F_a P_{a-x}}.$$

In discrete notation, we have the following approximation:

$$d(x,x+1)=\frac{D_{a,a+5;x,x+1}}{F_{a,a+5}P_{a+2-x}}. \tag{7}$$

As age $(a+2.5)$ is the mid point of the age group $(a,\ a+5)$, the children exposed to the risk of dying between x and $x+1$ are in fact those born before the mother reached an age between $(a+2.5-x)$ and $[a+2.5-(a+1)]$, or $(a+2-x)$ on average. In order to apply equation (7) to the Abidjan data, we also had to estimate the values of P_{a+2-x} by linear interpolation in the $(P_{a,a+5})$ series observed in the register. Also, for each group of mothers we redistributed proportionally all deaths with unknown age at death.

Table 10.9 presents the crude data, Table 10.10 displays the deaths estimated by applying equation (7), and Table 10.11 shows the survivors and probabilities of dying in the life tables constructed from these deaths. The number of observations is too small and the heaping at certain months (three and six) too high for life tables by individual months of age to be reliable. The findings seem to indicate that we can, however, distinguish between the neonatal and post-neonatal periods and then each year up to age 5. The infant mortality rate obtained hovers around 80 per 1000, with a significantly higher proportion of post-neonatal than neonatal deaths. Child mortality $({}_4q_1)$ remains high with a mean probability of 64 per 1000. These estimates are fairly close to those obtained using the first approach,

Table 10.9. *Distribution of reported deaths, by age at death and age group of mother, Abidjan, INSP register, 1980*

Age at death	Current age of mother								
	< 20	20–24	25–29	30–34	35–39	40–44	45–49	Unknown	Total
Months									
0	16	40	62	19	21	13	2	28	201
1	1	11	7	6	10	4	0	8	47
2	3	5	5	5	2	0	0	3	23
3	5	10	13	8	6	1	2	15	60
4	3	9	6	1	2	0	0	3	24
5	1	3	5	3	3	0	0	1	16
6	3	7	7	15	6	0	0	6	44
7	3	2	7	3	4	1	0	2	22
8	2	9	10	6	1	1	0	1	30
9	1	10	3	5	2	1	0	3	25
10	0	2	1	3	1	0	0	0	7
11	0	3	3	0	0	0	0	1	7
Subtotal	38	111	129	74	58	21	4	71	506
Years									
1	10	24	49	36	19	8	0	32	178
2	0	15	15	24	14	6	1	15	90
3	1	5	9	6	11	1	0	6	39
4	0	3	0	3	1	2	0	4	13
5–9	0	5	5	5	7	2	0	1	25
10+	0	0	2	1	2	2	0	2	9
Subtotal	49	163	209	149	112	42	5	131	860
Age unknown	5	25	50	42	27	5	0	39	193
TOTAL DEATHS	54	188	259	191	139	47	5	170	1053
Total mothers	1084	1204	765	339	155	54	8	501	4110
Total live-born	543	1842	2251	1495	862	342	45	1083	8463
Mean parity	0.50	1.53	2.94	4.41	5.56	6.33	(5.63)	2.16	2.06

where $l(2)$ was estimated to be around 901 (against an average of 896 here) and $l(5)$ around 878 (against 866 here). This convergence suggests that both methods are relatively sound as they are independent of one another and use different data from the same register. However, the method employed here seems more attractive since it yields more detailed mortality data and provides information on both age at death and the age cohort of the mother. For instance, early maternity is observed to influence the child's risk of dying: mortality, in particular neonatal mortality, is appreciably higher among children of women aged under 20. For mothers older than 20, mortality increases with age of mother. The only exception is neonatal mortality that displays no systematic pattern. This may be due to either the period when birth occurred or to a selection effect regarding mother's age. The former is likely to be important if childhood mortality has declined

Table 10.10. *No. of deaths per 10 000 new-born in generation life tables by age group of mother, Abidjan, INSP register, 1980*

Age x	Current age of mother				
	< 20	20–24	25–29	30–34	35–39
Months					
0	407	270	360	169	309
1	25	74	40	53	147
2	76	34	29	49	29
3	127	67	75	71	88
4	76	61	35	9	29
5	25	20	29	27	44
6	76	47	41	133	88
7	76	13	41	27	59
8	51	61	58	53	15
9	25	67	17	44	29
10	0	13	6	27	15
11	0	20	17	0	0
Years					
0	966	749	749	658	854
1	—	188	316	344	292
2	—	141	109	248	225
3	—	59	75	67	185
4	—	47	0	37	17
5	—	—	79	85	153
10	—	—	—	—	—

Table 10.11. *Mortality indicators from generation life tables by age group of mother, Abidjan, INSP register, 1980*

	Current age of mother				
	< 20	20–24	25–29	30–34	35–39
Survivors at age x ($l(x)$)					
Age					
0	1000	1000	1000	1000	1000
1 month	959	973	964	983	969
1 year	903	925	925	934	915
2 years	—	906	894	900	885
3 years	—	892	883	875	863
5 years	—	882	875	865	843
10 years	—	—	867	856	827
Conditional death probabilities (per 1000)					
q_{nn}	40.7	27.0	36.0	16.9	30.9
q_{pn}	58.3	49.2	40.4	49.7	56.2
${}_1q_0$	96.6	74.9	74.9	65.8	85.4
${}_4q_1$	—	47.0	54.0	74.5	78.6
${}_5q_5$	—	—	9.0	9.8	18.2

Notes: q_{nn} = probability of dying at 0–1 month; q_{pn} = probability of dying at 1–12 months.

steadily during the recent past. The latter effects should operate when the most fertile women, who are overselected beyond age 30 by the conditional sampling procedure, have given birth to children with a higher risk of dying. The proportion of children born in rural areas, or the frequency of omissions of dead children, might also be higher among older women. This suggests that the observed trend is therefore not a reliable sign of mortality decline and one should not attempt to date the estimates of mortality.

5. General Discussion

In African countries, where the available information on childhood mortality is patchy to say the least, any system adopted for observing mortality must aim at two targets. Firstly, to guide health intervention the planner must identify specific needs as they are reflected in levels of mortality. The methods of measurement applied to specific situations must be assessed according to three criteria: representativeness of the populations covered, reliability of the estimates of mortality levels, and the degree of detail with which mortality differentials may be assessed. Secondly, the planner must be able to measure the effectiveness of intervention in order to make decisions about its continuation. To do this he must consider two additional criteria: whether a method of observation is permanent or only occasional, and its responsiveness to short-term variations in mortality.

What role can the methods reviewed in this paper play in epidemiological observation and assessment of health in Africa? How do they stand in terms of the above conditions?

Taking advantage of medical intervention for observing mortality means being limited to the population involved in this intervention. By and large these populations are unrepresentative of larger aggregates. Moreover, as the aim of setting up a health unit is to reduce mortality, it may be assumed that, *ceteris paribus*, measuring mortality in a population where such a unit operates will lead to underestimation. However, this bias is limited by the retrospective nature of the data collected: current attendance of a pre-natal care centre or maternity hospital does not affect past births. As for unrepresentativeness, this is less important among urban populations than in the countryside where health coverage often remains very low. In many African towns, particularly the larger ones, pre-natal consultations and delivery in a maternity hospital have become common practice. Consequently, under-estimation of mortality may be very limited. In fact, the opposite may be true. For instance, in the case of the Abidjan register a study of the composition of the population by social strata reveals over-representation of the underprivileged classes so that the data we have analysed involve a sub-population that experiences mortality levels above the town's average.

Regarding the reliability of estimates of total and group differential mortality, the method presents the same drawbacks and advantages as retrospective survey data allowing the application of the 'classic' Brass methods. The estimation of mortality levels is largely unsatisfactory, suffering from a combination of observational errors and lack of robustness to violations of assumptions about mortality and fertility regimes. Yet, provided that the biases are uniform, the indices obtained clearly reflect the mortality differences between the sub-populations defined by the most commonly recorded variables: mother's educational status, ethnic group, mother's occupation, father's occupation, place of residence, mother's previous medical history, etc. Systematic processing of these registers could therefore provide an effective means of analysing factors affecting mortality risks.

Do retrospective methods such as those suggested improve continuous mortality monitoring? The answer is a resounding no. The proportions dead among children ever born for all birth orders combined and for women in a given age group are indicators that react too slowly to mortality changes to be adequate to calibrate ongoing adjustments required by health interventions. The only indicator that could perhaps serve as a barometer for measuring mortality before age 2 is the proportion dead among last-born. But a blend of two sources is an alternative solution to the impasse: vital registration can be useful for its immediate reaction to short-term variations and seasonal fluctuations, while maternity or pre-natal consultation registers could be used for correcting, albeit crudely, the serious deficiencies of vital records below age 5. This combination of two administrative sources provides a fairly detailed description of childhood mortality, its trends, and seasonal fluctuations (Fargues 1986). Currently, however, its applicability is feasible in towns only, since vital registration in rural areas remains largely theoretical. Although this method of observation is by no means a solution to all the problems hampering mortality analysis in Africa, it is too inexpensive relative to the services it offers to be dismissed: data collection for our Abidjan survey cost exactly 130 000 CFA francs, or $350.

References

Brass, W. (1969), 'Disciplining Demographic Data', *International Population Conference*, London, i. 183–204.

—— and Macrae, S. (1984), 'Childhood Mortality Estimates from Reports on Previous Births Given by Mother at the Time of a Maternity, i: Preceding Birth Technique', *Asian and Pacific Census Forum*, 11(2), 5–8.

—— —— (1985), 'Childhood Mortality Estimates from Reports on Previous Births Given by Mother at the Time of a Maternity, ii: Adapted Multiplying Factors Technique, *Asian and Pacific Census Forum*, 11(4), 5–9.

Fargues, P. (1985), 'L'Observation des grossesses, une occasion de mesurer indirectement la mortalité des enfants dans les pays à statistiques déficientes', *Population*, 40(6), 891–909.

—— (1986), 'La mortalité infantile et juvénile à Abidjan de 1973 à 1983', In *Estimation de la mortalité du jeune enfant (0–5 ans) pour guider les actions de santé dans les pays en développement* (Paris: Les Éditions de l' INSERM).

Hill, A. (1986), 'L'Enquête pilote sur la mortalité aux jeunes ages dans cinq maternités de la ville de Bamako', In *Estimation de la mortalité du jeune enfant (0–5 ans) pour guider les actions de santé dans les pays en développement* (Paris: Les Éditions de l'INSERM).

Macrae, S. (1982), *Childhood Mortality Estimates from Non Random Data (Using Maternity Histories Collected at Birth Registration)*, Cairo Demographic Centre Working Papers (Cairo: Cairo Demographic Centre).

11 Evaluation of Model Life Tables for East Africa

DOUGLAS C. EWBANK

Population Studies Center, University of Pennsylvania Philadelphia, Pennsylvania, USA

1. Introduction

In the study of African demography, even the trends in the most basic estimates of fertility and mortality are a matter of uncertainty in many areas. In the absence of complete registration of births and deaths in virtually all of the continent, estimates of fertility and mortality must be made using simple mathematical models and indicators of mortality like the proportion of persons with deceased parents. At the heart of these models is the assumption that we know the age pattern of mortality. Since we have little direct evidence of mortality rates, we have almost no direct evidence on the age pattern of mortality in Africa. Our estimates of levels and trends in mortality are therefore subject to an extra degree of uncertainty. For this reason, establishing the age pattern of mortality in Africa is a requirement for deriving more reliable estimates of the most basic demographic facts.

Most of the discussions of the age pattern of mortality in Africa have focused on three kinds of data: (1) several studies of child mortality in Senegal and Gambia, (2) maternity history data from fertility surveys (largely from the World Fertility Survey), and (3) reports of recent deaths in single-round surveys or of deaths between rounds of multiple-round surveys. All of these sources have severe limitations. Although the data from Senegal and Gambia are of very high quality, they are very limited geographically. We have little basis for judging the representativeness of their unusual age pattern of child mortality. Estimates of the age pattern of child mortality from maternity histories are often quite questionable because of the potential for biases in reported timing of births, ages at death, and selective underenumeration of early deaths. Retrospective multi-survey data sometimes appear to provide reliable data on adult mortality; however, the reported age pattern of mortality at the oldest ages (say over age 50) is very sensitive to the accuracy of reported ages.

In this paper I apply a simple model of the age pattern of mortality to indirect data on child and adult mortality to derive new information about the age pattern of mortality in East Africa. There are a number of mathematical models which might be used for this purpose. These vary in complexity from Brass's one- and two-parameter models (1975), to four-parameter models developed by Zaba (1979) and Ewbank *et al.* (1983), and eight-parameter models developed by Pollack and Heligman (1980) and Mode and Busby (1982). The larger the number of parameters used in the model, the greater the demands placed on the data.

Given that the data used in this paper provide only indirect evidence of the age pattern of mortality, it is not possible to use the more sophisticated models. Therefore, I have applied a modified version of Brass's two-parameter model which is sufficiently sophisticated to incorporate the maximum amount of information that can be extracted from the available data. The model applied here is basically Brass's two-parameter logit model except that the slope parameter, β, is allowed to differ over age 20 from the value used under age 20. Age 20 is selected for this purpose since (1) this is the highest age for which survival is generally estimated from child survival data, (2) it is reasonably close to the base ages generally used for adult mortality, and (3) mortality is relatively low at ages 10 to 25 so that any non-linearities in the logits will be imperceptible in the final life table.

The discussion begins with the age pattern of adult mortality since both the model and the results for adult mortality are simpler than the model and the results for child mortality.

2. The Age Pattern of Adult Mortality in East Africa

The most reliable data on adult mortality for much of Africa are reported proportions at each age with a surviving mother or a surviving father. Since these data rely on the reported age of the respondent rather than the reported age at death, they are less susceptible to the effects of age misreporting. They are also probably less susceptible to problems of under-reporting and events misplacement than data on recent deaths.

There are several variations on the basic model for estimating adult mortality from the proportions orphaned (e.g. United Nations 1983; Brass and Hill 1973), and Brass and Bamgboye (1981) have provided a technique for estimating the time period to which each estimate applies. These methods generally provide estimates of the proportion surviving from a base age, B, near the mean age of child-bearing to age $B+x$ where x is the mean age of the respondents. For example, the proportion with a surviving mother among respondents aged 30–4 might give an estimate of $l_{57.5}/l_{25}$. Similarly, the responses from persons aged 35–9 would yield an estimate of $l_{62.5}/l_{25}$. Although these are estimates of adult mortality, they are not commonly used indices such as life expectancy at age 20.

There are several approaches for the estimation of standard life table values from the survival ratios derived from orphanhood data. One approach is simply to compare these ratios with the same ratios calculated from different levels of the Coale and Demeny life tables. An alternative approach, suggested by Brass (1975), has the added benefit of producing an estimate of the age pattern of adult mortality. The method is based on the use of logit life tables. The logit of a proportion, p, is calculated as $-0.5*1n[p/(1-p)]$. If the logit of the life table value l_x is Y_x, then a life table can be compared to a standard life table with the same age pattern using the linear equation:

$$Y_x = \alpha + \beta Y_x^g. \qquad (1)$$

Brass recommends starting with value of l_B obtained from an estimate of survival to some age z from child survival, and using equation (1) with a value of β equal to 1. He recommends that z might be age 2; however, we will see that there are advantages to using an estimate of survival to age 20. This preliminary estimate of l_B is then multiplied by the proportion surviving from age B to age $x+B$ to get an estimate of survival to age $x+B$. Then we can estimate β from l_z, l_{x+B}, l_z^g, and $l_{x+B}{}^g$. Using this estimate of β we can re-estimate l_B from l_z. After two or three iterations, the estimates of β, l_B, and l_{x+B} stabilize.

In applying this technique we have used l_{20} as the second piece of data instead of the l_2 that is frequently recommended. There are two reasons for this. First, using l_2 gives a slope estimate that applies to ages 2 to about 70, whereas we are interested in a slope that applies only to the adult ages. Secondly, the estimates of $q(20)$ from child survival apply generally to the same time period as do the estimates of parental survival from the same data source. Therefore the estimation procedure is less affected by assumed time trends in child mortality. The actual estimate of l_{20} that was used for each estimate was from period life tables for child mortality described below.

Although the age pattern of mortality in the population being studied might be quite different from the standard used in the estimation procedure, the estimate of β gives a general impression of the way in which the population differs from the standard. In particular, a value of β greater than 1 means that mortality in the population is relatively high at the older ages. This does not provide a detailed picture of how the standard and the population differ (for example, it doesn't measure the amount of excess mortality among females during the reproductive ages), but does provide evidence of the overall pattern of adult mortality.

Table 11.1 displays estimates of the value of β for male and female adult mortality for the regions of Tanzania and some districts of Kenya. The standard is Coale and Demeny's North model life table level 12, which has a level of mortality similar to that in East Africa. The estimates for Tanzania for adult males and females are based on the 1978 census and for females from the 1973 National Demographic Survey.[1] Estimates for Kenya are

[1] The results for females from the 1973 and 1978 data sets are very similar, so that the estimate is generally an average of the two. In some cases, estimates are not available for 1973 because of changes in borders between the two dates.

Table 11.1. *Estimates of adult β values for areas of Tanzania and Kenya, according to Coale–Demeny patterns*

Area	Males	Females	Average
Tanzania			
Arusha	1.07	0.95	1.01
Coast	1.02	0.89	0.96
Dodoma	0.92	0.81	0.86
Iringa	1.06	0.91	0.98
Kigoma	—	1.05	—
Kilimanjaro	1.12	0.87	1.00
Lindi	0.94	0.79	0.87
Mara	—	0.73	—
Mbeya	0.90	0.76	0.83
Morogoro	0.99	0.85	0.92
Mtwara	0.99	0.75	0.87
Mwanza	—	0.94	—
Rukwa	—	0.88	—
Ruvuma	—	0.78	—
Shinyanga	1.10	0.95	1.03
Singida	1.00	0.97	0.99
Tabora	—	0.92	—
Tanga	0.93	0.86	0.90
West Lake	0.89	0.84	0.87
Kenya			
Busia	1.01	0.82	0.91
Bungoma	0.77	0.64	0.70
Embu	1.07	0.84	0.95
Kakamega	0.86	0.71	0.78
Kiambu	1.17	0.78	0.98
Kilifi	1.00	0.88	0.94
Kisii	1.23	0.84	1.03
Kisumu	1.00	0.77	0.89
Kwale	0.88	0.85	0.87
Lamu	1.06	0.88	0.97
Machakos	0.84	0.57	0.71
Meru	1.13	1.02	1.08
Muranga	1.04	0.72	0.88
S. Nyanza	0.94	0.78	0.86
Siaya	1.01	0.76	0.89

based on the 1979 census. For comparison, Table 11.2 includes the β derived by comparing other model life tables (Coale and Demeny's West, East, and South models and the United Nations models) and different levels of the North model (levels, 8, 10, 12, 14, and 16) to North level 12 for sexes combined. The βs are the logit slopes between ages 20 and 60.

The average of the β values for males and females range from about 0.85 to 1, values similar to those that relate East level 12, South level 12, the Latin American, and the UN General models to the North level 12. Figure 11.1

Table 11.2. *β values relating various Coale–Demeny and United Nations life tables to North model level 12*

Area	Child	Adult		
		Males	Females	Average
Coale–Demeny				
South 12	0.85	0.93	0.80	0.87
East 12	0.60	0.99	0.90	0.95
West 12	0.77	1.18	1.05	1.12
North 8	1.08	1.12	1.00	1.06
North 10	1.04	1.08	0.97	1.02
North 12	1.00	1.05	0.95	1.00
North 14	0.94	1.06	0.96	1.01
North 16	0.88	1.08	0.96	1.03
United Nations[a]				
Latin American	0.84	0.88	0.82	0.85
Chilean	0.51	1.07	0.90	0.98
South Asian	0.78	0.72	0.65	0.69
Far Eastern	0.76	1.34	1.22	1.28
General	0.83	1.01	0.94	0.97

[a] The United Nations life tables are those with an e_0 of 48 for males and 50 for females and a sex ratio at birth of 1.05.

shows that there is a very close relationship between the estimated values for males and females from the same area. The correlation between the two is 0.57. On average the female β values are about 0.17 lower than the male values (0.84 for females and 1.01 for males). This sex difference in the age pattern of adult mortality is somewhat greater than that found in the Coale and Demeny tables and the United Nations tables, with the exception of the Chilean pattern. In general, we might select a β of 1 for the males and 0.83 for the females. This would lead to an average β of about 0.91 for both sexes combined.

Examination of the estimates of the adult βs does not reveal any geographic areas which appear to have distinct age patterns of adult mortality.

3. The Age Pattern of Child Mortality

The estimation of child mortality from the reports of women on the survival of their children is accomplished by using any one of several variations on the basic model proposed by Brass. The estimates used here are derived from a computer program based on elaborations of Brass's original formulae. The most important of these elaborations is the inclusion of differences in infant mortality by maternal age. The program also calculates the cohort to which each estimate applies using similar elaborations of the equation provided by Palloni (1980). In addition, instead of using the

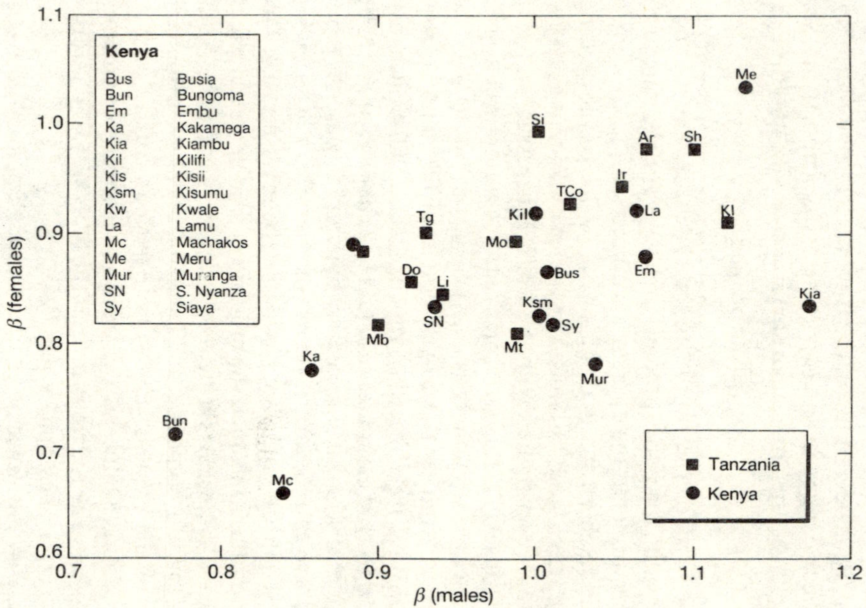

Fig. 11.1. Values of β for adult males and females, by region, Tanzania and Kenya

standard ages proposed by Brass for the mortality estimates (e.g. 1, 2, 3, 5, 10, 15, and 20) it selects the ages for which the required multiplier is closest to 1. The program also produces estimates based on the reports of women aged 50–4 and 55–9.

The child survival model provides one estimate of mortality from the reports of women in each age group. If we number the age groups of women from 15–19 equal to 1 to 55–59 equal to 9, then each age group provides an estimate of the proportion dying by some age, A_i. If mortality has been changing, then each estimate of mortality, $q(A_i)$, applies to a different cohort denoted by its year of birth, C_i. If we assume that mortality has been following a linear change in the α parameter for cohorts in equation (1), then we can describe the changes in survival over age and time as:

$$Y_i = \alpha_o + \beta\, Y_i^g + \tau C_i \tag{2}$$

where Y_i is the logit of the proportion surviving to age A_i and Y_i^g is the corresponding value from the standard life table. An estimate of β obtained by applying this model to the child survival estimates from women 15 to 49 or 20 to 59 will be an estimate of the average slope of the cohort life table between about ages 1 and 20. Since the values of Y_i^g and C_i from one data source are so highly correlated with the Y_i^g, it is necessary to use data from at least two censuses. In this case, the parameters of equation (2) can be estimated using regression analysis.

Even with data from two censuses, equation (2) must be applied very carefully. If the data are affected by misreporting, the application of equation (2) can lead to wild estimates of β and τ because of multicolinearity. In fact if there are errors in the reported proportions deceased it is possible for the estimate of β to turn negative. Less severe problems arise when the trend in mortality has not been constant. Two have been observed in East Africa. The first is populations in which the mortality trend began during the period covered by the estimates. In particular, it appears from the estimates that the mortality decline in some parts of Tanzania may have begun about 1950. In this situation, the value of C_i in the regression can be replaced by the maximum of C_i and the date of the start of the mortality trend. A similar adjustment can be made if it appears that the mortality trend stopped before the end of the series of estimates.

A second case of non-linearity in the trend is apparent in the data for a few regions. In this case there was apparently a temporary reversal in the trend in the middle of the period covered by the estimates. This leads to a very confusing picture since the estimates that are affected are based on the reports of women aged 15–29 in the first census and women aged 25–39 in the second census. The estimates based on the reports of young women are the average for a small number of years, while the estimates for older women are the average for a larger number of years. Therefore the estimates based on the reports of older women are less affected by temporary changes in the level of mortality. This problem can be reduced by weighting the regression by the number of children on which the proportion deceased is based. This reduces the weight given to the reports of the younger women.

In general, therefore, equation (2) must be applied with great caution and attempts should be made to compare the estimates with other sources and to check for consistency among the estimates for neighbouring regions. It is also important to check that the estimates are relatively stable, i.e. to make sure that the estimates do not change radically when one or more extreme points are excluded from the analysis. For these reasons, the estimation of equation (2) is very difficult. Each of the estimates provided below is the result of testing five to ten different regression models for each area. These generally include weighted and unweighted regressions excluding different outliers and where possible incorporating data from three or more censuses. The final estimates for each area are based on attempts to find a fit that is consistent with both the data and expectations based on estimates for neighbouring areas. The estimates of β are accurate to within about ± 0.1, although this is only an impressionistic estimate.[2]

The difficulties in estimating equation (2) are serious. Attempts to estimate β and τ using this procedure should be limited to cases where at least three data sets are available for some areas, where other data on the age pattern are available, or in cases in which it is judged safe to assume that mortality has been constant.

Table 11.3 presents estimates of β for cohorts of children from the regions of Tanzania and some districts of Kenya. It also presents estimates of β for period life tables. These period estimates are derived by using equation (2) to calculate life tables up to age 20 for single-year cohorts. A period life table was then constructed for 1975 from the ${}_1q_x$ values from the cohort tables. The period values show less variation than the cohort values. This is probably due to the fact that some of the uncertainty in the estimates of β that result from the colinearity of the Y^g_x values and the C_i estimates cancels out in the calculation of the period life tables.

The period-specific β's values range from about 0.75 to 1.1 (Tanga region of Tanzania shows a value of 1.26).[3] Although this range is relatively large, there is some evidence that it is not unreasonable. First, there are several cases in which reliable data are available for more than two censuses. In particular, we have combined several of the current regions of Tanzania to produce estimates for Lake region from 1948, 1957, 1967, and 1978 censuses. The data from all four censuses are consistent with a cohort β of

[2] It is not possible to calculate an exact confidence interval for these estimates since the main sources of error are reporting errors and errors in the model, rather than sampling errors. The sampling errors are very small because of the large sample sizes in the censuses (full counts in all but the 1962 and 1969 censuses of Kenya). This estimate of reliability comes from comparing the results of different regression fits (weighted and unweighted, deleting various outliers, etc.) with data from the same region.

[3] We have excluded estimates for areas where the data from the two sources are not consistent. In particular, several districts in Kenya showed values from the 1969 census that were all less than all of the estimates for 1979.

Table 11.3. *Estimates of childhood β values for areas of Tanzania and Kenya, according to Coale–Demeny patterns*

Area	Cohort values of β	*lx* values 1975		Period values of β
		$x = 1$	$x = 20$	
Tanzania				
Arusha	1.01	0.9085	0.7797	1.03
Coast	1.06	0.8886	0.7253	1.10
Dodoma	1.25	0.8874	0.6821	1.30
Iringa	1.00	0.8647	0.6967	1.02
Kilimanjaro	1.02	0.9342	0.8258	1.09
Lindi	1.07	0.8457	0.6515	1.07
Mara	0.74	0.8615	0.7326	0.82
Mbeya	0.74	0.8156	0.6698	0.78
Morogoro	1.05	0.8695	0.6885	1.10
Mtwara	1.13	0.8556	0.6554	1.13
Mwanza	0.71	0.8707	0.7573	0.77
Rukwa	0.86	0.8277	0.6625	0.89
Ruvuma	0.97	0.8509	0.6733	1.01
Shinyanga	1.04	0.8735	0.7057	1.05
Singida	0.72	0.8694	0.7459	0.82
Tanga	1.24	0.9075	0.7348	1.26
West Lake	0.88	0.8850	0.7415	0.98
Kenya				
Busia	0.97	0.8560	0.6768	1.04
Bungoma	0.93	0.9027	0.7731	1.00
Embu	0.85	0.9338	0.8408	0.98
Kakamega	0.95	0.8984	0.7589	1.03
Kiambu	0.79	0.9517	0.8848	0.94
Kilifi	1.00	0.8508	0.6759	1.00
Kisii	0.90	0.9248	0.8161	1.01
Kisumu	1.05	0.8636	0.6723	1.12
Kwale	0.99	0.8655	0.7026	1.00
Lamu	1.26	0.8837	0.6940	1.20
Machakos	0.84	0.8568	0.7154	0.86
Meru	0.98	0.9429	0.8464	1.09
Muranga	0.94	0.9502	0.8623	1.11
S. Nyanza	0.74	0.8380	0.6931	0.83
Siaya	0.79	0.8329	0.6800	0.85

about 0.85. For South Nyanza district in Kenya, the estimate of β of 0.74 is consistent with a linking of the 1962 and 1969 censuses as well as with a linking of the 1969 and 1979 censuses. If this slope were a result of differential completeness of reporting in one census or the other, it is unlikely that these two pairings would give similar estimates of β. Finally, the estimate for Machakos district in Kenya, 0.84, is even less extreme than the value of 0.66 between ages 1 and 15 from the data from the Machakos study.[4]

[4] This is based on life tables calculated from age-specific mortality rates given by van Ginneken *et al.* (1980).

Table 11.4. *Estimates of childhood β values for lake and coastal areas of Tanzania and Kenya*

Area	Cohort values of β	*lx* values 1975		Period values of β
		$x = 1$	$x = 20$	
Lake areas				
Tanzania				
Mara	0.74	0.8615	0.7326	0.82
Mwanza	0.71	0.8707	0.7573	0.77
West Lake	0.88	0.8850	0.7415	0.98
Kenya				
Kisumu	1.05	0.8636	0.6723	1.12
S. Nyanza	0.74	0.8380	0.6931	0.83
Siaya	0.79	0.8329	0.6800	0.85
Coastal areas				
Tanzania				
Coast	1.06	0.8886	0.7253	1.10
Lindi	1.07	0.8457	0.6515	1.07
Mtwara	1.13	0.8556	0.6554	1.13
Tanga	1.24	0.9075	0.7348	1.26
Kenya				
Kilifi	1.00	0.8508	0.6759	1.00
Kwale	0.99	0.8655	0.7026	1.00
Lamu	1.26	0.8837	0.6940	1.20

There are not as many opportunities to check the reliability of the high estimates of β. In most of the districts of Kenya that show high values of β the 1962 census data show gross under-reporting of deceased children and, therefore, the estimates cannot be based on more than two censuses. In addition, data from the Pare-Taveta study show slopes of 0.75 and 0.85 for two of the study areas, much lower than the census estimates for the neighbouring districts.[5] Since under-reporting of deceased children in the earlier census will bias upward the estimate of β, it is likely that the highest β estimates are a result of reporting errors.

In addition to these tests of consistency, we can examine the geographical pattern of the estimates. Since neighbouring parts of Tanzania and Kenya have very similar climates and cultural patterns, but have different census histories, geographic comparisons provide a useful test of whether the observed variations are real or the result of reporting errors. Table 11.4 summarizes the child survival data for six areas around Lake Victoria and seven areas along the coast. The data from the two countries suggest that the value of β for coastal areas is greater than the value around the lake. While the coastal values are all greater than about 1, the lake basin values are all less than 1. However, we cannot rule out the possibility that the high

[5] The third study area shows clear under-reporting of infant mortality (the IMR is 21).

estimates along the coast are due to under-reporting of deceased children in both the 1967 census of Tanzania and the 1969 census of Kenya.

A comparison of the estimates of β with the level of mortality, as measured by either the infant mortality rate or l_{20}, does not show the same relationship as found in the Coale and Demeny tables.

The value of β estimated from child survival data using equation (2) is an average slope between ages 1 and 20. It does not provide details on changes in the slope between these two ages. This can be quite important, since a two-parameter logit life table cannot describe all of the differences between two patterns. For example, the South and West model life tables (level 12) have similar values of β between ages 1 and 20 (0.85 and 0.77), while between ages 1 and 2 they have very different slopes (1.26 and 0.92). Therefore, this approach provides only a gross description of the age pattern of child mortality. It is important to note that this is sufficient to enable us to get more accurate estimates of the trends in the infant mortality rate from Brass-type estimates of the $q(x)$.

We have made some effort to test whether the North model is the appropriate one to use in East Africa when applying equation (2). We have rerun the analysis (i.e. re-estimated the $q(x)$ values as well as the regression fits) for Siaya and South Nyanza districts in Kenya. These two areas were selected because they both showed low values of β which suggest that the West or the South models might be more appropriate (see Table 11.2 for comparison values of β for the Coale–Demeny and United Nations patterns). Both areas also provided estimates from the 1969 and 1979 census that appear to be very consistent with the model described by equation (2). Table 11.5 presents the weighted R-squared values from fits to the different Coale–Demeny patterns. We fitted the data for each district first to all 16 points available from the two censuses[6] and then reran the regressions deleting the one point that had the largest deviation from the line.[7] For both

Table 11.5. *Comparison of fits (R^2 values) to various model life tables, Siaya and South Nyanza districts, Kenya*

District	Model	All points	Without 55–59 age group
Siaya	North	0.985	0.985
	South	0.975	0.975
	West	0.963	0.964
S. Nyanza	North	0.990	0.991
	South	0.982	0.984
	East	0.968	0.967
	West	0.979	0.980

[6] The 16 points come from the reports of women aged 15–19 to 45–9 in the 1969 census and women aged 15–19 to 55–9 in the 1979 census.

[7] For Siaya, we deleted the estimate from women aged 55–9 in the 1979 census, for South Nyanza we deleted the estimate from women aged 15–19 in the 1969 census.

districts it is clear that the North model level 12 fits the data better than the other models. Although this is not a definitive test it does suggest that the pattern is closest to a two-parameter transformation of the North pattern.

A second test of the choice of the appropriate Coale–Demeny pattern is to compare the mortality data from the maternity histories collected in the Kenya Fertility Survey of 1977–8 with the Coale–Demeny and UN tables. Table 11.6 presents an index of the age pattern of mortality under age 5, ${}_4q_1/{}_1q_o$, for the cohort born in Kenya in 1970–2 and the same index for the Coale–Demeny and UN patterns with similar levels of mortality. Two

Table 11.6. *Comparison of the age pattern under age 5 in Kenya with the patterns in model life tables*

	l_1	l_5	${}_4q_1/{}_1q_0$
Maternity histories and prospective studies			
Kenya A	0.8986	0.8303	0.75
B	0.8906	0.8303	0.62
Central	0.9391	0.9006	0.67
Coast	0.8794	0.8190	0.57
Nyanza	0.8799	0.7861	0.89
Rift	0.9298	0.9003	0.45
Western	0.8974	0.8068	0.98
Eastern and north-eastern	0.9203	0.8660	0.74
Nairobi	0.9246	0.8804	0.63
Machakos study	0.9459	0.9198	0.51
Tanzania	0.9288	0.8740	0.83
Lake regions	0.9353	0.8990	0.60
Coastal regions	0.8969	0.1680	0.71
Others	0.9370	0.8800	0.97
Pare-Taveta study (1962–66)			
Mamba	0.9785	0.9418	1.74
Mkomazi	0.8672	0.7894	0.68
Taveta	0.8799	0.8013	0.74
Patterns in model life tables			
Coale–Demeny patterns			
North 12	0.8726	0.7847	0.79
β = 0.9	0.8496	0.7620	0.69
β = 0.8	0.8233	0.7378	0.59
β = 0.7	0.7936	0.7120	0.50
South 12	0.8518	0.7541	0.77
East 12	0.8209	0.7573	0.43
West 12	0.8562	0.7850	0.58
UN model life tables (male e_0 = 48, females e_0 = 50)			
Latin American	0.8770	0.7990	0.72
Chilean	0.8526	0.8107	0.33
South Asian	0.8572	0.8198	0.31
Far Eastern	0.9058	0.8623	0.51
General	0.8790	0.7681	1.04

estimates of the IMR are presented for Kenya. The first estimate, 101.4, includes only those deaths that were reported at ages 0 to 11 months. The second estimate, 109.4, includes some adjustment for misreporting by including half of the deaths at 12 months and half of the deaths without a reported age at death as infant deaths. It is clear that the reported age pattern is very similar to the North level 12 adjusted with a logit β of 0.9 and to the UN Latin American model.

The ratios of ${}_4q_1$ to ${}_1q_0$ for the provinces of Kenya from the Kenya Fertility Survey are also given in Table 11.6. These ratios suggest quite extreme values of β and therefore provide little basis for evaluating the geographic differentials in the age pattern of mortality. It is likely that the sample sizes in the provinces are not large enough to get accurate estimates of the ratio of these two rates.

Table 11.6 also includes some data for Tanzania. The maternity histories from the 1973 National Demographic Survey give unreasonably low estimates of IMR for almost all of the regions. Despite this, the ratios of ${}_4q_1$ to IMR are reasonable for both the regions along the coast (Coast, Lindi, Mtwara, and Tanga regions) and those around Lake Victoria (Mara, Mwanza, and West Lake regions). Along the coast the values are consistent with a β of about 0.9, while around the lake a β of 0.8 is indicated. We also have mortality data from the Pare-Taveta study for the period of 1962–6. As noted above, these data are consistent with a β of about 0.9.

If we compare the life tables from the Machakos study with North level 12, we find that the slope between ages 1 and 5 is almost exactly the same as the slope between ages 5 and 15: both are about 0.66, lower than the 0.86 suggested by the census data.

These findings suggest that the East African pattern of child mortality is described rather well by transforming the North pattern with a β of about 0.9 to 0.95 (0.75 to 0.85 for areas around Lake Victoria) rather than by any of the other Coale and Demeny model life tables. Among the UN models, it is closest to the Latin American model. The evidence also suggests that child survival data from the 1969 census of Kenya and the 1967 census of Tanzania understate child mortality for the areas along the coast. If this is true, then the observed lack of mortality decline in these areas is also a result of under-reporting in the earlier censuses.

4. The Relationship between Child and Adult Mortality

Brass has suggested estimation of a two-parameter life table based on the child and adult survival estimates (Brass 1985). Although this approach has many advantages, it is useful to ask whether the evidence suggests that a two-parameter approach is adequate. In particular, we can ask whether independent estimates of the slopes under and over age 20 are similar.

Figure 11.2 relates the estimated slope for children with the estimated slope for adult females. There does not seem to be any relationship between

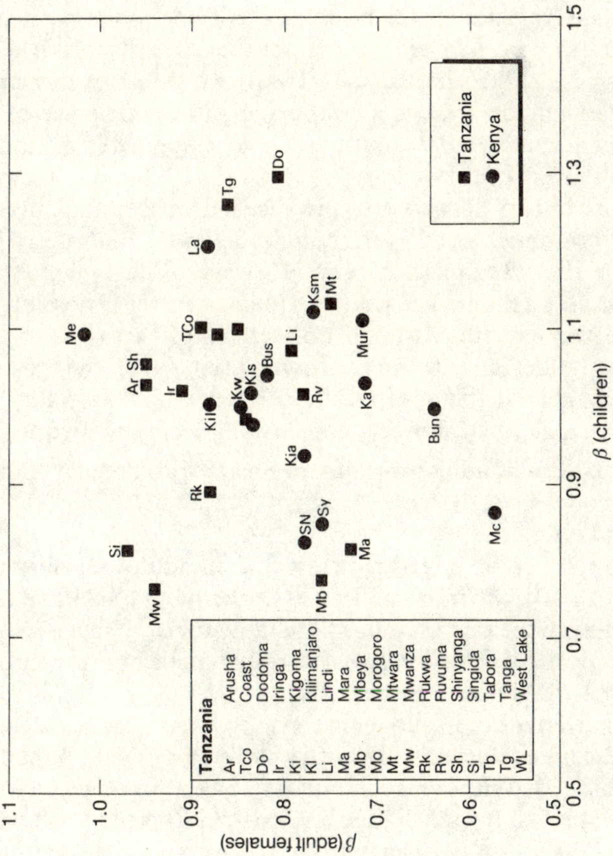

Fig. 11.2. Values of β for adult females and children, by region, Tanzania and Kenya

these estimates. However, the lake areas do tend to have values for both children and adults that are similar to the South model life table. For other areas, the average is close to 1 for children and 0.85 for adult females.

5. Conclusions about the Pattern of Mortality in East Africa

Although data on child survival and orphanhood are not designed for the study of age patterns of mortality, it is clear that they can provide useful evidence of the gross outlines of the age pattern. The model suggested for use with child survival data is the more difficult one to apply. It should be used only in areas where three or more sources of child survival data are available or where child survival data from two censuses can be supplemented by other reliable evidence on the age pattern. The use of parental survival data is probably more robust and can be applied in any area where these data are available.

The maternity history data seem to produce reasonable national estimates of the age pattern under age 5 even though the level of mortality is clearly understated in the Tanzanian survey. However, the regional data are apparently not very reliable for studying the age pattern, possibly because of the smaller (though still reasonably large) sample sizes.

Despite the limitation of the methods used here, they do provide reasonable estimates of the appropriate age pattern of mortality for East Africa. In particular, the following conclusions seem justified:

1. The age pattern of adult mortality in East Africa appears to be similar to the North model level 12, with a value of β of about 0.9 or 0.95 for both sexes combined.

2. There appears to be a larger differential in adult mortality by sex in East Africa than is built into the Coale–Demeny and UN life tables. In particular, if we use the same standard life table for both males and females (North model level 12), the β necessary to fit females must be about 0.15 smaller than the value used for the males.

3. The age pattern of mortality between ages 1 and 20 is best described by the North model life tables, possibly with a slope of β that is about 0.95. One exception to this is that the areas surrounding Lake Victoria seem to require a smaller slope, probably in the range 0.77 to 0.85 where the pattern is more similar to the UN Latin American pattern. It is also possible that the coastal areas require values of β greater than 1, although it is more likely that this is due to under-reporting of deceased children in the earlier censuses (i.e. 1967 in Tanzania and 1969 in Kenya).

4. Although both adult and child age patterns can be described on average by transforming the North model level 12 with a β of about 0.9 or 0.95, there is no evidence that within East Africa regional differences in adult and child mortality can be described by the same value of β. One possible exception to this is that areas surrounding Lake Victoria may require smaller slopes for both adults and children.

References

Brass, W. (1975), *Methods for Estimating Fertility and Mortality from Limited and Defective Data* (Chapel Hill, NC: Carolina Population Center).

—— (1985), *Advances in Methods for Estimating Fertility and Mortality from Limited and Defective Data* (London: Centre for Population Studies, London School of Hygiene and Tropical Medicine).

—— and Bamgboye, E.A. (1981), *The Time Location of Reports of Survivorship: Estimates for Maternal and Paternal Orphanhood and the Ever-Widowed*, Working Paper no. 81-1 (London: Centre for Population Studies, London School of Hygiene and Tropical Medicine).

—— and Hill, K. (1973), 'Estimating Adult Mortality from Orphanhood', *Proceedings of the International Population Conference, Liège, 1973*, iii. 111-23 (Liège: IUSSP).

Ewbank, D.C., Gomez de Leon, J.C., and Stoto, M.A. (1983), 'A Reducible Four-Parameter System of Model Life Tables', *Population Studies*, 37, 105-27.

Heligman, L., and Pollard, J.H. (1980), 'The Age Patterns of Mortality', *Journal of the Institute of Actuaries*, 107, pt. I, 49-80.

Mode, C.J., and Busby, R.C. (1982), 'An Eight-Parameter Model of Human Mortality: The Single Decrement Case', *Human Biology*, 44, 647-59.

Palloni, A. (1980), 'Estimating Infant and Child Mortality under Conditions of Changing Mortality', *Population Studies*, 34(1), 129-42.

UN (United Nations) (1983), *Manual X: Indirect Techniques for Demographic Estimation*, Population Study no. 81, Department of International Economic and Social Affairs, ST/EAS/Sera/81 (New York: UN).

Van Ginneken, J.K., Voorhoeve, A.M., and Omondi-Odhiambo (1980), 'Nachakos Project Studies: XVIII. Fertility, Mortality and Migration in 1975-1978', *Tropical and Geographical Medicine*, 32(2), 183-88.

Zaba, B. (1979), 'The Four-Parameter Logit Life Table System', *Population Studies*, 33, 79-100.

12 Assessing the Levels and Impact of Mortality in Crisis Situations

ALBERTO PALLONI

Center for Demography and Ecology, University of Wisconsin Madison, Wisconsin, USA

1. Introduction

Population crises characterized by sudden increases in mortality were a recurrent feature of pre-industrial populations. As a result, the study of their nature, intensity, causes, and impact on demographic and social structures has been largely the task of historical demographers. Long after their disappearance from the European scene, crises still reappear with some regularity in overburdened populations although with different intensity, regularity, and geographical scope. To the evidence revealing devastating catastrophes occurring during the end of the nineteenth century in India and China, information has been added about other crises of significant intensity and scope: the great Bengal famine of 1943, the Ethiopian famines of 1973 and 1974, the Russian famines in the late 1920s, the Bangladesh famines of 1971 and 1974, the China famine of 1959–61, and the Sahelian drought of 1970–2. In some cases the cause is the outbreak of political upheavals accompanied by pervasive and protracted warfare. In others, the culprit is the persistence of climatic conditions that lead to the collapse of normal production and distribution of food. With a few exceptions (for example the great malaria epidemic of 1934–5 in Sri Lanka), crises generated or dominated by the presence of epidemics have been absent. The spread of diseases, however, remains an important mechanism through which wars or adverse climatic conditions increase mortality, depress fertility, and alter the regional distribution of the population.

Population crises have long been thought to be a 'natural' mechanism through which a homeostatic equilibrium is regained once populations experience more or less prolonged drifts away from it. Malthus saw them as a court of last resort, the archetype of the positive checks preventing population growth beyond the carrying capacity. In the past two decades or so, the debates about the relations between population and socio-economic well-being have recast the role that population crises play in the absence of deliberate population control. They are seen either as the ultimate consequence of failure to bring down the rates of population growth (Meadows *et al.* 1974) or as exaggerations, the product of misreading of population and economic statistics (Simon 1980). The wave of renewed interest has also prompted efforts to reorder our notions about the role of population crises in the pre-industrial period (Watkins and van de Walle 1983). Partial evidence suggests that, at least in the long run, they could not have provided sufficient relief from population pressure build-up (Watkins 1985). The recent rapid world-wide spread of a deadly AIDS epidemic could alter the terms of these debates, as the death toll in certain areas could mount to unforeseen levels and, if unchecked, upset the functioning of entire economic, political, and social systems.

The aim of this paper is to study the nature of population crises, to describe the techniques for measuring their intensity, and to assess the potential role that they play as a check on population growth. In Section 2, I present a model of the mechanisms leading to population crises. This model is a generalization: one suggested by Chen and Chowdhury (1977) for the study of famines. It will be used to formulate a typology of crises and to define the expected behaviour of common demographic indicators. In Section 3, I discuss the typical trajectory of demographic indicators immediately before, during, and immediately after a crisis; allowance is made for variations due to the nature, magnitude, and duration of crises. In Section 4, I present an assessment of alternative measures of the intensity of population crises. I show that proper formulation and application of these measures would benefit from a prior examination of the mechanisms that trigger the crises. In Section 5, I deal with the impact of population crises on prospective population growth and on selected aspects of the social structure. Here I contest the argument according to which the 'effectiveness' of crises as mechanisms of population control is only of minimum significance.

2. The Nature of Population Crises

2.1. Basic Variables and Relations

Although the causal chain that leads to the outbreak of population crises is complicated and variable, a few universal features can be highlighted. First, there is always a triggering event that, under certain conditions, will lead to responses of mortality, fertility, and migration. The triggering event may be exogenous or endogenous. Thus, the occurrence of flooding or droughts or other vagaries of climatic nature is an exogenous event since it is not directly induced by social design or population conditions. It may well be that population pressure on land and patterns of land utilization and settlements could, in the long term, affect weather patterns and, through soil degradation, increase the sensitivity of output to changes in weather. In these cases, however, the role of population as a cause is remote and can be safely disregarded. On the other hand, political unrest, accompanied by more or less systematic warfare or the implementation of economic policies affecting food production or its distribution, shall be considered endogenous. In most cases the outbreaks of epidemics are part of a chain initiated by an exogenous or endogenous triggering event. However, there are situations when they are the result of long-term changes in a delicate balance between agents of disease and human hosts. In these cases, they shall be considered exogenous events. This distinction is introduced here to dispose of the idea that all crises are endogenously produced, the result of excessive pressure on resources. The connection between the occurrence of the crises and population pressure should not be postulated a priori but left to investigation. It may be that the triggering event and the chain of reactions that follows it would not have taken place in the absence of rapid population growth or unbalanced population distribution. Yet this is a proposition for falsification, not an axiom on which interpretations should be based. Secondly, the triggering event may lead to two alternative and possibly complementary responses. The first is the production of food shortages and the second is the disturbance of the patterns of entitlement to goods and services for all or part of the population. The distinction between these two mechanisms was first introduced by Sen (1980) and later used in other research on the effects of crises (Chen and Chowdhury 1977; Bongaarts and Cain 1981; Watkins and van de Walle 1983; Watkins and Menken 1985). Sen argued that in many cases where famines had been documented there was no evidence of a collapse of food production and a consequent food shortage. Rather, what led to starvation and its sequelae was the deterioration of terms of trade for some groups of the population. For groups whose claim over goods and services is based on self-subsistence, famines *may* occur as an outcome of adverse conditions affecting production if and when the capability to command resources outside the boundaries of self-subsistence is also compromised. By and large, this takes place when there is a general collapse of food availability. It may also occur, however, when total food production is unaltered but the institutional mechanisms to make it accessible to the self-subsistent population are lacking. What leads to famines is the inability of the self-subsistent groups to find adequate alternatives. Groups whose claims over goods and services depend on exchange of products or factors of production face adverse conditions when relative prices and market conditions are altered even though production of basic foods remains unchanged. It follows that the use of food production or availability per caput as an indicator of actual availability is meaningless inasmuch as it masks transformations in the system of entitlements. Serious disruptions of food production may lead to changes in the systems of entitlements for groups other than those relying on self-subsistence and, vice versa, alterations in entitlements may eventually cause a drop in food production. But, although the two mechanisms are related, their effects have to be considered separately. More importantly, the triggering event may affect only one of them without necessarily affecting the other.

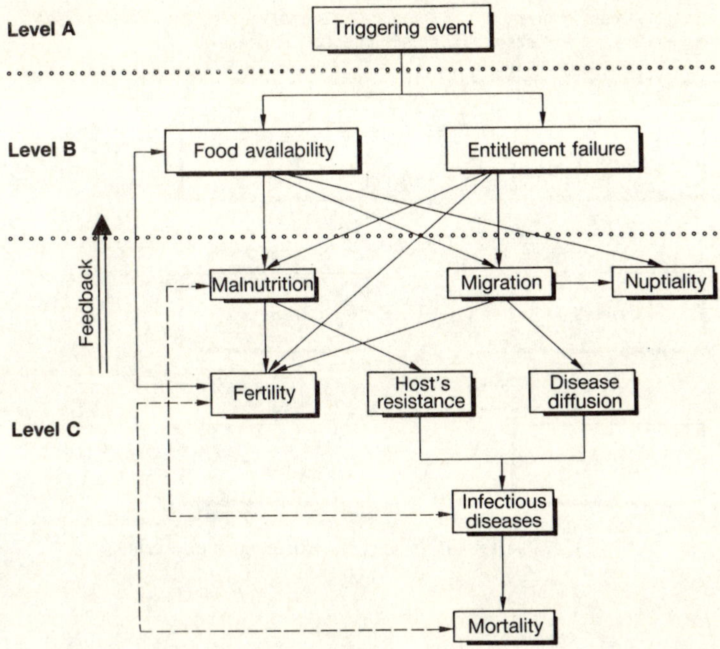

Fig. 12.1. Relations between main variables in population crises

The relations just described are illustrated on the top levels, A and B, of Figure 12.1. They represent what we may consider the 'background' variables affecting population responses. The third level of Figure 12.1, level C, illustrates the relations between selected 'intermediate' variables and the demographic responses. Whether due to food shortage or disruption of entitlements, a first impact is felt on the levels of nutrition of the population affected. Two additional responses are highly likely: a decrease in nuptiality and an increase in emigration. The levels of marital fertility are affected directly by nutrition (Bongaarts and Delgado 1977), by spouse separation due to emigration, or by decreased frequency of sexual intercourse. Finally, mortality increases due to the incidence of infectious diseases (facilitated by lowered resistance of hosts), the spread of diseases due to higher population mobility, aggravation of overcrowding, and, lastly, direct starvation. Feedback mechanisms are quite important in this representation. First, an increased incidence of diseases aggravates malnutrition and reinforces emigration. Secondly, an increase in infant mortality disrupts patterns of breast-feeding, reducing the length of postpartum amenorrhoea and lowering intervals between conceptions. Thirdly, increased mortality and emigration disrupt production and distribution of goods. The inclusion of feedback mechanisms is useful for the study of crises with effects that are dampened over time as well as of those that have a tendency to self-reproduce, aggravating the impact of initial effects. The

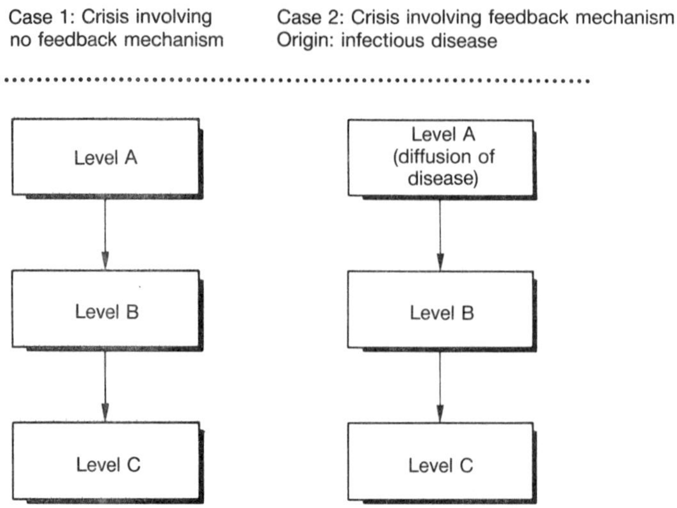

Fig. 12.2. Alternative sequence of events in a crisis

representation is general enough to apply to situations where the triggering event is the presence of a disease, as was the case during pre-industrial times, rather than the vagaries of weather, natural disasters, or military actions. Figure 12.2 represents the two alternative types of mechanism leading to population crises.

2.2. Types of Crisis

The distinction just made permits us to define two main types of crisis: one in which the triggering mechanism operates directly through level B and the other in which the first impact of the triggering mechanism is on variables in level C. Two additional types of population crisis may be distinguished depending on the importance of food shortages relative to changes in entitlement systems. In each of these cases, two further types can be defined according to the most typical demographic response. While in some cases emigration will dominate over changes in the patterns of nuptiality and marital fertility, in others the most frequent response will be postponement of marriages, separations, and delayed child-bearing. Finally, within each of these types one could further distinguish crises according to the nature of the mortality response. Although crises may result in age- and sex-invariant (approximately neutral) mortality increases, it is more common to find strong age and sex differentials in mortality excesses.

Table 12.1. *Classification of some recorded population crises according to selected characteristics*

Event*	Triggering event	Losses of entitlements versus food production	Emigration relative to nuptiality, fertility	Nature of mortality response	Presence of feedback
Bangladesh (1970–71)	War	Food production	Both	Age-selective	Epidemics, food production
Bangladesh (1974–75)	Flooding	Both	Nuptiality, fertility	Age-selective	No
Bengal (1943)	War	Entitlements	Both	Neutral	Epidemics, food production
China (1958–62)	Drop in grain output (agricultural policies)	Food production	Nuptiality, fertility	Age-selective	No
India (1870–80)	Weather patterns (drought, flooding)	Both	Stronger response of nuptiality	Neutral	Several epidemics, food production
Russia (1920–39)	Agricultural policies, war	Both	Nuptiality, fertility	Age-selective	?

* Dates correspond to time period bounding the duration of the crisis.

With no further elaboration beyond the simple sequence of dichotomous types that I have established before, there are a total of 16 types of crisis. A final set, depending on the intensity of the mortality response and the duration of the crisis, could be added (Charbonneau and Larose 1979). Since the dimensions used to construct these types are correlated, it is likely that only a small subset will be of any empirical relevance. In Table 12.1 I suggest a classification of a handful of population crises according to a small number of criteria.

The distinctions highlighted by the typology are useful to define the expected behaviour of various demographic indicators immediately before, during, and immediately after the crisis. In turn, the latter can be used as evidence to infer the type and characteristics of the crisis.

3. Demographic Responses and Their Interactions

In Section 2 I made no mention of specific demographic indicators and instead identified demographic responses in general. This is because the trajectory of demographic indicators depends on which ones are used. Typically one relies on the easiest ones to compute, the crude rates (Chen and Chowdhury 1977; Appleby 1978; Livi-Bacci 1978; Hollingsworth 1979; Bongaarts and Cain 1981; Anderson and Silver 1985). These, however, are affected by changes in the age structure of the population which occur as a result of the sudden variation of vital events. For example, in the absence of strong selection effects, a large increase in infant mortality will set the stage for future increases in the crude death rates even if the adverse conditions were not altered over a period of several years. This suggests that one should first trace the response in age-specific rates and then attempt to infer the trajectory of more conventional demographic indicators.

3.1. Mortality

3.1.1. Crises due to relative food shortages When the triggering event affects directly the levels of food production or system of entitlements, the age-specific rates should rise rather slowly for an interval of time after food shortages become evident. This is because at the onset of the crisis the population may trade goods or resort to liquidation of assets in exchange for access to food as a means of coping with relative scarcity. Unless the pre-crisis levels of nutrition are too low to begin with, the increase in mortality accelerates only after the scarcity of food leads to widespread undernourishment, chronic deficiencies in the immune system, and increases in crowding. As food production and distribution return to normal levels, the force of mortality progressively decreases. Infant mortality is expected to rise due to malnutrition among children older than 1 month and, to a lesser extent, due to decreases in birth weight affecting the force of neonatal mortality. In most societies, however, the practice of breast-feeding provides a protection for all infants and makes them less vulnerable to sudden shifts in food availability unless there is pervasive malnutrition among mothers. Mortality in the age group 1–5 is highly sensitive to crises triggered by relative shortages of food. Young children depend on solid foods and their immune system is not completely developed. Unlike infants, they do not enjoy the double protection offered by breast-feeding. The combination of food scarcity and spread of infectious diseases has had devastating consequences within this age group in almost all documented cases (Curlin *et al.* 1971; Chen 1973; Chen and Chowdhury 1977; Turpeinen 1979; Kane 1984; Ashton *et al.* 1985). Sex differentials in mortality are exacerbated within this age group. Over-mortality of females has been reported even in cases where this is not a feature of pre-crisis mortality patterns (Kane 1984; Ashton *et al.* 1985). Mortality above age 60 is also increased more than proportionately. This occurs as a result of higher susceptibility to diseases as well as to shifts in the distribution of food and services favouring young adults. Unfavourable distribution of resources within the household aggravated by the stress of lactation or child-bearing is at the root of female excess mortality within the reproductive period. In most documented cases, female mortality increases are higher and the absolute levels may reach higher values than for males (Curlin *et al.* 1971).

3.1.2. Crises due to wars The onset of the rise in the force of mortality arrives sooner and is steeper when the initial stage of the crisis is characterized by warfare, as it should reflect the excess deaths due to military casualties. As the latter account for a substantial amount of excess deaths, the cessation of hostilities brings a return to normal levels of mortality at a faster pace than in other types of crisis. This should hold at least for the age-specific death rates at ages 15–50. The age pattern of mortality excesses shows a higher relative increase for males in the age group 15–50 than in other types of crisis. However, since wars bring disruption of food production and distribution, both the levels and the age–sex pattern of increases in mortality may eventually be indistinguishable from those precipitated by food shortages. Differences will be discerned only during the initial stages and in the tail-end of the crisis, as mortality declines less rapidly when the main triggering event is war.

3.1.3. Crises due to epidemics In these cases the onset of mortality increases depends on the speed with which the disease is diffused in the population. The pace of its decline depends on the relative accommodation of host and agents of disease, the proportion of individuals who are not susceptible, and the degree of isolation of the infected population. At least during the initial stages of the epidemic, the levels and age–sex patterns of mortality increases are dependent on its virulence and age- and sex-specific incidence. Thus, for example, before the introduction of the vaccine, 50% of the excess deaths caused by a smallpox epidemic was accounted for by children younger than 5. On the other hand, the age patterns of mortality excesses during a cholera epidemic are strongly biased toward adults and old people (Salvini 1977). As occurs with population crises triggered by wars, those triggered by disease may be just one stage of a protracted period of crisis during which the initial effects are exacerbated by disruptions in food production and distribution. If this occurs, the absolute levels and the age- and sex-specific patterns of mortality excesses will eventually follow the profile described earlier: to the pattern of excesses due to disease alone is superimposed another one due to the feedback mechanism.

In all three types of crisis, there is an underlying selection process which will affect the post-crisis levels of age-specific mortality. On the one hand, when the selection process leads to an elimination of the most frail, post-crisis population composition should be biased toward the least frail groups. The discrepancies should be higher at ages where the selection is stronger. As a consequence, the levels of mortality will be somewhat lower for a time after the crisis has run its course. On the other hand, when the selection process disproportionately affects those who are least frail, one would expect, at least in the short run, higher levels of mortality during the

Table 12.2. *Relations between timing, levels, and age–sex patterns of over-mortality and type of crisis*

Type of crisis	Levels	Age pattern	Sex pattern	Most favoured in selection
Triggered by war with no feedback	Variable; immediate increases	Concentrated in 15–50	Male excesses	Most frail
Triggered by relative food shortages	Variable; gradual increases to a maximum	Group 1–10 most affected	Generally female excesses	Least frail
Triggered by disease but no feedback	Variable; timing follows diffusion curve	Variable with disease	No differentiation	Least frail

post-crisis period than during the pre-crisis period. If the triggering event leads to food shortages or loss of entitlements, we would expect to have heavy selection of the *least* frail among the youngest. If the triggering event is warfare, we would expect heavy selection of the *most* frail at ages when recruitment into the military is most likely.

Table 12.2 summarizes the relations between levels, age–sex patterns, and timing of mortality excesses on the one hand, and types of crisis on the other. One caveat is necessary here. So far I have assumed a homogeneous population, one with no social inequalities. In reality, pre-crisis mortality *differentials* are exacerbated during and immediately after a crisis. The historical record shows that when the latter occur due to the diffusion of diseases, the wealthier groups stand better chances of surviving (Livi-Bacci 1978). Yet the contrasts are not nearly as high as those experienced in contemporary developing societies (Alamgir 1980). Regional differentials will be consistent with the regional patterns of food shortages and with regional differences in access to markets. In general, however, it is the urban areas that are least affected.

3.2. Fertility

Fertility responses to crises have two main components. The first corresponds to reactions of the nuptiality pattern. The second corresponds to effects on frequency of intercourse, length of postpartum amenorrhoea, and fecundity.

3.2.1. Nuptiality The sensitivity of nuptiality patterns to crisis situations is enhanced whenever marriages are dissociated from menarche. This was the case in societies characterized by the 'Western European marriage pattern' (Watkins and van de Walle 1983). The flexibility of this mechanism is less so in contemporary developing societies where the association between marriage and menarche is stronger. Yet all documented cases show evidence of significant changes in nuptiality. Transaction costs, negative future expectations, and actual physical displacement of potential spouses all contribute to reducing the probabilities of marriage in the prime ages. As indications of serious disruption of food production and distribution of food become evident, households reallocate scarce resources to meet expected shortages. As a result, marriage contracts depending on dowry payments are reduced. Similarly, increases in the costs of adopting a spouse as a new household member or drastic changes in the prospective costs of setting up a separate household lead to systematic avoidance of new marriages (Bongaarts and Cain 1981; Caldwell *et al.* 1984). When wars are the triggering event or when other triggering events lead to generalized emigration, the normal course of nuptiality is, in addition, interrupted by sheer contraction of the marriage market. Depending on the intensity of the crisis, the onset of the nuptiality response can be swift and massive, and frequently becomes an early warning of the impending crisis (Kane 1984). By the same token, post-crisis levels of nuptiality may rise very rapidly, as marriages postponed over a rather extended period take place within a short interval of time. Such response is either attenuated or altogether prevented when the crisis has more or less permanent effects on marriage markets. This occurs with wars and wholesale emigration. As in the case of mortality, nuptiality responses may vary by social groups. Available evidence shows that some groups use postponement of marriages as a strategy for survival, whereas in others marriages experience little if any changes (Caldwell *et al.* 1984).

3.2.2. Marital fertility The response of marital fertility is somewhat swifter and depends on three factors. First, frequency of intercourse is reduced significantly. This occurs as a result of stress and poor physical condition, physical separation due to war and emigration, or deliberate abstinence. Secondly, fecundity is diminished as stress or malnutrition impair normal amenorrhoea and anovulatory cycles. Thirdly, higher levels of infant mortality lead to shorter postpartum amenorrhoea periods (via reductions in the length of breast-feeding) thus offsetting somewhat the trend towards lower fecundity. Its strongest impact will be felt one year or so after the outbreak of the crisis but it is likely to be dwarfed by effects with opposite signs associated with the other two factors. At any rate, the downswing in fertility will be of lower magnitude than when breast-feeding is rarely practised. Unlike the onset of the response in nuptiality which precedes the outbreak of the crisis, the earliest indication of marital fertility reductions will take place several months into the crisis. Similarly, whereas the post-crisis nuptiality levels may experience an immediate boost, restoration of the pre-crisis levels of marital fertility may take as long as nine months after the peak of the crisis is passed. The post-crisis rate of increase

Table 12.3. *Associations between type of crisis, timing, and intensity of the response of nuptiality and effects on marital fertility*

Type of crisis	Nuptiality	Marital fertility		
		Fecundity	Intercourse	Amenorrhoea
Triggered by war with no feedback	Pervasive postponement	—	Affected by separation	—
Triggered by relative food shortages	Pervasive postponement	Ovulation affected by nutrition	Affected by separation in case of emigration, and by stress	Affected by reduced breast-feeding
Triggered by disease but no feedback	Minor reduction in the absence of migration	—	Affected by conditions produced by the disease itself	Affected by reduced breast-feeding

of marital fertility will be a function of the effects that the crisis had on marital disruption. If they are unimportant, the increase may be substantial even in the absence of any intentional behaviour to 'replace' the lost population. This is because the composition of the female population by susceptibility status will be unusually favourable to higher fertility (Bongaarts and Cain 1981).

Table 12.3 summarizes the most important associations between types of crisis, timing, and intensity of the response of nuptiality, and effects on marital fertility.

3.3. Emigration and Return Migration

Intense and sudden flows of migrants out of the areas most stricken by the crises has been well documented in pre-industrial and contemporaneous populations (Caldwell 1975; Chen and Chowdhury 1977; Livi-Bacci 1978; Meng-Try 1981; Kane 1984; Ashton *et al.* 1985). These flows may permanently alter the regional distribution of population (no return migration) or they may produce temporary relief for as long as the crisis lasts. Apart from the fact that emigration itself can lead to higher mortality among those who migrate (dislocation, inability to adapt to new environment) and in the areas of destination (diffusion of diseases), it is unclear just what effect they actually have on the conditions at the origin. It has been argued that outflows relieve pressure on resources and, hence, lead to a more rapid recovery than when only food shortages are the cause of the crisis. However, it is also the case that, like mortality, migration is selective and may involve the most active and productive members of the sending population. If this is the case and no substantial return migration takes place, the sending areas should deteriorate even further and their post-crisis mortality should remain at levels higher than the pre-crisis levels as those who remain are the most frail. Massive displacements of population are an invariable characteristic of crisis triggered by warfare (e.g. the cases of Central America, Eritrea, Afghanistan, Kampuchea). Movements of such magnitudes are regularly accompanied by dismal health conditions and extraordinarily high levels of mortality at all ages. This is why protracted warfare, even though localized, may end up having a devastating cumulative effect on the demography of a population. Massive displacements of population not only lead to deterioration of health conditions; they also have a deep impact on fertility and, more generally, on household and family organization. In extreme cases, the latter disintegrates and is replaced by alternative institutional forms (Montes 1985). On the other hand, although fertility is initially depressed, it may experience important and sustained surges as pre-crisis restrictions in cohabitation and sexual intercourse break down. Reproduction may begin to take place among couples not in regular unions with the consequent disarray in the system of child care and child rearing. This reinforces the conditions keeping high levels of mortality.

Migration has also been the response to epidemics (Livi-Bacci 1978; Charbonneau and Larose 1979) and appears as an early warning of relative food shortages (Caldwell 1975; Lardinois 1982; Caldwell *et al.* 1984; Ashton *et al.* 1985). In both cases, return migration is more likely to occur than when wars are the triggering event since, unlike wars, epidemics and food shortages in general do not involve destruction of the environment. Similarly, the age and sex pattern of migration is likely to be more selective than in the case of wars.

3.4. The Behaviour of Demographic Indicators

The foregoing considerations lead to some inferences about the most likely course of selected demographic indicators. Consider the case of a crisis triggered by food shortages with no feedback mechanisms. I will take the probability of dying before age 5 (*Q*5) and the CDR as indicators of mortality. As indicators of fertility I will use the TFR and the CBR. Assume that generalized food shortages appear at time T_0 and that they last about a year. Immediately before T_0 there will be a slight increase in *Q*5 and CDR as important segments of the population experience gradually worsening conditions due to a lack of trade goods and funds for protection. During the initial phase of the crisis, CDR will increase rapidly as the force of mortality at all ages increases. A similar trajectory will be observed for *Q*5. Within the year after the end of the crisis, the CDR will begin to decline, reproducing the decline in age-specific rates. However, its level may dip further than *Q*5 (and lower than the pre-crisis levels) as a consequence of

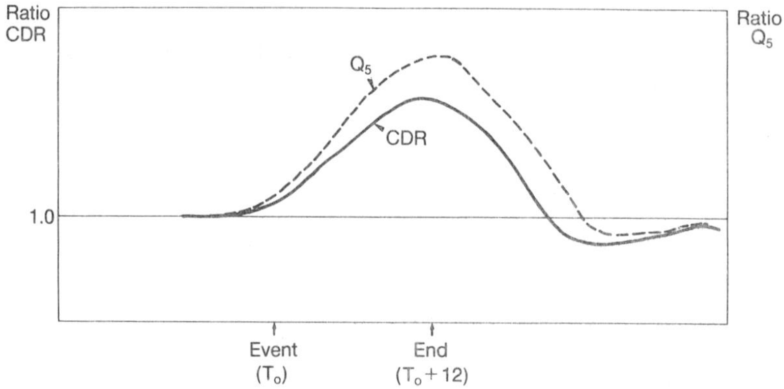

Fig. 12.3. Ratios of mortality measurements during crises to mortality measurements during normal periods, CDR (crude death rate), Q_5 (probability of dying before age 5)

the changed composition of the population. First, as a result of short-term reductions of fertility and higher levels of mortality in infancy, the age structure of the population becomes more 'favourable' to a lower CDR. Secondly, the composition of the population by frailty is biased toward those least frail, and the age-specific mortality rates ought to be on the whole somewhat lower than the pre-crisis levels. Figure 12.3 displays the trajectories corresponding to CDR and *Q5*. As a result either of an increase of emigration or of a tighter nuptiality regime, both CBR and TFR begin to decline *before* the onset of the crisis. Nine months into the crisis, they reach their 'crisis levels'. This occurs as a result of lowered conceptions during the previous nine months. The fall in CBR is slower than in TFR since the former also reflects the changed age composition which is more favourable towards population in reproductive ages. If return migration takes place and the level of marital disruption caused by adult mortality is insignificant, then the restoration of pre-crisis levels of nuptiality and marital fertility occurs in two stages. First, there is a sudden surge of nuptiality immediately after the end of the crisis. The CBR and TFR remain at low levels until about a year past the end of the crisis. The second stage begins with increases in CBR and TFR as a consequence of the surge of marriages and of resumed marital relations. Both indicators will increase to levels exceeding those observed during the pre-crisis period. The 'excess' in TFR will be due to the unusual composition of the female population by susceptibility status. The increase in the CBR will be the direct consequence of increased TFR as well as of an age structure which has higher proportions in the child-bearing ages. Figure 12.4 displays the trajectory of TFR and CBR. Figures 12.3 and 12.4 should be contrasted with Figure 12.5, where I have plotted the trajectory for selected indicators of mortality

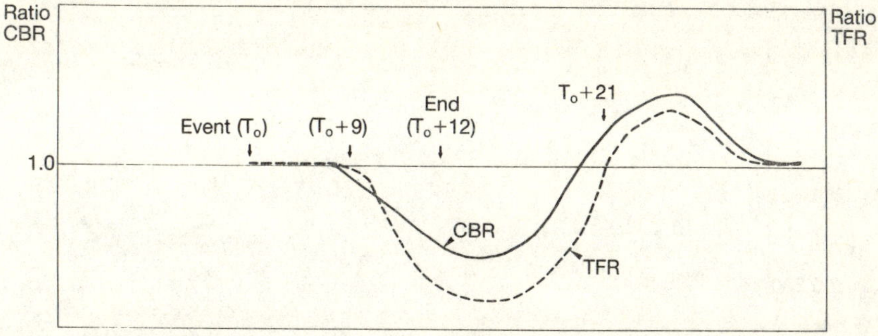

Fig. 12.4. Ratios of fertility measurements during crises to mortality measurements during normal periods, CBR (crude birth rate), TFR (total fertility rate)

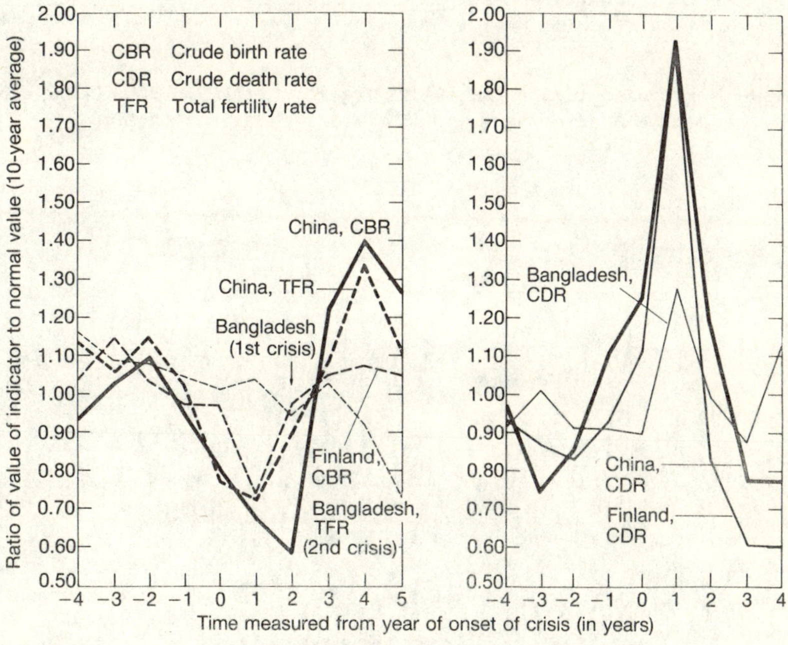

Fig. 12.5. Trajectory of demographic indicators in three selected countries. *Source*: see Appendix 12.1

and fertility in Bangladesh, China, and Finland during periods bounded by and containing crises (see Appendix 12.1 for actual listing of quantities plotted). Finally, Figures 12.3, 12.4, and 12.5 should be compared to Figure 12.6, which displays crude rates for simulated regimes of crises.

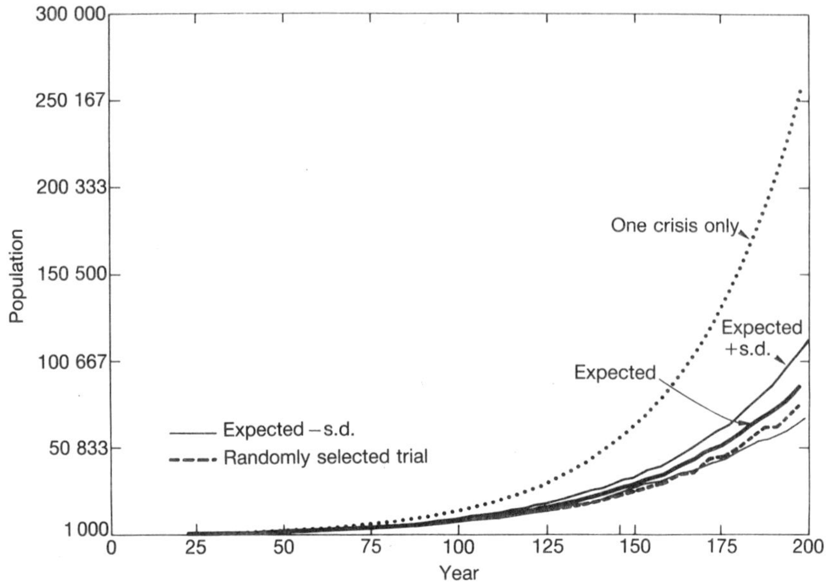

Fig. 12.6(*a*). Projected populations. Crisis regime = HHS. Years 1–200. Initial population: growing at r – 0.027. (s.d. = standard deviation)

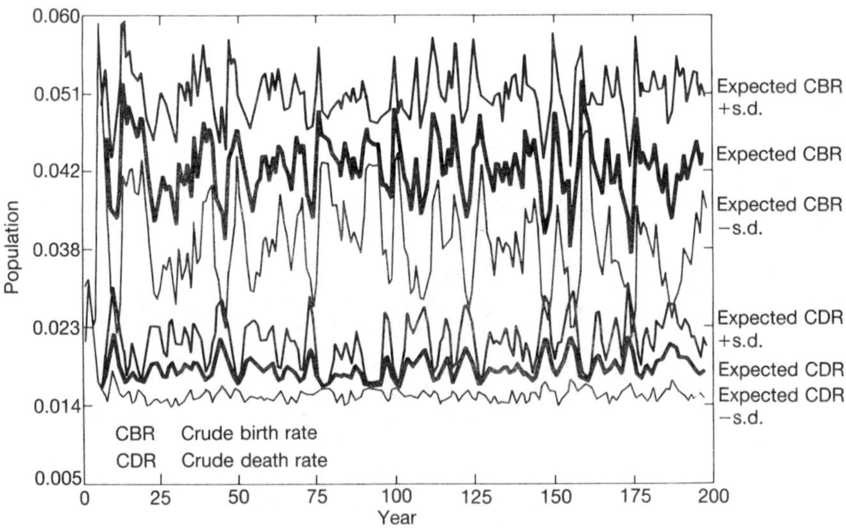

Fig. 12.6(*b*). Vital rates. Crisis regime = HHS. Years 1–200. Initial population: growing at r – 0.027. (s.d. = standard deviation)

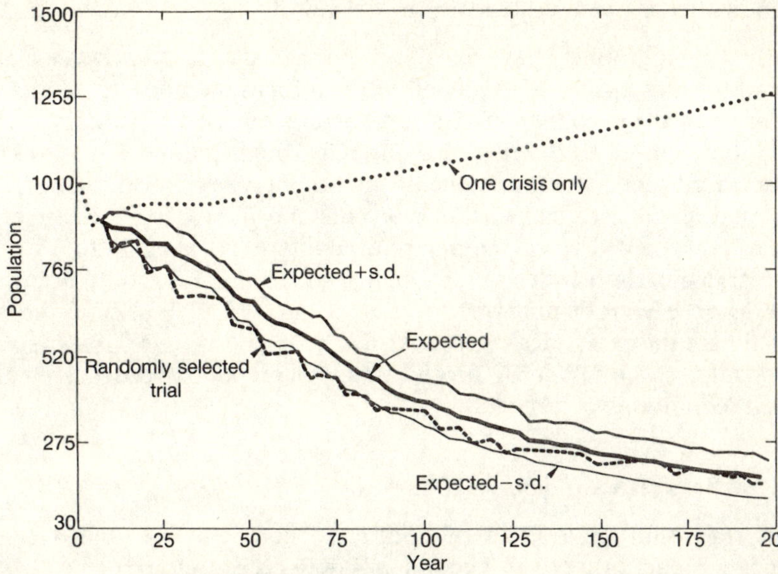

Fig. 12.6(*c*). Projected populations. Crisis regime = HHS. Years 1–200. Initial population: nearly stationary. (s.d. = standard deviation)

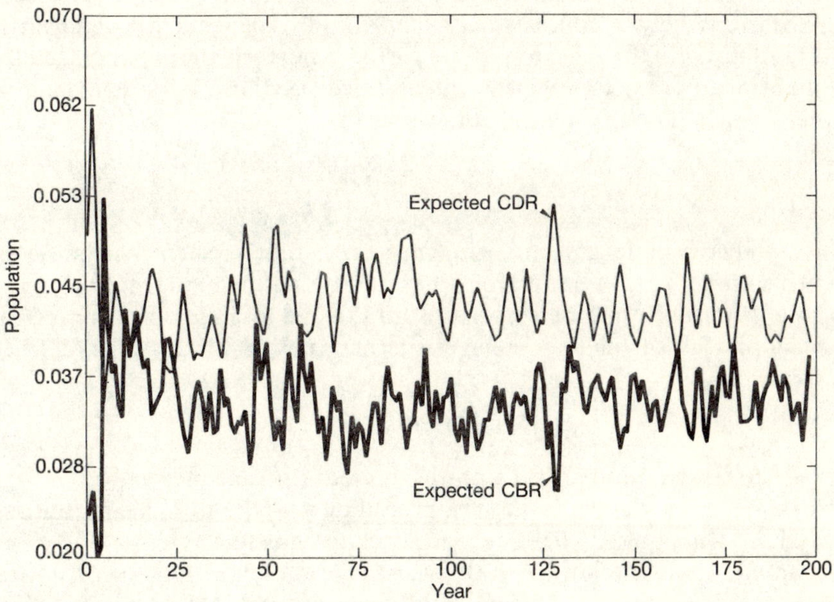

Fig. 12.6(*d*). Expected crude birth rate (CBR) and crude death rate (CDR). Crisis regime = HHS. Years 1–200. Initial population: nearly stationary. (s.d. = standard deviation)

4. Measurement of the Intensity of Crises

Historical demographers have developed a long list of alternative measurements of crisis intensity. Virtually all of them focus on mortality alone and most are based on counts of deaths that are not tied to counts of the population exposed to the risk of dying (Charbonneau and Larose 1979). An important preoccupation is the distinction between mortality increases or fluctuations due to extraordinary events and increases or fluctuations that can be expected from sampling variability. This is explicable by the historical demographers' focus on small populations and should not be of great relevance when dealing with large aggregates. In all these cases the idea is to estimate the excess deaths that are due to the occurrence of a crisis or, equivalently, the number of deaths that would not have occurred had 'normal' conditions persisted.

4.1. Hollingsworth's Index

This is the result of a series of criteria to highlight the objective and subjective impact of a crisis. The index is defined as follows:

$$\mathrm{I} = Q/(1-Q) \cdot (n)^{2/3} \cdot (t)^{-1/3}, \tag{1}$$

where Q is the proportion of the population who dies within period of length t, and n is an estimate of the total population. The index has been applied in a variety of contexts but its magnitude is dependent on and varies substantially with the selection of the period t. Further, the exponents in the formula are parameters whose values have been fixed for convenience rather than estimated directly from the data.

4.2. Dupaquier's Index

This closely resembles a normalizing measure in that it relates the number of deaths within a year (or month or other suitable defined time interval) to an average ('normal') number of deaths and to the average deviation of the number of deaths from their mean during such periods. The index is defined as follows:

$$\mathrm{II} = [M(t) - U]/S, \tag{2}$$

where $M(t)$ is the number of deaths in the year (or time interval) t, U is the mean number of deaths in the ten preceding years, and S their standard deviation. Dupaquier (1979) suggested classifying the intensity of crises depending on whether or not the value of index II fell within one of several ranges that were defined a priori. Unlike index I, index II does not incorporate a subjective dimension. The only 'subjective' item here is the selection of the number of years for the period of reference. It is likely that the value of the index varies considerably depending on the choice of the reference period.

An alternative to index II is to fit a functional form to a subset of deaths observed within a period. The intervals of crises are then determined in much the same way as one would identify conspicuous residuals in general linear models (Madai 1979; Guttman 1980). However, this procedure also has limitations. First, the choice of functional form is not trivial, although for most purposes a linear form would suffice. Secondly, the selection of the interval of time within which the function is fitted may have important consequences for the identification of large deviations. Finally, identification of large residuals is closely dependent on the method used to estimate the functional form. Least-squares techniques are too sensitive to outliers and hence do not permit a clear-cut differentiation between normal and non-normal periods. In most cases, it would be preferable to use one of a variety of resistant procedures that are robust to outliers. For example, one could first divide the set of observations corresponding to a period of say *n* years into two groups. Then one could estimate the slope and intercept of a line using the median values in the two groups. Finally, the deviation of each observed value from the estimated line yields a measure of its closeness to a 'normal' year.

The various modifications suggested to these procedures attenuate the influence of random fluctuations and of the levels of mortality created by the crisis itself. Although their degree of success is somewhat different, all of these procedures are quite efficient in identifying the years of crisis. But none of them is good enough to measure the actual intensity of the crisis. Using 800 simulated populations experiencing crises with different frequencies, durations, and intensity during a period of 200 years, it was found that Dupaquier's index, the index based on fitting a line through ordinary least squares and, lastly, the one based on fitting a line using resistant procedures successfully identified the crisis at least eight out of ten times. Failures occurred more often in simulations combining a pattern of crises with the highest rate of occurrence and the longest durations. However, in only a handful of cases (1 out of 80) was any index successful in estimating the intensity of the crisis (measured by the excess of crude death rates in relation to a normal year). In most cases there was a downward bias in the estimate of mortality excesses. This occurs because none of the procedures takes into account the reduction in the crude death rates that occurs as a consequence of lower birth rates immediately before and during a crisis. It should be noted that their performance in this regard would have been worse had the simulations incorporated the effects of population heterogeneity.

The foregoing limitation is all the more relevant when estimates of excess deaths are obtained from two population censuses bracketing the period during which the crisis occurred. By and large, however, the estimated excesses are too high (rather than too low) since the estimated over-mortality assumes unchanged fertility during the period of crisis. The sensitivity of these estimates to the choice of assumptions about normal mortality and fertility is just too high to place any confidence in them (Anderson and Silver 1985). However, if supported by careful internal consistency checks, they may result in good approximations (Kane 1984; Ashton *et al.* 1985).

4.3. Indices of Replacement of Population Lost

An alternative to conventional indices of intensity of crisis is to incorporate mortality jointly with a measure of the capability of the population to recover the losses. This suggestion was first made in connection with the study of crisis in Toscana (Del Panta and Livi-Bacci 1979) but has not received much attention. I present here a pair of elaborations and extensions of the original measure.

Assume that the normal female fertility rate in the age interval $x, x+dx$, is given by $f(x)dx$, that maximum potential fertility in the small interval $x, x+dx$ is given by $(M^*f(x)dx)$, and that the normal probabilities of surviving from birth to exact age x are $p(x)$. The net reproduction rate is then given by:

$$\text{NRR} = \int_{o}^{\beta} p(x) f(x)\, dx. \tag{3}$$

If a crisis of one-year duration increases the force of mortality at age x, $x+dx$ by a fraction τ, then the cohort aged x at the time of the crisis will require an increase F in fertility to restore the pre-crisis value of the ratio between generations. F is calculated as:

$$F = [S(x)]^{-\tau}, \tag{4}$$

where $S(x)$ is the probability of surviving from age $x-1$ to x. If the population's normal rates of increase hover around zero, the value of F is that required to guarantee replacement of the cohort affected by the crisis. Table 12.4*a* shows the value of F for the cohort born during the year preceding the crisis $(x-1)$. The values are high only for crises with intensity representing doubling or quadrupling of pre-crisis mortality levels $(\tau = 1, 3)$. In none of the contemporary documented crises has the intensity reached such levels. All other values of F are easily attained through post-crisis rebounds of the levels of fertility. Naturally, these values are harder to attain by populations that have higher levels of fertility, e.g. their pre-crisis levels of fertility are closer to a maximum. The ratios of F to the maximum possible increases for a set of combinations of fertility and mortality regimes are displayed in Table 12.4*b*.

Assume now that a cohort just born is exposed more or less continuously

Table 12.4*a*. *Proportionate increases in fertility required to maintain net reproductive rate (one-year crisis, cohort born during year preceding crisis)*

Proportionate increases in mortality due to crisis	Probability of surviving first year of life		
	0.80	0.85	0.90
0.10	1.02	1.02	1.01
0.50	1.12	1.08	1.05
1.00	1.25	1.18	1.18
3.00	1.95	1.63	1.63

Table 12.4*b*. *Ratios of the increase in fertility necessary to maintain pe-crisis net reproduction rate (NRR) to the maximum possible increase (GRR = 7.5) (one-year crisis, cohort born during year preceding crisis)*

Proportionate increases in mortality due to crisis	Probability of surviving first year of life								
	0.80			0.85			0.90		
	NRR*			NRR*			NRR*		
	2.36	1.77	1.00	1.36	1.77	1.00	2.36	1.77	1.00
0.10	0.54	0.41	0.23	0.54	0.41	0.23	0.54	4.0	23.00
0.50	0.59	0.45	0.25	0.58	0.43	0.24	0.56	0.42	0.24
1.00	0.67	0.50	0.28	0.63	0.47	0.27	0.63	0.47	0.27
3.00	1.04	0.80	0.44	0.87	0.65	0.37	0.87	0.65	0.37

* The values of NRR correspond to values of GRR equal to 4.00, 3.00, and 1.69, respectively.

to excess mortality due to crises during its first 15 years of life and that the average increase in the force of mortality between ages 0 and 15 (the beginning of the reproductive period) is τ, then F is calculated as:

$$F = [p(15)]^{-\tau}, \tag{5}$$

where, for simplicity, I have assumed that no mortality increases occur above age 15. To evaluate the implications of (5) I have calculated the minimum increase in mortality that would be necessary for the cohort to be unable to replace itself. This will occur whenever the fertility multiplying factor, F, is higher than the minimum value that would be required to bring the current fertility up to the maximum fertility observed in human societies (corresponding to a GRR of about 7.5). Table 12.5 displays some results for selected combinations of fertility and mortality regimes. In a society with high fertility (GRR = 4) and high mortality ($\mathring{e}_0 = 35$), replacement of the cohort would be attainable if the crisis to which the cohort is exposed up to age 15 leads to average increases of mortality equivalent to 4.25 times the levels prevailing in normal times. In a society with high fertility but lower mortality ($\mathring{e}_0 = 50$), the crisis should increase mortality eightfold for

Table 12.5. *Minimum proportional increase in mortality (values of* T*) that would require the maximum possible increase in fertility (up to GRR = 7.5) to ensure replacement (cumulated effects of crisis evaluated at age 15)*

Life expectancy at birth ($\mathring{e}_0$)	Gross reproduction rate	
	4.00	1.75*
35	3.25	3.26
50	7.74	7.78

* The combination GRR = 1.75 and $\mathring{e}_0$ = 35 corresponds to a stationary population.

replacement to be impossible. Note that the dimensions of the crisis would have to be truly catastrophic for the affected cohorts to be unable to replace themselves. In fact, on average it would require that only 17% of the original cohort attained age 15. Just from this information alone one may conclude that, unless they are recurrent and severe, population crises are not the most efficient devices to check population growth.

A diagnostic of the severity of the crisis using the above-described measure is, of course, strongly dependent on what one considers to be the maximum reproductive potential of the population. Since the latter is not unique, this measure of a crisis's intensity will vary due to factors other than those strictly related to the intensity of the crisis.

5. Measuring the Impact of a Population Crisis

5.1. Existing Approaches to Measurement

An interesting question is whether or not population crises can be taken as effective checks on population growth. This, after all, is the assumption that was at the core of Malthus's catalogue of positive checks. It also appears from time to time in contemporary debates as a strategic, if seldom documented, premiss. The answer to the question depends largely on the measurement instruments designed to assess quantitatively the magnitude of the effects of a crisis. I will review three different strategies, each having merits and disadvantages. It turns out that the underlying model for determining how crises are generated is essential for drawing conclusions. Quantitative assessments that rest on a static conception of crises lead to the conclusion that crises cannot be seriously taken as efficient checks on population growth. A model that explicitly accounts for periodicity, intensity, and duration leads one to expect a considerable reduction of population pressure in the medium to long term. The degree to which crises actually check population growth depends on a stochastic mechanism which simultaneously determines their rate of occurrence, intensity, and duration.

5.1.1. Calculation of the time for recovery In a series of papers (Watkins and van de Walle 1983; Watkins 1985; Watkins and Menken 1985) it is shown that a population with characteristics similar to those of pre-industrial societies will take only a short time to reach the pre-crisis population size after having experienced a crisis of high intensity. In fact, the time for recovery has been calculated to be between six and 50 years, depending on the initial rate of growth and the severity of the crisis. This range is considered to be sufficiently low to make only a minor dent in the population pressure experienced by the society. However, equally compelling is the calculation of the relative sizes of two populations, one experiencing and one not experiencing crises. Using the same data presented by Watkins (1985) I calculated that the ratio of the absolute size of the latter to that of the former 20 years after the crisis would be 1.06 if the crisis were minor and 1.26 if the crisis were major. The ratios would be 1.08 and 1.32 respectively 95 years after the crisis. These figures are invariant with the original rates of growth of the population. The slow-down of growth implied by these ratios is relevant if the initial size of the population—the level achieved immediately before the crisis strikes—is below sustaining capacity. Calculations based on simulations to be presented later show that if a crisis of major proportions took place, on the average, every 30 to 40 years, the size of the population unaffected by crises could become nine times larger than the expected size of a population subject to crises in the relatively short interval of 200 years. This is hardly a token reduction in population growth. Thus, in addition to the calculation of the time to recovery, one needs to consider the population growth that was foregone as a consequence of the crisis. The only case in which the latter is irrelevant is when the initial size of the population was already the equilibrium size.

5.1.2. The foregone population growth under stability The suggestion stated before can be better and more elegantly implemented using some regularities that hold in stable populations. These have been used elsewhere to study the long-term impact of emigration (Keyfitz 1977) and have been applied to examine the long-term impact of some types of population crisis (Le Bras 1969; Salvini 1977). It is well known that under conditions experienced by contemporary developing societies, a policy of year-after-year emigration would require that about 30% of the population at the peak ages of child-bearing migrate in order to keep the population from growing at more than 0.02 per year. The requisite outflow is evidently higher if the target was to keep the population at stationary levels. This suggests from the start that a one-time shock induced by a population crisis will be even less effective to check population growth in the absence of reductions of fertility. In fact, if the losses of population due to a crisis were equivalent to a fraction D of all members between the ages 0 and 50, the ratio of the ultimate population (affected by the crisis) to the ultimate population (not affected by the crisis) would be $1-D$. The contribution of each age in the range 0–50 to this total loss t years after the crisis is given by the following quantity:

$$(D/\nu)\cdot N(x)\cdot \nu(x)\cdot \exp(r\cdot t)/b, \tag{6}$$

where a is the mean age at child-bearing in the stable population, $N(x)$ is the absolute number of the population aged x at the time the crisis strikes, $v(x)$ is the reproductive value at age x, r is the rate of growth, and b the intrinsic birth rate. The values of r, b, and $v(x)$ are those associated with the patterns of mortality and fertility in the pre- and post-crisis situation. Since $v(x)$ peaks at ages 15 or so, the losses experienced at this age weigh more heavily on the calculation of the ultimate losses due to the population crisis. Thus, crises in which the population 0–15 is more affected are the ones with the highest potential effectiveness for reducing ultimate population growth. In order to generate a standard formula for the calculation of ultimate losses to be attributed to excess mortality in the most strategic age group, 0–15, I derived the following formula:

$$\text{Ultimate loss} = j\cdot[\exp(r\cdot t)/(b\cdot \nu)]\cdot C(15)\cdot B(15) \tag{7}$$

where j is an indicator of mortality increases induced by the crisis (equals 1 if mortality is doubled, 2 if mortality is tripled, etc.), r, b, and a are defined as before, $C(15)$ is a weighted average of the reproductive values within the age range 0–15, and $B(15)$ is the number of deaths that would take place in the population in the ages 0–15 exposed to the crisis if subjected to the pre-crisis force of mortality. The weights are given by:

$$w(x) - \left[N(x)\cdot(x)\right] \Big/ \left[\int_0^{15} N(x)\cdot \mu(x)dx\right], \tag{8}$$

where $\mu(x)$ is the force of mortality at x in the pre-crisis pattern of mortality and $N(x)$ is defined as before. Formulae (7) and (8) are valid if mortality in the age segment 0–15 follows a Gompertz curve. The latter has been shown to be a quite reasonable approximation to a disparate set of mortality patterns. To illustrate the magnitudes attained by (7), I have assumed a stable population with a rate of growth of 0.01 per year, a life expectancy of about 50 (implying a crude birth rate equal to 0.029), and a mean age at child-bearing of 26. If the crisis lasts between 1 and 2 years and mortality increases to a level twice as high as the pre-crisis levels, the quantity $B(15)$ is about 24 per 1000 of the original population, and the value of $C(15)$ is 1.5. Thus the amount of the loss after 20 years is close to 58.32 per 1000 of the original population. After 50 years the loss is in the neighbourhood of 79 per 1000 of the original population. Since in these calculations I have not accounted for the post-crisis upswing in births, these figures should be taken as upper limits. As they stand, however, they do not appear to be insignificant. They are small enough to shatter the idea that a crisis may be any more effective than emigration to keep populations from growing above certain targets. However, they are significant enough to make a recurrent crisis if not a court of last resort at least a highly disruptive occurrence in the trajectory of a population.

5.2. An Integrated Framework for Measuring the Impact of a Crisis

The two alternative forms of measurement presented above rest on models of generation of crises that disregard: (*a*) the repetitive nature of crises, and (*b*) the trajectory of vital events resulting from a crisis (see Section 3). To assess the quantitative impact of crises, it is not sufficient to follow the consequences of *one* of them in the absence of any repetitions. Recurrence of crises is a fundamental characteristic of societies that have been vulnerable to them. Similarly, full and accurate estimation of the impact of a regime of crises requires appropriate modelling of the trajectory of vital events. As indicated in Section 3, these trajectories depend on the nature of the crises themselves. In order to eliminate these two deficiencies, I proceeded to simulate the trajectories of several populations exposed to several regimes of fertility, mortality, migration, and crises. As indicated below, the parameters for each of these regimes were estimated from known records of population crises in the pre-industrial period and during the twentieth century. The results that I obtained indicate that some crisis regimes that are likely to affect contemporary populations can be devastating if sustained for relatively short periods of time. The simulation models provide a richer and more complete picture than models resting on the assumption of a single shock.

5.2.1. A random mechanism for the occurrence of crises To incorporate the possibility of multiple crises in an interval of time, it is necessary to postulate a mechanism of recurrence. In the absence of any theory about it, I assumed that crises are generated by a Poisson process. Correspondingly, the intercrises waiting times are independent and exponentially distributed. Estimates of the rate for the exponential distribution were obtained from several historical records of crises (Charbonneau and Larose 1979; Del Panta 1980) and from summaries of crises in contemporary developing countries (see Appendix 12.2). Although it would have been possible to estimate rates specific to type of crisis, country (region), and historical period, I thought it sufficient to estimate a range. A low frequency of crises corresponds to a regime with a rate equal to 30 per 1000 years (or a waiting time for a crisis equal to 33 years). A regime with high frequency is characterized by a rate equal to 100 per 1000 years (or a waiting time for a crisis equal to ten years). It should be noted that the lower bound is by all accounts too low to represent the regime of crises in the pre-industrial period. On the other hand, the upper bound is too low to represent the experience of some developing countries such as Bangladesh during what has elapsed of the twentieth century. Thus, the range used here is quite conservative for applications to both pre-industrial societies and contemporary developing societies.

5.2.2. Definition of types of crisis A regime of crises is characterized by two variables. The first is the duration and the second the intensity. The duration of a crisis was arbitrarily set equal to two or five years. This was done to make results comparable with those obtained by other researchers. There was no compelling reason to select either of them although it could have been possible to estimate empirically the average durations by type of crisis, by country (region), and historical period. However, this elaboration is superfluous in the absence of a theory linking type of crisis and durations. The intensity of a crisis depends on two correlated components: the levels of over-mortality and the levels of fertility reduction. I assumed that minor crises led to minor reactions in both fertility and mortality and that major crises led to major changes in both fertility and mortality. Post-crisis mortality was assumed to go no lower than pre-crisis levels of mortality. Thus, I disregard completely the effects of selectivity. Post-crisis levels of fertility were characterized by a one-year rebound that took them to levels higher than during the pre-crisis period. Consistent with the discussion in Section 3, the levels of mortality and fertility for the year immediately preceding the crisis are subject to a mild increase and decrease respectively. The magnitude of these changes depends on the intensity of the crisis.

Age-specific fertility is affected by the same proportionate amount so that a crisis does not bring about a change in the age pattern of fertility. A high-intensity crisis led to a decline of GRR of about 50% and a rebound of about 20% for the year following the end of the crisis. During the year before the onset of crisis levels, fertility declines gradually to 75% of its pre-crisis level, to 63% of its original value within the first year of the crisis, and is halved for a period beginning nine months after the onset of the crisis until the end of the crisis. A low-intensity crisis leads to an overall decrease of 10% in the GRR and to a post-crisis rebound of 5%.

Over-mortality was assumed to follow a pattern by age that was obtained as an average of various experiences in the pre-industrial period and during the twentieth century: increases are heavier among younger children and older people. The pattern is displayed in Appendix 12.2. A high-intensity crisis was defined as one that, given the age pattern of over-mortality, would increase the CDR by about 80% in a population with a GRR equal to 3 and a life expectancy equal to 50. The increase begins one year before the onset of the crisis—during which the levels of mortality are increased by an amount equivalent to 15% of the CDR—and terminates at the end of the duration of the crisis. A low-intensity crisis leads to increases in the force of mortality that, given the age pattern of over-mortality, translates into an increase of the CDR of about 15% in a population with the same characteristics mentioned above.

The values given to the parameters in the cases of highest and lowest intensity are bracketed by those observed in the following cases: China (1950–60), Bangladesh (1971–5), Finland (1750–1900), Tuscany (fifteenth century), and various other regions' and countries' experiences recorded by Charbonneau and Larose. The only feature that I have left out is migration, which can be approximately captured by increasing the impact of mortality. This is an approximation inasmuch as the age pattern and the timing of emigration are not necessarily the same as those of over-mortality.

5.2.3. Simulation of crises Two initial populations were selected. The first was nearly stationary, having a life expectancy of about 27.5 (East model) and a GRR equal to 2.5. The second had a life expectancy equal to 50 (West model) and a GRR equal to 3, with an intrinsic rate of growth of 0.027. The first population resembles the Chinese farmers studied by Barclay and the second one is very similar to Bangladesh during 1976–9. Having selected an initial population and a crisis regime a pseudo-random number generator was utilized to determine the timing of crises within a period of 200 years. In all cases, a crisis at the beginning of the projection period was imposed. Randomness entered in the determination of all crises following the first one. The population was then projected forward year by year for a period of 200 years. The simulation process was repeated 100 times and then a new crisis regime was selected and applied in analogous fashion. Altogether 800 simulations were carried out with the second initial population and 400 with the first initial population.

5.2.4. Main results Tables 12.6 and 12.7 summarize selected results for both initial populations. Each table displays the expected population in four different crisis regimes, the population experiencing only one crisis, and the population experiencing no crisis after 10, 50, 100, 150, and 200 years from the initiation of the projection. The crisis regimes were chosen to provide the widest possible range of results. Whenever randomness entered into the determination of a regime, I used the expected population (average over 100 simulations) even though the corresponding figures represent no particular experience. To enhance the reliability of the figures, I also show the estimated standard deviations.

Figures 12.6*a* to 12.6*d* display the trajectory of crude rates and populations in selected regimes of crises. Expected rates and populations are shown and each series is enclosed within a band determined by one standard deviation on each side of the expected values. These figures should be compared with Figures 12.3, 12.4, and 12.5. The most striking feature of Tables 12.6 and 12.7 is the divergence between the trajectories of the randomized crises and the one corresponding to the one-crisis, one-shock regime. Notice that the combination of an initially stationary population with a relatively severe crisis regime leads to population extinction. Here, there is no place for the calculation of time to recovery. This is a result that would simply not be obtained in a one-crisis, one-shock regime. In other,

Table 12.6. *Population sizes of initial population with* r = *0.0022 in six different situations*

Year	Crisis regime[a]								One crisis[b]	No crisis
	1 (HHS)		2 (HHL)		3 (LLL)		4 (LLS)			
	EV[c]	Standard error	EV	Standard error	EV	Standard error	EV	Standard error		
0	1000		1000		1000		1000		1000	1000
10	891	(42.9)	831	(0.000)	977	(0.000)	991	(7.6)	977	1017
50	715	(71.5)	629	(110.0)	989	(24.8)	1032	(21.1)	1036	1098
100	483	(52.3)	369	(151.0)	1008	(47.6)	1080	(36.1)	1130	1203
150	317	(51.1)	184	(80.7)	997	(71.6)	1133	(50.3)	1238	1318
200	220	(40.9)	84	(40.5)	1027	(86.8)	1176	(65.2)	1357	1443

[a] The nature of each regime is defined by a triplet where the symbols refer to (in order) low (L) or high (H) frequency, low (L) or high (H) intensity, and short (S) or long (L) duration.

[b] One crisis at the beginning of the projection period lasting five years and having low intensity.

[c] EV = expected value over 100 simulations.

Table 12.7. *Population sizes of initial population with* r = *0.027 in six different situations*

Year	Crisis regime[a]								One crisis[b]	No crisis
	1 (HHS)		2 (HHL)		3 (LLL)		4 (LLS)			
	EV[c]	Standard error	EV	Standard error	EV	Standard error	EV	Standard error		
0	1000		1000		1000		1000		1000	1000
10	1176	31.9	1094	0.0001	1246	0.0002	1264	8.6	1192	1290
50	2805	278.6	2357	230.1	3735	73.9	3882	82.2	3682	4024
100	8596	1167.3	5615	1242.3	14903	545.0	15918	494.7	15368	16858
150	25722	3490.1	13156	4747.1	59496	3098.0	65223	2149.3	64368	70619
200	73356	11106.9	33680	19205.0	240699	13960.0	266826	11542.0	269645	295836

[a] The nature of each regime is defined by a triplet where the symbols refer to (in order) low (L) or high (H) frequency, low (L) or high (H) intensity, and short (S) or long (L) duration.

[b] One crisis at the beginning of the projection period lasting five years and having high intensity.

[c] EV = expected value over 100 simulations.

less severe cases, the crisis regime leads to losses amounting to 32% of the size achieved at the end of 200 years by the population experiencing only one crisis and 40% of the size achieved in the absence of crisis. Naturally, the potential for recovery is much greater in the population initially growing at a fast rate (Table 12.7). Although severe crisis regimes do not appear to lead to extinction, they imply massive losses in a relatively short time. Note that after 200 years the ratio of absolute sizes in the population experiencing no crisis to the populations affected by the two most serious regimes are 4 and 9 (first and second columns). Even the less severe crisis regimes produce losses amounting to at least 10% of the final size in the society free of crisis. In the short run (first 100 years) the differentials are greater in regimes with less severe crises and smaller in regimes with high-intensity crises. Thus, even with populations growing rapidly, a randomized crisis regime inflicts losses that cannot be gauged using a one-crisis model. This holds regardless of the time horizon that one selects to use for the projection exercise.

Table 12.8 displays information on the ratio of the expected populations in various crisis regimes to the population in the absence of crises at $\tau = 100$ and $\tau = 200$ years after the initiation of the projection. The information is represented in terms equivalent to rates of growth: the figures are the values of the difference between the rates of growth in two populations that would lead to the observed ratios after 100 and 200 years. For example, at the end of 100 years, a population exposed to no crises is expected to be about 1.96 times larger than a population with frequent crises of high intensity but

Table 12.8. *Estimated rates of growth consistent with the ratio of population with crises to population without crisis at two values of* τ

τ		Comparison of population with more than one crisis to population with no crisis				Comparison of population with only one crisis to population with no crisis			
	Frequency	Intensity				Intensity			
		Low		High		Low		High	
		Short	Long	Short	Long	Short	Long	Short	Long
100									
	Low	0.0006	0.0012	0.0031	0.0048	0.0001	0.0002	0.0005	0.0009
		(2.1)[a]	(4.5)	(11.2)	(17.2)	(0.4)	(0.7)	(1.7)	(3.3)
	High	0.0013	0.0023	0.0067	0.0110				
		(4.7)	(8.2)	(24.2)	(39.7)				
200									
	Low	0.0005	0.0010	0.0027	0.0046	0.0001	0.0002	0.0005	0.0009
		(1.8)	(3.7)	(10.0)	(16.6)	(0.4)	(0.7)	(1.7)	(3.3)
	High	0.0013	0.0022	0.0070	0.0109				
		(4.7)	(7.9)	(25.3)	(39.4)				

[a] The percentage differences relative to the original rate of growth appear in parentheses (0.0277).

short durations. This is equivalent to having two populations growing at rates differing by about 0.0067 (or 24.2% of the initial rate of growth). Table 12.8 only shows results for the fast-growing initial population. Note that the differences are substantial. Randomized crisis regimes lead to larger differentials than the one-crisis, one-shock model regardless of the type of crisis. Also, randomized crisis regimes are strongly responsive to changes in intensity and frequency and less so to changes in duration. However, there are powerful interaction effects. Thus, shifting frequencies has a larger impact when the crises are long than when they are short, and when they are of high rather than low intensity.

All in all, these results confirm the conjecture made at the outset: provided that crises are recurrent, even a rather benign regime would end up producing substantial losses of total population. Under certain conditions, recurrent crises lead to long-term extinction; under more general conditions, a pattern of population crisis can provide significant relief from excessive growth, much more than implied by the one-crisis, one-shock model.

5.2.5. Limitations of the simulations The assumptions of independence of successive crises, the lack of built-in correlation between duration and types of crisis, the absence of combination of different types of crisis within the same projection routine are all pitfalls that make the procedure implemented here a somewhat inaccurate and unrealistic one. However, it provides a better assessment of the quantitative impact of crises than those that rest on the assumption of only one crisis during the period over which calculations are carried out. Further refinements of this procedure are possible. Moreover, the procedure can be related to the formal theory of populations subjected to cyclical fluctuations (Cohen 1976, 1977; Frauenthal and Swick 1983).

There are, however, limitations that go beyond those due to lack of acceptable measurement. In fact, the simulations are carried out in a context where neither social differentiation nor regional inequalities are taken into account. Crises do not affect equally all members of the population. A realistic representation of the effects of population crises should predict the differential impact on the various social strata. In so far as social groups are endowed with different resources, the balance between demographic and non-demographic responses will vary from case to case. Only in very extreme cases will all groups be affected equally. Neglecting social heterogeneity inevitably creates distortions that are masked by the general character of population projections.

Most crises are local since they affect certain geographic areas rather than entire nations. This effectively limits their scope and their potential impact if only because exchanges (of persons or resources) between regions will enter as an adaptive response (Caldwell 1975). Except for the allowance made for emigration and return migration, the projection exercise completely neglected this. Thus, the projections portray an artificial state of affairs since they are only valid within a bounded (and perhaps small) region and only when the crisis is allowed to run its course without the beneficial influence of outside relief. Since social and geographical heterogeneity are pervasive, the results of a simulation exercise that has neglected them should only be used to gain a rough sense of the magnitudes of effects involved.

5.3. Other Consequences of Population Crises

An intriguing idea emerges from the combined studies of population crises and of the conditions sustaining high levels of fertility (Bongaarts and Cain 1981; Caldwell *et al.* 1984). The idea is that, in contrast to what has been conventionally thought, recurrent crises may lead to the development of mechanisms that support high fertility as a form of risk avoidance. This should occur whenever institutionalized systems to minimize risks, such as social security, health services, transportation networks, are absent. If this were in fact the case, the occurrence of a crisis itself reinforces sustained growth in the aftermath of the crisis. This is a paradoxical consequence of the positive checks. The relation between fertility and crisis will of course vary across subgroups in the population and will depend on the latter's position to minimize risks. This conjecture can be neatly connected with Sen's hypothesis about entitlements to yield predictions about what subgroups in the population are subject to higher risks and hence to potentially higher fertility and adverse behaviour toward fertility limitations. By the same token, since a crisis rarely leaves untouched the structure of families and households and since the latter are the fabric of rules of couples' behaviour, the study of fertility cannot overlook the frequency and intensity of crises, quite apart from the impact that these may have on the size of a population. In fact, the study of one cannot be carried out without an understanding of the mechanisms determining the other. From this point of view the analysis of population crises is far from an esoteric exercise. As it occurs with the study of physical materials subjected to extreme conditions, the analysis of population crises should reveal the vulnerability of a society's adaptive response and of its resilience.

6. Conclusion

This chapter has reviewed the state of the art in the analysis of population crises. I showed that there are models that guide the understanding of a crisis's origin and development. These models enable us to formulate typical trajectories for the most important vital phenomena. This provides an opportunity for indirect measurement in so far as hypothetical trajectories could be contrasted with observed data to infer characteristics about the crisis.

Although affected by numerous limitations, alternative measurement techniques to quantify the intensity of crises have been developed; they should be used to create a map of their frequency and chronology. In addition, available methods allow us, with variable success, to quantify the magnitude of the impact of crises and assess their role as checks on population growth. These methods can be designed to represent the stochastic nature of the process and preserve its recurrent character. These are essential characteristics of crisis regimes which, if neglected, can lead to misleading conclusions. Thus, for example, I showed that a regime with randomly generated crises can lead to significant losses that would not have been discovered by more conservative modelling. Extensions of these designs may be possible if they are associated with conventional advances in modelling of populations subject to cyclical fluctuations.

The study of crises is inseparable from the study of fertility and family and household structures: not only do population crises affect family organization but the latter may be designed to minimize the expected losses in high-risk situations leading to recurrent crises. Thus, population crises ought to receive more than the token attention paid to them by those who use them to create or exorcise the evils of population growth.

References

Alamgir, M. (1980), *Famine in South Asia: Political Economy of Mass Starvation* (Cambridge: Cambridge University Press).

Anderson, B., and Silver, B. (1985), *Demographic Analysis and Population Catastrophes in the USSR*, Michigan Working Papers in Soviet and East European Studies.

Appleby, A. B. (1978), *Famine in Tudor and Stuart England* (Palo Alto, Calif.: Stanford University Press).

Ashton, B., Hill, K., Piazza, A., and Zeitz, R. (1985), 'Famine in China, 1958–61', *Population and Development Review*, 10, 613–45.

Bongaarts, J., and Cain, M. (1981), *Demographic Responses to Famine*, Working Paper no. 77 (New York: The Population Council, Center for Policy Studies).

—— and Delgado (1977), *Effects of Nutritional Status on Fertility in Rural Guatemala*, Working Paper no. 7 (New York: The Population Council, Center for Policy Studies).

Caldwell, J. (1975), *The Sahelian Drought and Its Demographic Implications*, Overseas Liaison Committee Paper, no. 8 (American Council on Education).

—— Reddy, P. H., and Caldwell, P. (1984), 'Periodic Risk as a Cause of Mortality Decline: Survival Strategies in the 1980–83 South Indian Drought', Paper presented at the *Seminar on Social and Biological Correlates of Mortality, IUSSP*, Tokyo, Nov.

Charbonneau, H., and Larose, A. (eds.) (1979), *The Great Mortalities: Methodological Studies of Demographic Crises in the Past* (Liège: Ordina Éditions).

Chen, L. C. (1973), *Disaster in Bangladesh: Health Crises in a Developing Nation* (New York: Oxford University Press).

—— and Chowdhury, A. K. M. A. (1977), 'The Dynamics of Contemporary

Famine', *International Population Conference*, Mexico, 8–13 August 1977, i. 409–26 (Liège: IUSSP).

Cohen, J. (1976), 'Ergodicity of Age Structure in Populations with Markovian Vital Rates, i: Countable States', *Journal of the American Mathematical Association*, 71, 335–9.

—— (1977), 'Ergodicity of Age Structure in Populations with Markovian Vital Rates, ii: General States', *Advances in Applied Probabilities*, 9, 18–37.

Curlin, G.T., Chen, L., and Hussain, B. (1971), 'Demographic Crisis: The Impact of the Bangladesh Civil War on Births and Deaths in a Rural Area of Bangladesh', *Population Studies*, 30, 87–105.

Del Panta, L. (1980), *Le Epidemie nella Storia Demografica Italiana: Secoli XIV–XIX* (Turin: Casa Editice Loescher).

—— and Livi-Bacci, M. (1979), 'Chronology, Intensity, and Diffusion of Mortality in Italy: 1600–1850', in Charbonneau and Larose (1979).

Dupaquier, J. (1979), 'L'Analyse statistique des crises de mortalité', in Charbonneau and Larose (1979).

Frauenthal, J., and Swick, K. (1983), 'Limit Cycle Oscillations of Human Populations', *Demography*, 20, 285–98.

Guttman, M.P. (1980), *War and Rural Life in the Early Modern Low Countries* (Assen, Netherlands: Van Gorcum).

Hollingsworth, T. (1979), 'A Preliminary Suggestion for the Measurement of Mortality Crisis', in Charbonneau and Larose (1979).

Kane, P. (1984), 'Famine in China 1959–1961: Demographic and Social Implications', Paper presented at the *Seminar on Social and Biological Correlates of Mortality, IUSSP*, Tokyo, Nov.

Keyfitz, N. (1977), *Applied Mathematical Demography* (New York: Wiley Interscience).

Lardinois, R. (1982), 'Une conjoncture de crise démographique en Inde du sud au XIXe siècle: La famine de 1876–1878', *Population*, 20, 371–404.

Le Bras, H. (1969), 'Retour d'une population à l'état stable après une catastrophe', *Population*, 24, 861–96.

Livi-Bacci, M. (1978), *La société italienne devant les crises de mortalité* (Florence: Dipartimento Statistico).

Madai, L. (1979), 'Les crises de mortalité en Europe dans le deuxième moitié du XIXe siècle', in Charbonneau and Larose (1979).

Meadows, D.L., *et al.* (1974), *Dynamics of Growth in a Finite World* (Cambridge: Wright Allen).

Meng-Try, E.A. (1981), 'Kampuchea: A Country Adrift', *Population and Development Review*, 7, 209–28.

Miranda, A. (1982), *The Demography of Bangladesh*, The Chr. Michelsen Institute (Bergen: Derap).

Montes, J. (1985), Personal communication about the results of a pilot project on refugee camps in El Salvador.

Salvini, S. (1977), 'Un Metodo di Misura delle Crisis di Mortalità', *Genus*, 33, 119–42.

Sen, A. (1980), 'Famines', *World Development*, 8, 613–21.

Simon, J. (1980), 'Resources, Population, and Environment: An Oversupply of Bad News', *Science*, 208, 1431–6.

Turpeinen, O. (1979), 'Fertility and Mortality in Finland since 1750', *Population Studies*, 3(1).

Watkins, S.C. (1985), 'A Skeptical View of the Demography of Famines: Results from a Simulation Model', *International Population Conference*, Florence, June, iv. 339–54 (Liège: IUSSP, Ordina Éditions).

—— and Menken, J. (1985), 'Famines in Historical Perspective', *Population and Development Review*, 11, 647–76.

—— and van de Walle, E. (1983), 'Nutrition, Mortality and Population Size: Malthus' Court of Last Resort', *Journal of Interdisciplinary History*, 13(2), 205–26.

Appendix 12.1

Table 12.A1. *Data to construct Figure 12.5*

	CBR (per 1000)	CDR (per 1000)	TFR (per 1000)	Infant mortality rate
Bangladesh				
1966–1967	47.1	15.0	6.7	110.7
1967–1968	45.4	16.6	6.3	125.4
1968–1969	46.6	15.0	6.3	123.8
1969–1970	45.3	14.9	6.1	127.5
1970–1971	43.5	14.8	5.9	131.3
1971–1972	44.5	21.0	6.0	146.6
1972–1973	41.8	16.2	5.5	129.2
1973–1974	45.6	14.2	6.0	128.8
1974–1975	40.1	20.0	5.4	167.2
1975–1976	27.6	18.2	4.3	150.4
China				
1955	32.6	12.3	6.3	177
1956	31.9	11.4	5.9	167
1957	34.0	10.8	6.4	156
1958	29.2	12.0	5.7	144
1959	24.8	14.6	4.3	236
1960	20.9	25.4	4.0	205
1961	18.0	14.2	3.3	132
1962	37.0	10.0	6.0	236
1963	43.4	10.0	7.5	88
1964	39.1	11.5	6.2	96
Finland				
1863	n/a	29.8	4.8	211
1864	n/a	22.6	5.3	144
1865	n/a	25.9	4.8	193
1866	n/a	33.6	4.5	217
1867	n/a	38.1	4.5	223
1868	n/a	77.5	3.4	391
1869	n/a	25.2	4.5	141
1870	n/a	18.2	4.9	136
1871	n/a	17.9	5.0	141
1872	n/a	19.7	4.9	173

Note: n/a = not available. The entire set of vital rates on which the construction of Figure 12.5 is based appears in this appendix.

Sources: Turpeinen (1979) for Finland; Curlin *et al*. (1971); Chen and Chowdhury (1977); Miranda (1982) for Bangladesh; Charbonneau and Larose (1979) for London.

Appendix 12.2 Basic Information Used to Construct Patterns of Fertility and Mortality during Crises

1. Pattern of Over-mortality

Table 12.A2. *Pattern of over-mortality expressed as ratios of increases in the age group 0–1 for some actual crisis and pattern used in the model*

Age	Used	Bangladesh 1971	Finland 1832–1833	Finland 1867–1868	London plague 1603
0	1.00	1.00	1.00	1.00	1.00
1–4	3.74	2.26	3.03	2.27	4.34
5–9	3.94	12.01	4.89	3.78	15.02
10–14	3.82	1.85	3.83	3.47	16.30
15–44	1.45	1.35	2.39	2.62	6.20
45–64	1.66	1.93	1.64	2.31	1.44
65+	1.91	3.09	0.39	1.00	0.37

Sources: as for Table 12.A1.

The recorded percentage increases in CDR used as references are the following:

Bangladesh	(war 1971–2)	28
Bangladesh	(famine 1974)	24
China	(famine 1959–62)	70
Finland	(famine 1867–8)	158

These percentages were calculated using as baseline the average of five years preceding the crisis (excluding observations corresponding to other crises).

2. Pattern of Decline in Fertility

The average percentage decrease in TFR for China, Bangladesh, and Finland during the various crises are 48, 33, and 28 respectively. The rebound over the year following the crisis was characterized by increases of 19% and 5% in China and Finland respectively. The sources for these figures are the same as for Table 12.A1.

3. Frequency of Crises

(*a*) Note that in China there have been two documented crises over a period of 70 years and that in Bangladesh there have been three over the same period. If no other crises occur, these frequencies represent rates of 20 per 1000 years and 30 per 1000 years.

(*b*) Del Panta (1980) documents rates in a range of 13.3 per 1000–100 per 1000 in Italy as a whole and of 40.8 per 1000–218 per 1000 in Tuscany during the sixteenth century.

(*c*) Charbonneau and Larose (1979) document a very broad set of frequency rates for various periods and geographical regions. The average that could be established for crisis of medium intensity is 33 per 1000. However, the range of frequency rates that they can document is 24 per 1000–230 per 1000 for crises of low to high (but not highest) intensity.

13 The Influence of Cohort Effects on Differentials and Trends in Mortality

GRAZIELLA CASELLI

Dipartimento di Scienze Demografiche, Università di Roma 'La Sapienza' Rome, Italy

1. Introduction

Recent mortality changes in Europe, particularly in the more developed countries, have triggered a resurgence of interest in the study of cohort-specific mortality patterns. Two recent findings are worthy of note. First, it has been shown that male cohorts which participated in World War I subsequently experienced higher mortality than adjacent cohorts who were not involved in the conflict (Vallin 1973; Horiuchi 1983; Caselli *et al.* 1985). Secondly, the same excess mortality has been detected among the cohorts born or adolescent during 1914–18 (Horiuchi 1983).

The rapid evolution of European mortality since the turn of the century makes it difficult to estimate cohort effects. Pure cohort effects (or at least what is commonly defined as such) only appear to be marginal, largely obscured by general mortality declines. The deceleration of this decline in recent years makes the mortality histories of specific cohorts and their influence on overall mortality trends easier to observe.

The rapid improvement that took place in the earlier period accounts for the strongly heterogeneous histories of the 100 or so cohorts making up the present-day population. The cohorts which experienced the dramatic events of World War I are a striking example, but even those born more recently are also highly differentiated. In Italy, for instance, the cohorts now aged between 30 and 80 experienced very different patterns of childhood mortality. The probability of survival at age 15, which was only 68% for males born in 1918, increased to 77% for those born in 1921 and to 81% and 82% respectively for those born in 1931 and 1935 (Natale and Bernassola 1973). One may ask whether the cohorts exposed during childhood to the highest risks of death were subsequently at an advantage, because of the early loss of their weakest elements (selection effect), or at a disadvantage, because their health in later life was made more fragile by the difficult context of their first years (weakening effect).

As early as the beginning of the century, Mortara (1912) and Gini (1914) showed that in Sweden the cohorts which experienced the highest levels of infant mortality were those which subsequently experienced the highest risks of death in old age. In the 1930s, a more refined study of Kermach *et al.* (1934) also led to the conclusion that the trend in adult mortality in the different cohorts depends on the trend in their infant mortality. More recently, Livi-Bacci (1962), Preston and van de Walle (1978), and Caselli and Capocaccia (1989) have outlined the contours of a positive relation between infant and child mortality and adult or old-age mortality. Other authors, on the other hand, have stressed the positive influence of selection among the cohorts affected by higher infant mortality, frequently excluding the weakening effect from their analysis (Manton *et al.* 1981; Stato *et al.* 1984; MacMillen and Nam 1985).

In this chapter we do not attempt to review all the models which may be used to represent the combined effects of period and generation. Instead we aim at illustrating the present stage of research in this field by introducing three different approaches. The first approach is based on the comparison of period data for four different time periods whereas the other two consider a complete set of data by year of age and year of birth since the turn of the century. In the presentation of these approaches, we will successively attempt

- to determine the long-term effects of the two world wars by using an empirical model based on the mortality variations from one age group to the next;
- to analyse the role of cohort effects in the mortality variations between different countries and between the sexes by using contour maps of mortality surfaces;
- to measure the importance of cohort effects in the decline of adult mortality by means of a mathematical model separating age, period, and cohort effects.

2. An Empirical Model for Analysing Age-Specific Mortality Variations: The Long-Term Effects of the Two World Wars

As mentioned before, certain authors have observed the existence of a higher vulnerability among male adults involved in World War I in various European countries. The data bases and methods of analysis used by these authors vary widely and this may in part account for some of the differences in their findings. In France, for instance, the most deeply affected cohorts seem to be those actively engaged in the war (Vallin 1973), whereas for the Federal Republic of Germany and Italy this vulnerability appears to be shared also by those who were adolescents during the war years (Horiuchi 1983; Caselli and Capocaccia 1986). It is of interest then to examine whether the differences observed are real or the result of the diversity in data and methods, and to see whether they are confirmed when we apply the same model to the same data for different countries.

To test this, we chose to compare Italy and France, using mortality data on males and females aged over 40 for four time periods: 1950–2, 1960–2, 1970–2 and 1977–9. This analysis is thus based entirely on period data. Earlier studies had shown that differences among cohorts were well estimated using a measure of relative variation between one age group and the next (Mortara 1908). The measurement is expressed as follows:

$$K(x) = \frac{d \log \mu(x)}{dx}, \tag{1}$$

where $\mu(x)$ is the force of mortality, or instantaneous death rate, at age x. Assuming that 1_x (the probability of survival between x and $x+1$), is linear implies that the single-year probabilities of dying are equal to the instantaneous mortality rates. Thus:

$$K_x^* = \log\ q_{x+1} - \log q_x. \tag{2}$$

The use of death probabilities by individual years of age makes the value of K_x^* all the more prone to strong random fluctuations. It was smoothed by a method proposed by Horiuchi and Coale (1985) which involves nine-point moving averages weighted by a triangular distribution of weights over the nine points:

$$\hat{K}_x = \sum_{n=-4}^{4} \frac{5-|n|}{25} K_{x+n}^*. \tag{3}$$

This measure is very useful for identifying the age-specific mortality variations related to differences in cohort behaviour. After age 40, the risk of dying increases monotonically with age. Therefore, if in a given period-specific life table we find at age x a cohort with relatively high mortality, and at age $x+1$ a cohort with relatively low mortality, the increase in the risk of dying from the younger age to the older one is small and $\hat{K}_x$ should assume low values. In the opposite case, $\hat{K}_x$ should approach very high values. Thus, the peaks or troughs of the $\hat{K}_x$ curve obtained from the period data make it possible to detect cohorts affected by excess or unusually low mortality above and beyond the levels expected from purely random fluctuation.

Figure 13.1 illustrates the findings for French and Italian males for the first three time periods considered. Both countries show evidence of the negative effects of World War I on the mortality of the cohorts directly engaged in the conflict. This is clear from the three hollows observed around ages 55 (for 1950–2), 65 (for 1960–2), and 75 (for 1970–2). Regarding the cohorts whose members were adolescents during World War I, in particular those born between 1900 and 1905, the effects of the war on their mortality levels are less clear-cut, but not altogether absent. By and large, however, the variations observed are much stronger between the different Italian cohorts than between their French counterparts. This could indicate that the health of the former was more deeply affected by the war than the latter, in particular for the younger cohorts. This is hardly surprising. In 1917–18, a truly dramatic period for Italy, children as young as 15 were being sent to the front. In addition, living conditions were extremely difficult for those less directly engaged in the conflict due to widespread food shortages and inadequate diet, poor general health conditions and hygiene, and the outbreaks of epidemics (e.g. Spanish influenza).

The children who were born during the war years were also affected by these poor living conditions. This is clearly visible on the $\hat{K}_x$ curve, where a hollow is observed at ages 40–5 for 1960–2, for females as well as males (Figures 13.1 and 13.2). Thus, the common assumption that male cohorts born during World War I were only penalized at older ages seems inadequate here.

Figure 13.3 illustrates the situation for 1977–9. For males (Figure 13.3*a*), the effects of World War I tend to disappear as the cohorts age. On the other hand, an irregularity appears for the cohorts born around 1930: for males who were adolescents in 1940–5 and youths directly involved in the conflict mortality passes from a peak to a hollow. A similar feature has been observed for the same cohorts in Japan and the Federal Republic of Germany (Okoro 1981; Horiuchi 1983). By contrast, no such irregularity appears in the female data (see Figure 13.3*b*), suggesting that the male irregularity should be interpreted more as a result of their active participation in the war at a very young age rather than as the effect of generally poorer health conditions for these adolescents. Naturally, we cannot exclude the possibility that the lack of World War II specific effects on female mortality levels could also be due to biological and social factors, with the former easing and accelerating recovery for women and the latter providing some protection from the effects of post-war industrialization.

3. The Role of Cohort Effects in Mortality Variation between Different Countries and between Sexes

A full comparison of mortality trends in two countries or for males and females within the same country requires more than a series of observations at various time periods since the permanent influence of a cohort's past history on its present mortality must also be taken into account. This requires a long, continuous series of data by age and year of observation for each country (or sex), within which the differences (between countries or sexes) attributable to a context shared at one point in time by all the cohorts (period component) must be distinguished from the differences (between countries or sexes) due to the past history of each cohort (cohort component).

To simplify the presentation of such data, Vaupel *et al.* (1985) have generalized an idea originally suggested by Delaporte (1941) and have proposed a method of contour line representation which is greatly simplified by computerization. These figures are constructed by projecting mortality onto a plane whose axes are age and calendar year and drawing

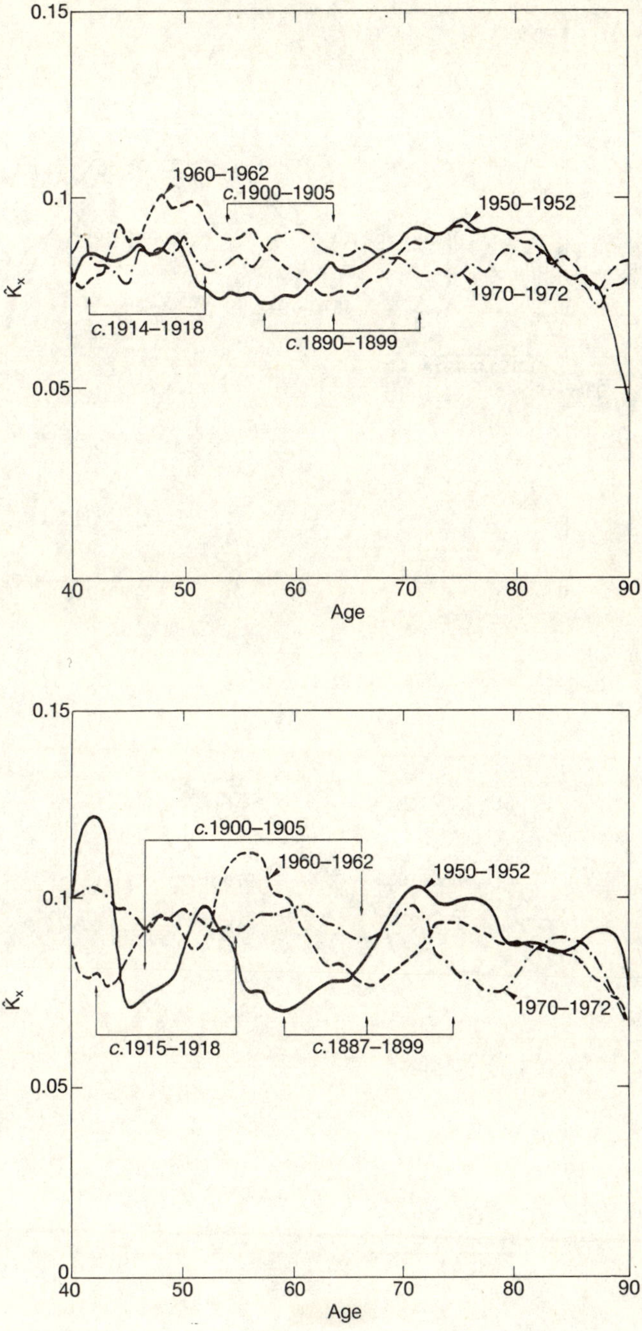

Fig. 13.1. Age distribution of $\hat{K}_x$: comparison of three periods, France and Italy, males
(*a*) France
(*b*) Italy

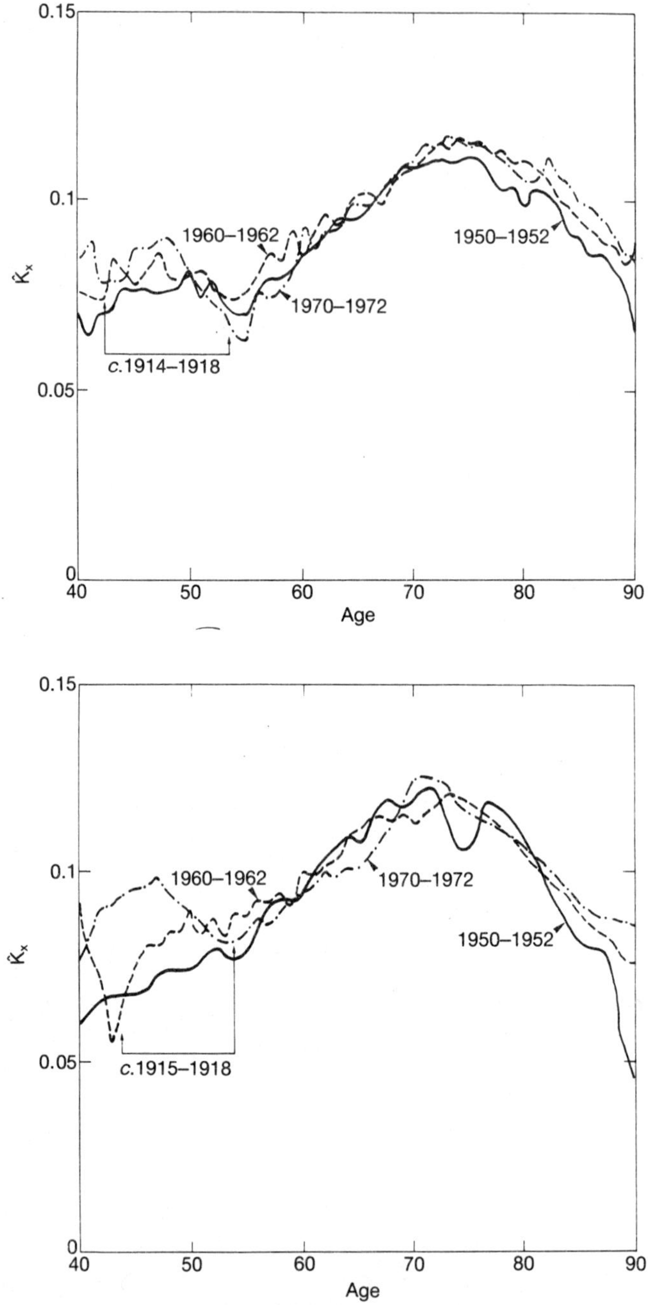

Fig. 13.2. Age distribution of $\hat{K}_x$: comparison of three periods, France and Italy, females
(*a*) France
(*b*) Italy

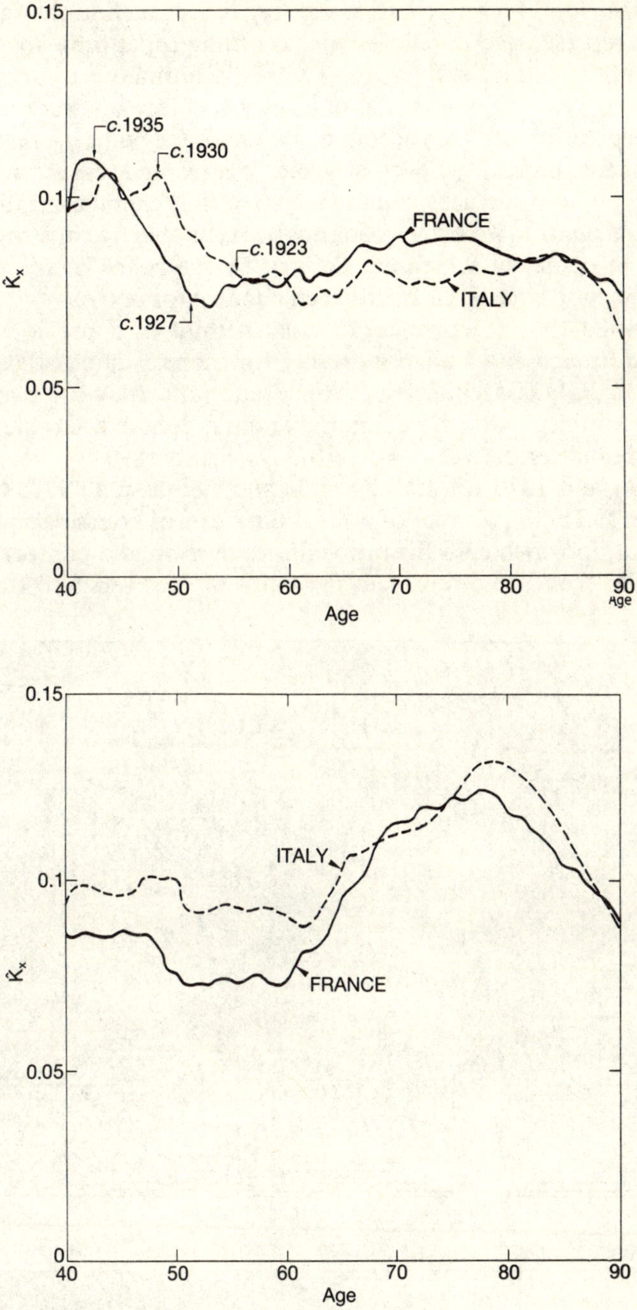

Fig. 13.3. Age distribution of $\hat{K}_x$ in 1977–1979: comparison between France and Italy, males and females
(*a*) France
(*b*) Italy

contour lines representing these mortality levels, much in the same way as altitude is represented on relief maps according to latitude and longitude. As the death probabilities $q_{(x,x+1),(y,y+1)}$ are discontinuous by age, $(x, x+1)$, and calendar year, $(y, y+1)$, the mortality surfaces $q_{x,y}$ were obtained by linear interpolation between adjoining points. Each point $q_{x,y}$ on the surface represents the mortality level at exact age x and exact time y. The comparison of two surfaces, and therefore of the general mortality features of two countries (or two sexes), can thus be facilitated by constructing a new surface, the points of which are defined by the ratios or the differences between the mortality levels $q_{x,y}$ of each country (or sex).

We verified the effectiveness of this method in a previous study by comparing French and Italian mortality for each sex and male and female mortality in Italy (Caselli *et al.* 1986). France and Italy are two countries that have complete, directly comparable data: cohort life tables based on death probabilities between two birthdays, from 1899 for France (Vallin 1973, 1984) and 1870 for Italy (Natale and Bernassola 1973; Caselli and Greco 1983). These two sets of cohort data can of course also be read as period data, in which case the probabilities overlap two consecutive years.

Figures 13.4 and 13.5 represent the ratios of the French to Italian death

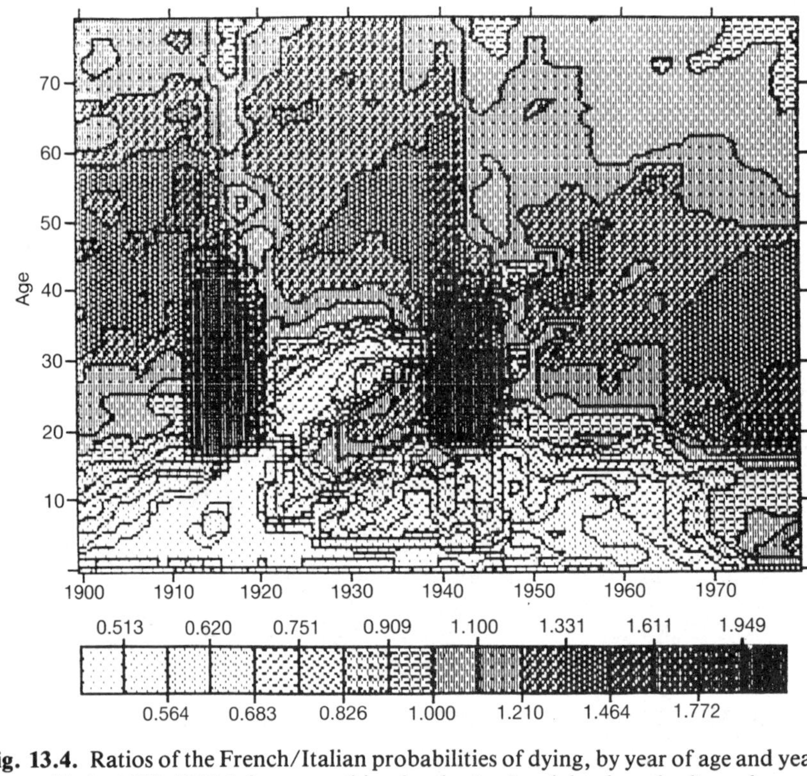

Fig. 13.4. Ratios of the French/Italian probabilities of dying, by year of age and year of birth, 1899–1979 (after smoothing by the 5 × 5 weighted method), males

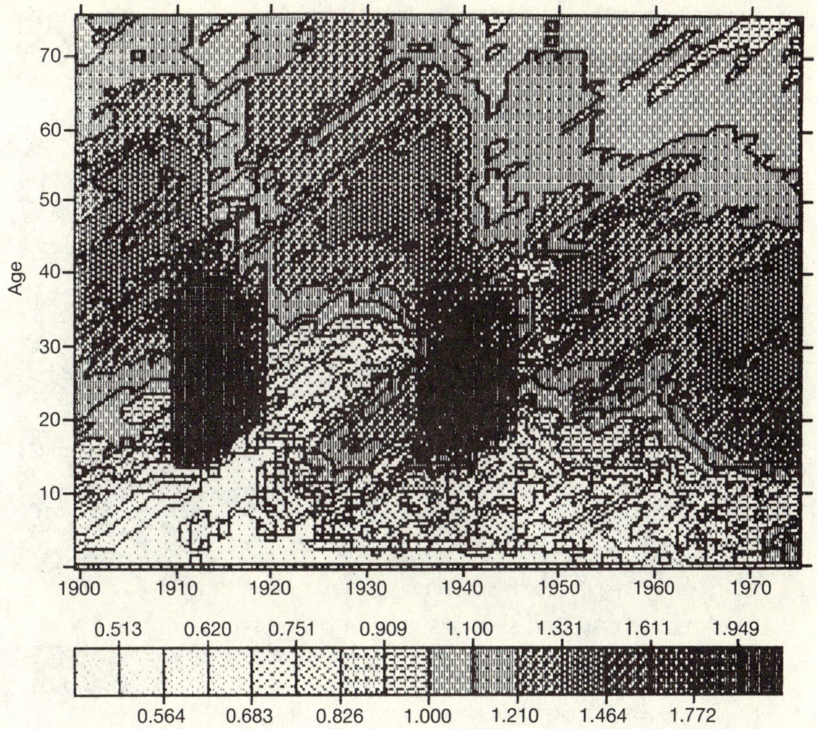

Fig. 13.5. Ratios of the French/Italian five-year probabilities of dying, by cohort, males

probabilities for males aged 0 to 79, from 1899 to 1979. Figures 13.6 and 13.7 illustrate the ratios of male to female probabilities for Italy at the same ages from 1870 to 1979. In both cases, the first figure (Figures 13.4 and 13.6) was based on the probabilities by individual year of age, but was smoothed to make it easier to read (Vaupel *et al.* 1985). The smoothing procedure eliminates random fluctuations but also obscures real patterns. To preserve the effects due to each cohort, we constructed in each case a second figure (Figures 13.5 and 13.7) based on five-year cohort probabilities computed for each year of age and calendar year. This procedure naturally accentuates the cohort effects. In Figures 13.4 and 13.5 the darker hatching represents excess French mortality, while the lighter hatching represents excess Italian mortality. The differences in excess mortality according to age for a period can be read from bottom to top. From left to right, we can read for each age the evolution of this excess mortality through time and, finally, diagonally we can observe the evolution of the different cohorts. It is the capability for this triple representation that makes this type of figure so appealing.

The French–Italian mortality pattern reflects a complex history marked

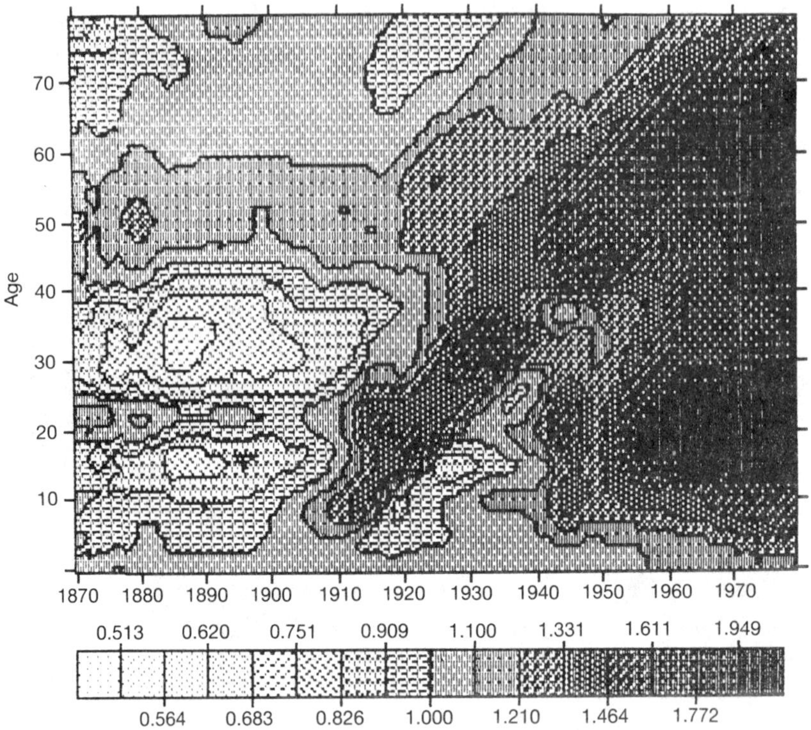

Fig. 13.6. Ratios of the male/female probabilities of dying, by year of age and year of birth, 1870–1979 (after smoothing by the 5 × 5 weighted surfaces method), Italy

by important period fluctuations. The most visible of these appears to be the result of the two world wars, which cause a very strong rise in excess French mortality at adult ages. More generally, the main features of the period under study are: pronounced excess Italian mortality at young ages; excess French mortality among adults; excess French mortality among the elderly at the beginning of the period, followed by excess older-age mortality among Italians during the two wars and at the very end of the period.

Figures 13.4 and 13.5 also bring out the influence of cohort effects on the mortality differentials between the two countries. These cohort effects may explain the distinctive patterns of mortality excesses observed between 1920 and 1940 and then between 1950 and 1970. The cohorts implicated are those which we identified in our preceding analysis, namely, those directly involved in World War I and those who were adolescents during the war years. By enabling us to consider the complete mortality histories of the different cohorts, the method of contour line representation makes it easier to understand the mechanisms behind the evolution observed in each country. The French cohorts aged between 30 and 40 at the end of World

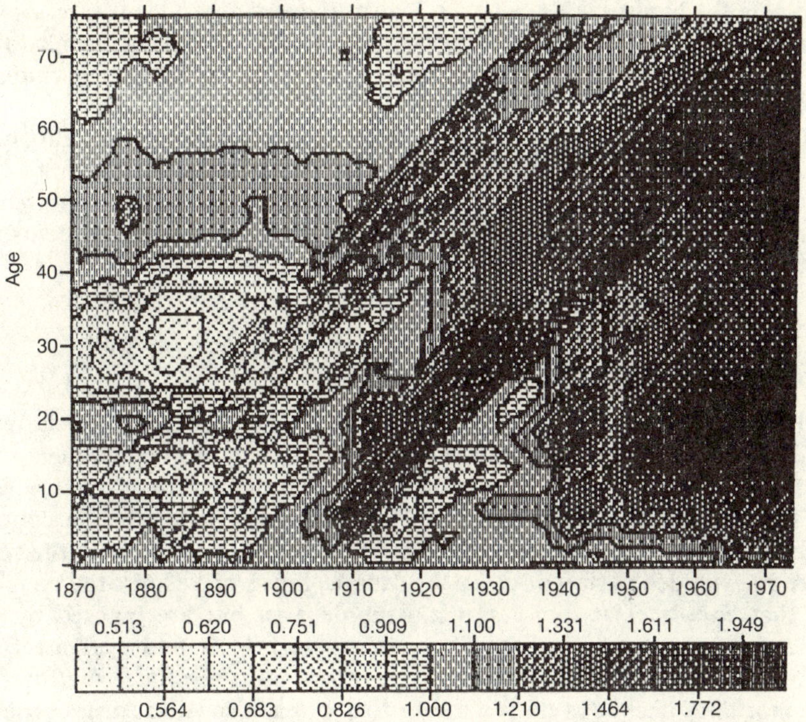

Fig. 13.7. Ratios of the male/female five-year probabilities of dying, by cohort, Italy

War I subsequently experience a higher mortality than the corresponding Italian cohorts at all ages. On the other hand, the Italian males aged around 20 at the end of the war experience higher mortality than their French counterparts at all ages after 20. This phenomenon, which also affects the cohorts born during World War I, continues beyond World War II until the 1950s, before giving way to higher French mortality.

Contour line representation illustrates even more clearly the contrasts between male and female probabilities of death in Italy (Figures 13.6 and 13.7). The negative effects of World War I on the health of the male cohorts, more directly involved than the females, broaden the gap between the sexes. The dark band running diagonally across the figures reflects these unequal scars left by World War I, which persist over time. Events such as World War II have been added to such differentials but the exceptionally high levels of excess male mortality observed in the late 1950s and early 1960s at ages 55–65 still reflect a major cohort effect. These cohorts are the ones born around 1900, therefore aged between 10 and 20 during World War I. Very high levels of excess male mortality are also found in the late 1960s for cohorts born during or just before the war (aged around 55 in the mid-1960s). Figures 13.6 and 13.7 also suggest a growing excess male mortality for the cohorts affected by World War II during their adolescence, those aged 40–50 in the mid-1970s. This confirms the findings revealed by the first approach.

In summary, the strong increase observed in excess male mortality over the last 20 years can be partly explained by cohort factors. These are not, however, limited to past history since current factors such as the recent sharp increase in mortality among young males due to road accidents are also contributing to the increase.

4. A Statistical Model for Interpreting the Evolution of Adult Mortality

With the application of the first two approaches we have identified the most evident cohort effects, but also shown that the long-term consequences of the history of specific cohorts gradually fade away with time as the survivors of the original group grow older. It would seem that only the cohorts directly affected by traumas experienced during their youth bear the scars over an extended period of time. We cannot, however, dismiss the possibility that cohort effects of lesser magnitude exist but are masked by the general decline in mortality. To reveal these effects and evaluate their influence on age-specific mortality trends, we must estimate what part of the mortality decline is due to the history of each specific cohort (cohort effect) and what part is due to changes in health and social conditions through time (period effect). The problem is not an easy one, as it means estimating the effects of components which are already difficult to distinguish and identify. Hobcraft *et al.* (1982) and Hobcraft and Gilks (1984) have stressed the limitations of certain models based on questionable assumptions.

The model presented here is based on a working assumption for which we cannot claim generalized validity. The aim is not to measure precisely the weight of the cohort effects, but essentially to establish their existence and their role in the age-structure of mortality and its evolution. The method consists in determining the influence of the cohort component by estimating the differences between the mortality levels actually observed and those which would have been reached had the cohort component been equal for all the cohorts studied. The levels of mortality of the different cohorts between ages 25 and 80 are estimated after controlling for the influence of mortality in the first 15 years of life. The cohort component retained here is therefore what emerges from the relation existing between overall infant–adolescent mortality and the age-specific risks of death after age 25; the remaining variability is attributed to the period component. Naturally, such a simplification prevents us from observing in full the course of the different cohort histories and from clarifying how, at each age, the mortality levels bear the mark of events experienced at earlier ages. However, this method seems the most adequate for assessing the relation between adult mortality and living conditions during childhood and adolescence.

In an earlier study, a statistical model for separating age, period, and cohort effects proved quite suitable for answering this type of question (Caselli and Capocaccia 1989). It is a log-linear model of the death probabilities of the cohorts born between 1882 and 1953, in which the cohort component is a function of mortality between 0 and 15 years. The model is expressed as:

$$\text{logit}\ [q_{i,k}({}_{15}q_0)] = \alpha_i + \beta_k + \delta_i \cdot {}_{15}q_0 + \text{constant}, \tag{4}$$

where:

- α_i represents the age effects for each age *i* between 25 and 80 years;
- β_k represents the period effects for each calendar year *k* between 1907 and 1979;
- $\delta_i \cdot {}_{15}q_0$, the global cohort effects (equal to the product of the death probabilities between 0 and 15 years for each cohort and the δ_i coefficients characterizing the cohort effects). δ_i corresponds to the logarithm of the relative risk of dying associated with a unit increase in the independent variable for each age studied.

The vectors α, β, and δ were estimated using maximum likelihood procedures on formula (4) applied to the number of deaths and number of individuals exposed to the risk in each cell corresponding to the combination of the three factors. An analysis of residuals showed no indications of systematic deviation between the estimated and observed values. The standardized residuals were distributed more or less normally, with a variance independent of the factors under study. The correlation coefficient between the observed and expected death probabilities can be taken as a global indicator of the goodness of fit. The value obtained ($R^2 = 0.996$) indicates that the fit is quite satisfactory.

The coefficients α, β, and δ can be interpreted as relative risks of death for each age, period, and cohort. We should note that, in our model, age effects are independent of period (and therefore applicable to each calendar year), but dependent, through coefficient δ_i, on mortality between ages 0 and 15. Period effects are independent of age (and therefore applicable to all ages between 25 and 80) and of early mortality. The model expresses the relation between early and adult mortality, eliminating the influence of variations in the period effects.

To illustrate the descriptive capacity of the model we present in Figure 13.8 the observed and estimated values of the death probabilities between ages 25 and 80 for two periods, 1960–1 (the first period for which we have the complete history of the cohorts studied) and 1978–9. In order to determine the association between infant–adolescent mortality, on the one hand, and mortality at later ages, on the other, we first assess the relation between adult mortality differentials (measured as relative risks, *RR*) and mortality

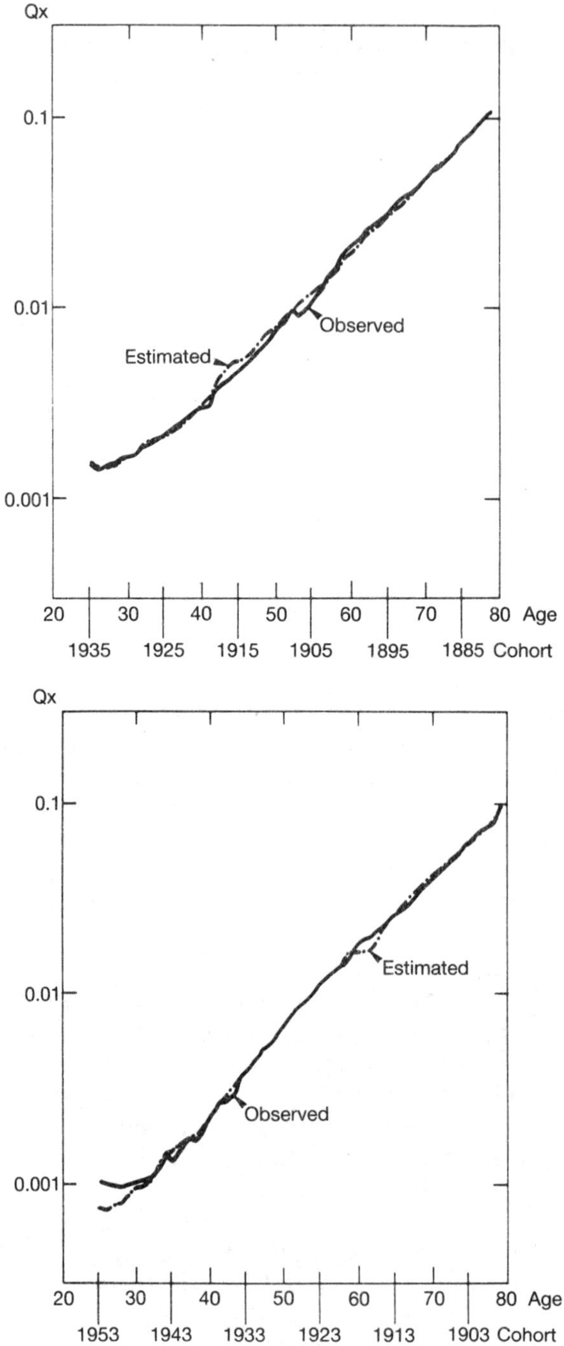

Fig. 13.8. Age-specific mortality in 1960–1961 and 1978–1979: observed and estimated probabilities, Italy, males
(*a*) 1960–1961
(*b*) 1978–1979

differentials between ages 0 and 15 for all the cohorts considered. We use the following approximation:

$$RR(\Delta_{15}q_0) = \exp(\delta_i \Delta_{15}q_0), \qquad (5)$$

where $\Delta_{15}q_0$ is assigned a value equal to the standard deviation of $\Delta_{15}q_0$.

Figure 13.9 displays the age structure of the risks corresponding to a standard deviation of the probabilities of death between ages 0 and 15. In the case of Italy, the mean probability of death between ages 0 and 15 for the cohorts born between 1882 and 1953 is equal to 296.2 per 1000, with a standard deviation of 79.2 per 1000. The findings confirm the existence of a link between mortality early in life and mortality in subsequent years: a higher mortality between ages 0 and 15 goes hand in hand with a higher mortality between ages 25 and 45 and a lower mortality afterwards. The effects of infant–adolescent mortality are, however, stronger on mortality between ages 25 and 45 than between ages 45 and 80. The older cohorts born before the end of World War I, who experienced very high infant and adolescent mortality, are disadvantaged during the first part of adult life amongst the more recent cohorts, who experienced a much lower early mortality. Conversely, their situation is more favourable after age 45. Once we eliminate the effect of mortality in early life by applying the observed mean (296.2 per 1000) to all cohorts, we can evaluate the mortality differences between cohorts according to the model's other components, age and period. Thus, for all the cohorts including individuals between 25 and 80, we have a theoretical mortality level by age which depends exclusively on the period component.

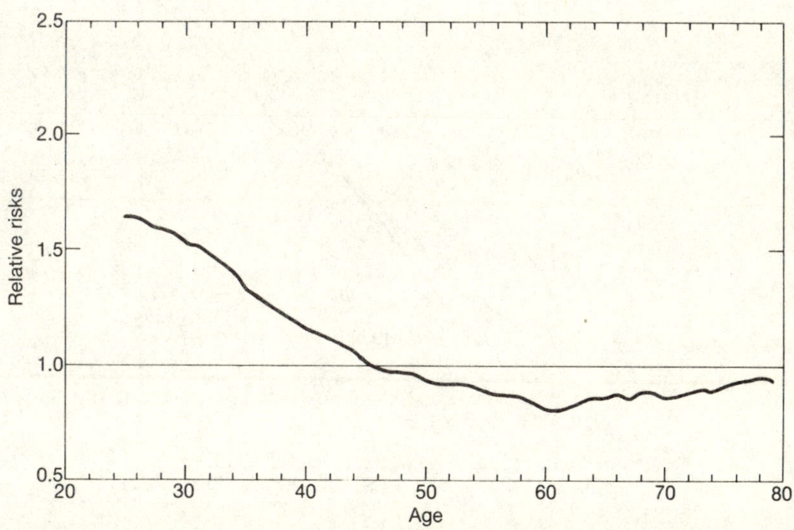

Fig. 13.9. Relative risks corresponding to a standard deviation of the death probabilities between 0 and 15 years ($\Delta_{15}q_0$) = standard deviation)

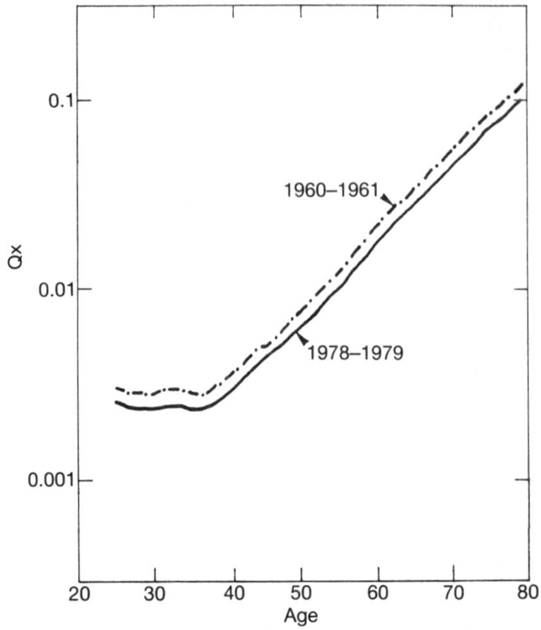

Fig. 13.10. 'Theoretical' death probabilities, by age (without cohort effects), 1960–1961 and 1978–1979, Italy, males

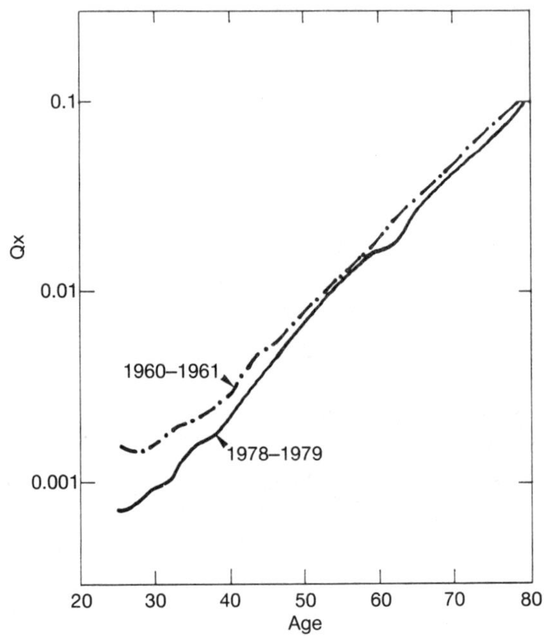

Fig. 13.11. 'Real' death probabilities, by age, 1960–1961 and 1978–1979, Italy, males

Figure 13.10 facilitates the cross-sectional comparison of these results. It displays the mortality levels that would be observed in 1960–1 and 1978–9 if the period component had been the only one to affect the mortality of the cohorts considered. The gap between these two curves is an indicator of the effects of the period component on mortality change during this time interval. The agreement between observed and estimated probabilities shown in Figure 13.8 means that the latter can be used as an expression of real mortality and compared to the theoretical values illustrated in Figure 13.10. A comparison of Figures 13.10 and 13.11 reveals the importance of cohort effects on adult mortality levels and age patterns. At each age, the difference between the gap separating the 'real' curves (Figure 13.11) and the gap separating the 'theoretical' curves (Figure 13.10) is an indicator of the influence of heterogeneity of the early mortality levels of the cohorts involved. To simplify our interpretation, we should bear in mind that the real mortality levels between ages 25 and 40 in 1960–1 involves cohorts born between 1920 and 1935, whereas those for 1978–9 involve cohorts born between 1938 and 1953, who experienced a much lower early mortality than the preceding ones. Figure 13.9 shows indications of the positive contribution to mortality below age 45 of those cohorts with an experience of low infant and adolescent mortality. Figure 13.12 adds support to this argument. Theoretical and observed probabilities are compared for each of the two calendar years. In this case the gap between the curves expresses the effect of the cohort component alone (limited, by assumption, to the cohort effects attributable to mortality before age 15) on the evolution of age-specific mortality. The comparison of these two curves above age 45 yields results that differ according to the period considered. In 1960–1, the most favourable levels for these ages continue to be those of the observed probabilities. Indeed, during 1960–1 the cohorts attaining ages above 45 are those that experienced above-average infan–adolescent mortality and that should, according to the model, be in an advantageous position at these ages. Instead in 1978–9 the situation is different: the results are reversed between ages 45 and 55, which are then attained by cohorts who have experienced below-average mortality in the first part of their life. Thus, after age 45 counterbalancing forces seem to come into play, with a downward trend induced by period effects that is partially offset by cohort effects.

The results obtained with the third approach allow us to make some considerations of a general nature. Firstly, as other studies have shown, these findings confirm certain associations between the risks experienced during the first part of life and mortality at later ages. This can lead us to speculate that better living conditions not only allow more individuals to survive, but also make the survivors more resistant to death until around age 50. It is only after this age that some indicators suggest that a reduction of the selection effect by early mortality may have a negative effect on mortality at the older ages, because a greater proportion of more fragile

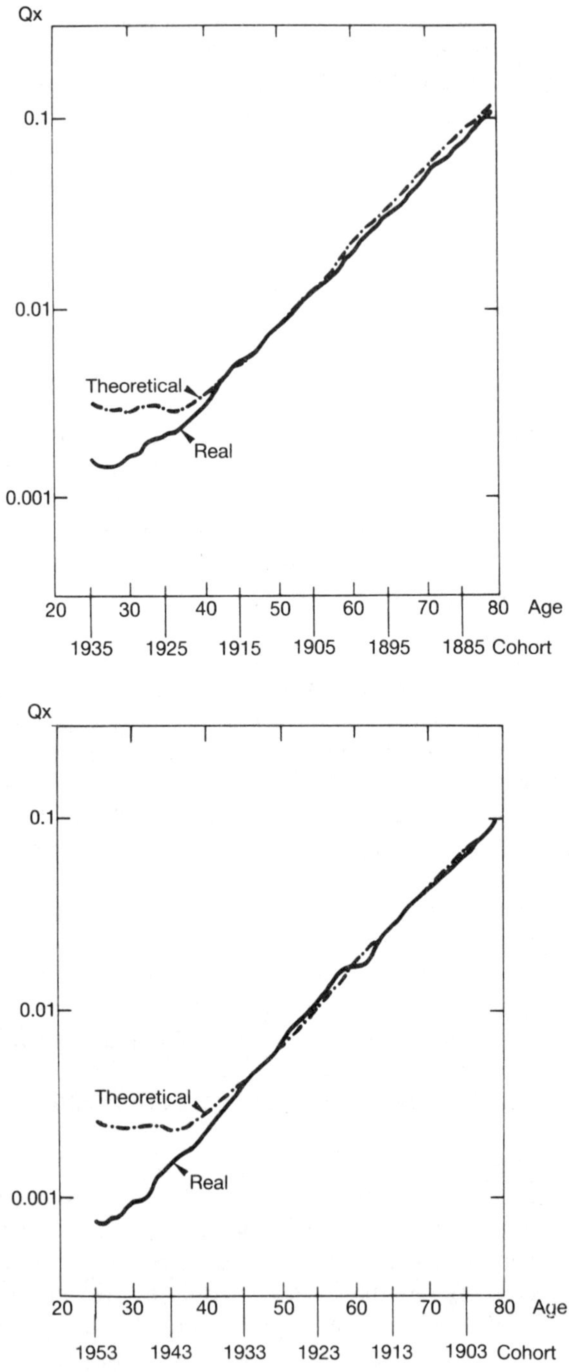

Fig. 13.12. 'Real' death probabilities and those estimated after elimination of the cohort effect ('theoretical'), by age, 1960–1961 and 1978–1979, Italy, males
(*a*) 1960–1961
(*b*) 1978–1979

individuals has survived. The assumption on which the model is based has also allowed us to provide a measure of the influence of the cohort component on the evolution of adult mortality. Although the measure is based on an unverified assumption, it permits the utilization of a cohort approach to link mortality at certain ages and mortality later in life.

5. Conclusions

The three approaches presented here have provided evidence showing the influence of the past history of the different cohorts on mortality trends and differentials between countries or between the sexes. We have shown that French and Italian male mortality since 1950 and 1970 has been affected by the traumas experienced by cohorts whose members were children or adolescents during the war years or directly involved in one of the two World Wars (first approach). These same cohorts contribute substantially to the differences in adult mortality observed either between the two countries or, for Italy, between the two sexes (second approach). Important cohort effects (based on events which have deeply affected certain cohorts at a given point in their history) can adequately be revealed by the first two approaches. The choice between one or the other depends on the data available and the level of detail required. While the first approach is straightforward to use, it has the drawback of being limited to the analysis of variations in adult mortality. It cannot provide any information on cohort effects that affect the first part of life. On the other hand, the contour line representation of mortality surfaces that traces entire cohort histories offers a more complete picture. This type of figure makes it possible to capture graphically the processes which lead to the diversity of experiences observed. As different cohort histories and their chronology are considered, assumptions for interpreting the observations can be more readily formulated. For instance, this approach led us to infer that the evolution of the cohorts born around 1900 in France and Italy suggests the action of selective forces acting alternately in the two countries during the lives of these cohorts. Its only drawback, of course, is that this approach requires long, very detailed data series by age and year of observation. However, such data are necessary if one is to follow age-specific mortality patterns through time and from one cohort to the next. When such data series are available the range of possibilities is very broad, and many types of hypothesis can be tested. With our third approach we were able to compare the evolutions of early mortality and adult mortality for different cohorts, revealing the contributions of age, period, and cohort effects. This procedure enabled us to distinguish what part of recent trends in mortality between ages 25 and 80 is due to economic, social, and health changes (period effect) and what part is due to the earlier life of each cohort (cohort effect). But, however well adapted a model may be, it obviously has certain limits. In our case the most important is that it assumes that there is only one cohort-specific factor of relevance (namely, the experience during the age group 0–14). Reality is no doubt more complex: at each age, mortality is surely influenced by all the events experienced at all the preceding ages. The findings provided by this type of model must therefore be compared to those obtained using different methods.

Important insights into the mechanisms explaining trends in adult and old-age mortality could be gained if one were to consider not only the past mortality experience of the different cohorts, but also their past health and social contexts. A great deal of old-age mortality may, for example, be accounted for by exposure at adult ages to certain risks related to working life or, more generally, by the consequences or late developments of morbid processes contracted earlier (neoplasms, ischaemic heart diseases, etc.). Within this perspective, longitudinal studies such as those conducted in several countries, and primarily England, are valuable as they make it possible to investigate the relation between past mortality and past experiences and, at the same time, provide useful material for interpreting the relation.

References

Caselli, G., and Capocaccia, R. (1989), 'Age, Period, Cohort, and Early Mortality: An Analysis of Adult Mortality in Italy', *Population Studies*, 43(1), 133–54.

—— and Greco, B. (1983), *Aggiornamento delle Tavole di Mortalità per Generazione di Natale e Bernassola, Periodo 1965–1979* (Rome: Università la Sapienza, Dipartimento di Scienze Demografiche).

—— Vallin, J., Vaupel, J., and Yashin, A. (1986), *L'évolution de la structure par âge de la mortalité en Italie et en France depuis 1900: Effets de période et effets de génération*, Dossiers et Recherches, no. 4 (Paris: INED).

—— Vaupel, J., and Yashin A. (1985), 'Mortality in Italy: Contours of a Century of Evolution', *Genus*, 16, 1–2.

Delaporte, P. (1941), *Évolution de la mortalité en Europe depuis l'origine des statistiques de l'état civil (tables de mortalité de générations)* (Paris: Imprimerie Nationale).

Gini, C. (1914), 'Sull' Utilità delle Rappresentazioni Grafiche', *Giornale degli Economisti e Rivista di Statistica*, Feb.

Hobcraft, J., and Gilks, W. (1984), 'Age, Period and Cohort Analysis in Mortality Studies', in J. Vallin, J. Pollard, and L. Heligman (eds.), *Methodologies for the Collection and Analysis of Mortality Data* (Liège: IUSSP, Ordina Éditions).

—— Menken, J., and Preston, S. (1982), 'Age, Period and Cohort Effects in Demography: A Review', *Population Index*, 48(1), 3–43.

Horiuchi, S. (1983), 'The Long Term Impact of War on Mortality: Old Age Mortality of First World War Survivors in the Federal Republic of Germany', *United Nations Population Bulletin*, vol. 5.

—— and Coale, A. (1985), 'Age Patterns of Mortality for Older Women: An Analysis Using the Age Specific Rate of Mortality Change with Age', spontaneous paper presented at the *International Population Conference*, Florence, June.

Kermach, W. O., MacKendrick, A. G., and MacKinley, P. L. (1934), 'Death Rates in Great Britain and Sweden: Some General Regularities and Their Significance', *The Lancet*.

Livi Bacci, M. (1962), 'Alcune Considerazioni sulle Tendenze della Mortalità Senile e sull 'Eventuale Influenza Selettiva della Mortalità Infantile', *Rivista Italiana di Economia, Demografia e Statistica*, vols. 3–4.

MacMillen, M. M., and Nam, C. B. (1985), 'Mortality Crossovers by Causes of Death and Race in the US, in the 1970s', spontaneous paper presented at the *International Population Conference*, Florence, June.

Manton, K. G., Stallard, E., and Vaupel, J. W. (1981), 'Methods for Comparing the Mortality Experience of Heterogeneous Populations', *Demography*, 18.

—— —— —— 'Alternative Estimates of the Heterogeneity of Mortality Risks among the Aged', unpublished manuscript.

Mortara, G. (1908), *La Mortalità Secondo l'Età e la Durata della Vita Economicamente Produttiva* (Rome: Cooperativa Tipografica Manuzio).

—— (1912), 'Le Variazioni della Mortalità da Generazione a Generazione in Svezia', *Annali di Statistica*, series 5, vol. 4.

Natale, M., and Bernassola, A. (1973), *La Mortalità per Causa nelle Regioni Italiane: Tavole per Contemporanei 1965–1966 e per Generazione 1790–1964* (Rome: Istituto di Demografia, Università di Roma).

Okoro, S. (1981), *Increase in Mortality of Middle Aged Males in Japan*, Nupri Research Paper, series 3 (Tokyo: Nupri).

Preston, S., and van de Walle, E. (1978), 'Urban French Mortality in the Nineteenth Century', *Population Studies*, 32(2), 275–98.

Stato, R., Zeckman, R., and Rizzo, J. (1984), 'Hidden Heterogeneity in Risk: Evidence from Japanese Mortality', unpublished manuscript.

Vallin, J. (1973), *La Mortalité par génération en France depuis 1899*, Travaux et Documents, Cahier no. 63 (Paris: INED, PUF).

—— (1984), *Tables de mortalité du moment et par génération 1899–1981: Mise à jour provisoire des tables annexées du cahier 63* (Paris: INED).

Vaupel, J. W., Gambill, B. T., and Yashin, A. I. (1985), *Contour Maps of Population Surfaces*, Working Paper (Laxenburg, Austria: IIASA)

14 Classification of Developed Countries according to Cause-of-Death Patterns: A Test of Robustness during the Period 1968–1974

NICOLAS BROUARD

Institut National D'Études Démographiques (INED), Paris, France

1. Introduction

The development of statistical theory originated in the study of mortality and measurement of death risk. Both were of interest to insurance companies and bankers. Indeed, a correct estimate of mortality by age makes it possible to wager on human life along insurance principles, and the service rendered by this insurance is a source of profit. Unlike literature on causes of death, the literature on estimates and modelling of mortality by age is abundant. For actuaries, it is less important to know what a client is going to die of than his date of death. However, knowledge of causes of death interests another sector, that of health organizations. There is little doubt that developments on modelling of mortality by causes of death are scarcer and less penetrating than the advances in modelling of mortality by age. This may be due to the combination of paucity and unreliability of the data and the intrinsic difficulties of the subject. We have already learnt, using the model tables by Ledermann (1959) and Coale and Demeny (1966), that two countries can have the same life expectancy and yet experience two very different age-specific death risk patterns. With causes of death the analysis becomes more complicated since two countries which have identical death rates at given ages, such as Japan and Sweden, may have radically opposed causal profiles.

In a recent paper (Brouard and Lopez 1985) we defined a typology of countries with low mortality according to causes of death. This typology was established on the basis of a cross-sectional analysis of 45 countries around 1980. One of the results of the analysis is that Sweden is characterized by a pattern dominated by cardiovascular diseases. This is a rather paradoxical fact since the death rate by cardiovascular disease in Sweden is presently one of the lowest in the world. But this is only because Sweden has the second-highest observed life expectancy in the world. When the bulk of countries reach comparable levels of life expectancy, Sweden will appear as a country where mortality from cardiovascular diseases is most significant.

In this chapter I will attempt to confirm the validity of the typology elaborated before by testing its robustness on a set of developed countries during and after 1968. The study uses all available WHO mortality data for developed countries where the deaths by cause are classified according to the Eighth and Ninth Revisions of the International Classification of Disease. The first year is usually 1968 and the last is 1984 and thus the study covers a period of roughly 15 years. I shall also try to assess the robustness of the method by analysing geographical variations within one country. Before presenting the results, I will briefly summarize the methodology and the nature of the data, and state the main hypotheses.

2. Nature of the Data and Method

2.1. Data

It should be remembered that only the initial cause of death is itemized in WHO statistics: the process, the multiple causes, the circumstances which led to the death are totally unknown to us. This is an important limitation of the study. We focus on 35 developed countries which regularly send their statistics to WHO (this excludes the USSR). While the statistics are available according to a list of 150 causes of death, we restrict our study to 16 specific causes (see Appendix 14.1). This reduction allows us to maximize simplicity: the intercorrelation between causes is complex enough to require minimization of the number of parameters to be estimated. The reduced list of causes of death also achieves parsimony in the sense that we preserve relevant information whenever desired. Thus, for example, in the case of tumours we gain simplicity by retaining only those items which are either quite frequent or of relevance to health. organizations, physicians, and epidemiologists. We gain in parsimony by not aggregating those items with frequencies that vary considerably across countries. Thus, tumours are broken up into four rubrics: lung, respiratory and digestive, breast (for women) and prostate (for men), and stomach.

2.2. The Statistical Method

In accord with the general goal of model tables, we are interested in the proportional distribution of deaths by causes. We want to show that countries have a cause-of-death profile that, without being completely invariant, remains very characteristic in the sense that the difference between two profiles of one country at two different dates is less marked than that of two different countries at the same date. Moreover, since it is too complicated to introduce the full age detail of these cause-specific profiles (Duchene 1980) we have chosen three broad age groups: 15–24,

45–54, and 65–74 years. Mortality in the first age group does not involve the same causes as does mortality in the other two and is dealt with separately. Causes of death in this group are reduced to four: (*a*) road accidents, (*b*) suicides, (*c*) other accidents, and (*d*) other causes. In the other age groups all 16 causes play an important role (see Figure 14.1). The average level of the death rate in the age group 45–54 for all 35 developed countries is six to seven times lower than that in the age group 65–74 and its cause-specific distribution is very different: in particular, cardiovascular diseases play a much more important role whereas accidents play a lesser one. Figure 14.1 also shows that there are marked differences in the distribution of causes of death by sex. The distributions calculated at the beginning of the observation period (1968–9) and at the end of the period (1983–4) show that, if the average level of mortality has decreased (from 475 per 100 000 to 432 for males in the age group 65–74, for example), the structure has not greatly changed. Admittedly, these average distributions hide significant variability between countries but we can show that age- and sex-specific peculiarities in the distribution of cause of death tend to be reproduced for other ages and the other sex.

We now describe the statistical method applied to understand the relation between the results of the longitudinal analysis and those of the cross-section. In Appendix 14.2 we give details on the calculations as well as additional numerical results. Consider a table where countries are placed in rows and the 16 chosen causes of deaths appear in the columns. Each cell on the table may represent the death rate in country *i* by cause *j* for a given sex and age. The sum of the cells in a row gives the total death rate. Dividing each of the cell values into the total death rate, yields a cause-of-death profile or the cause-of-death distribution. If we then define a distance between two profiles, calculate a table of distances between countries, and apply an automatic classification procedure to the table of distances, we will obtain a typology of countries. Admittedly, the procedure has limitations since: (*a*) there is no unique definition of distance, (*b*) there are multiple classification algorithms, and (*c*) the number of groups desired in the typology is arbitrary. Nevertheless, we have been able to assess the robustness of the typology by changing the definition of distance and altering the classification algorithm. Although the proximity between groups varies according to the methods used, the typology suggested here is robust and, without being optimal, is in all cases quite acceptable. Instead of summarizing the different typologies obtained, I will show the details of the method used so that its results can be replicated more easily.

As indicated in Appendix 14.2, instead of presenting two separate analyses, one for the age group 45–54 and the other for the age group 65–74, I have juxtaposed the tables corresponding to each of them. This was done only after having verified that the resulting typologies were consistent with each other. Thus the total for a row yields now the sum of the rates in the two age groups. To assign weight to causes of lesser significance and to

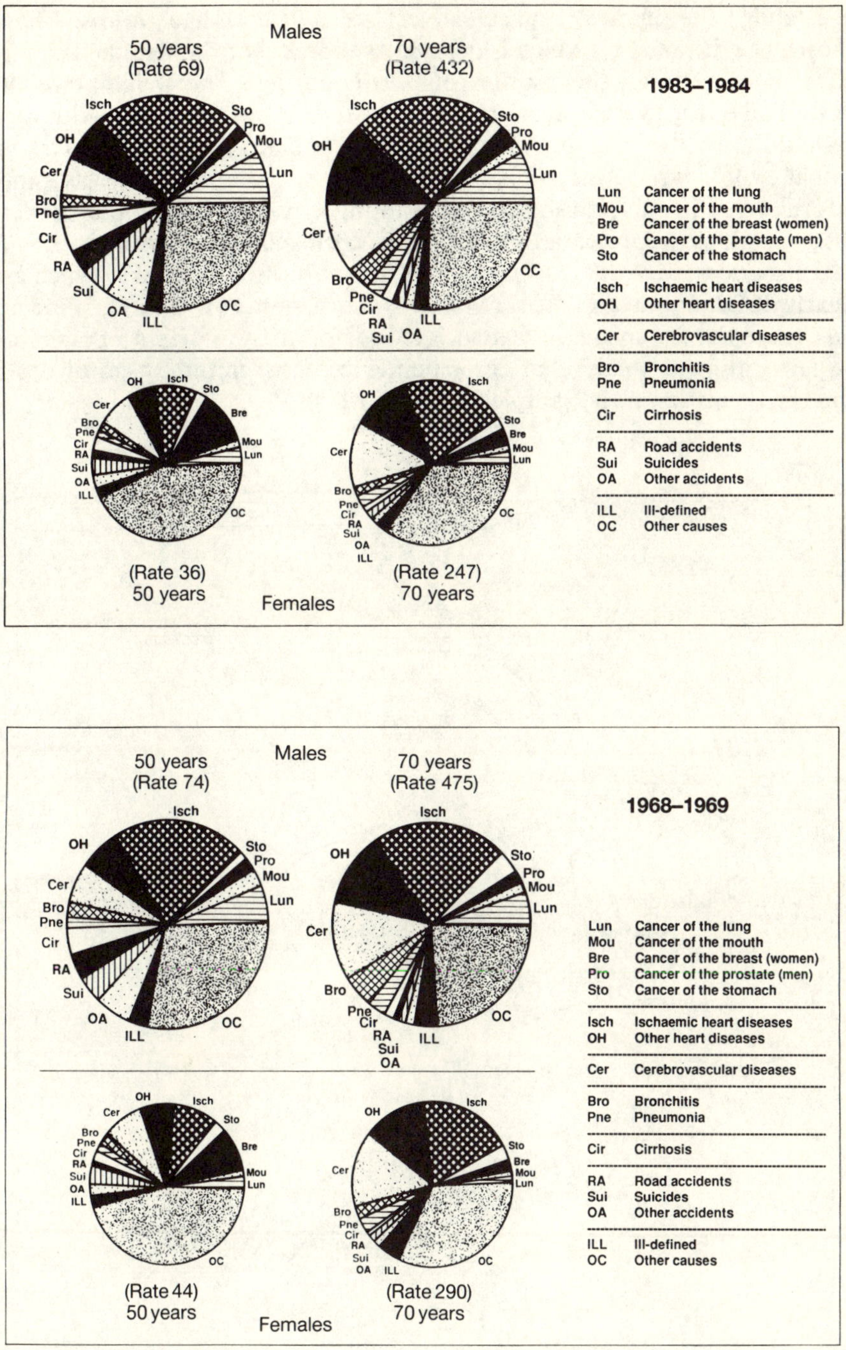

Fig. 14.1. Distribution of deaths by cause at 45–54 years and 65–74 years for males and females. Average from 35 countries in 1983–1984 and 1968–1969

re-establish the relative balance between rates in both age groups, I have chosen the distance x^2, which is defined as the total of the weighted square differences between the profiles of two countries. The weights are the reciprocal of the average death rate by cause (see Appendix 14.2). Although it would have also been possible to combine female and male tables into a global typology, it was thought that a sex-specific analysis was more informative. In fact, although the resulting typologies are quite similar, their graphical representation and their correspondence with causes of death are very different (this is especially true for breast cancers, which are clearly not distributed in the same way as cancers of the prostate). If indeed we choose the distance x^2, the associated factor analysis offers a possibility of performing an analysis of correspondence between the causes of death and the countries in the sample (Benzecri 1973).

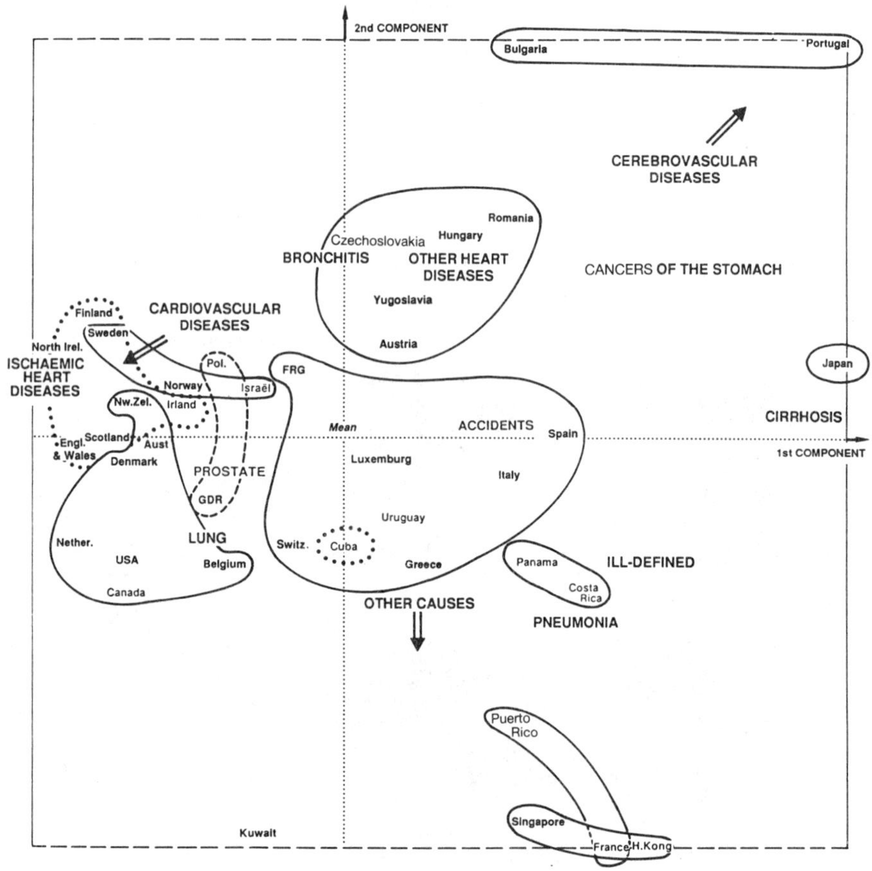

Fig. 14.2. Classification of countries according to their death-by-cause profile at 45–54 years and 65–74 years, men, first and second components

The typology consists of 11 groups. The developing countries which are listed (in parentheses) have been removed from subsequent analysis since our study is only concerned with developed countries. The list of countries in each group is presented below. The representation in the form of contours on two factorial planes intersecting components 1 and 2 and 1 and 3 appear in Figures 14.2 and 14.3 respectively. The groups are:

A: Japan;
B: Singapore, Hong Kong;
C: France, (Puerto Rico);
D: Portugal, Bulgaria;

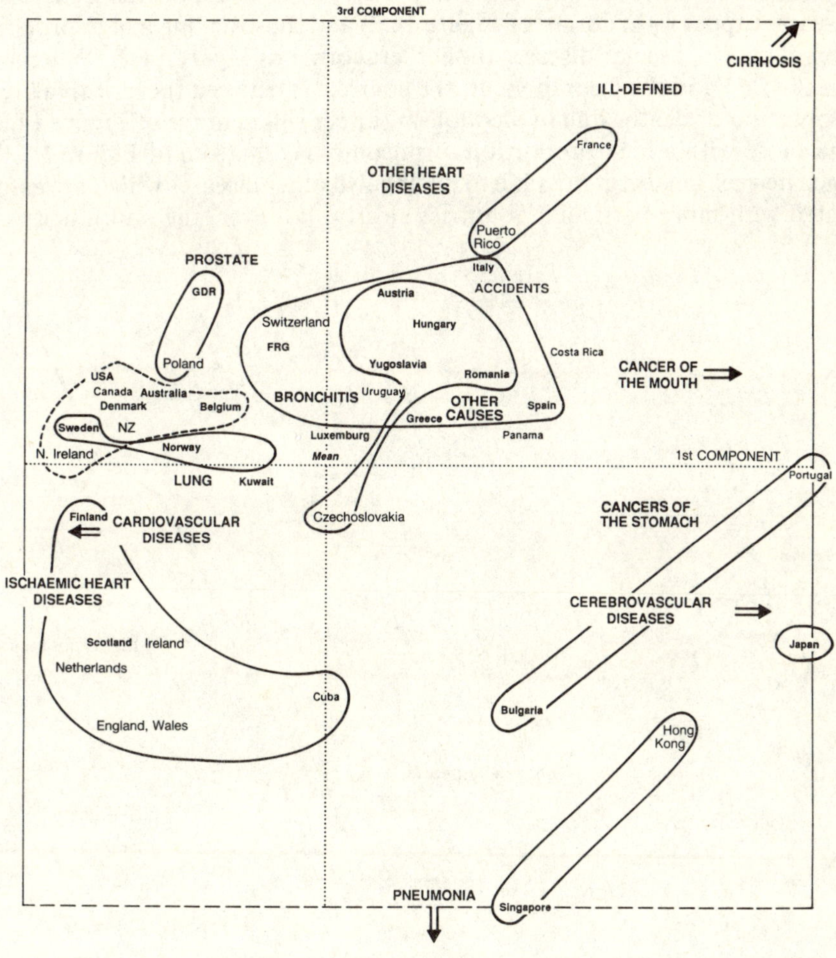

Fig. 14.3. Classification of countries according to their death-by-cause profile at 45–54 years and 65–74 years, men, first and third components

E: (Panama, Costa Rica);
F: Spain, Italy, Greece, Uruguay, Federal Republic of Germany (FRG), Switzerland;
G: Romania, Yugoslavia, Hungary, Austria, Czechoslovakia;
H: Poland, German Democratic Republic (GDR);
I: United States, Canada, Australia, New Zealand, The Netherlands, Denmark, Belgium;
J: England and Wales, Scotland, Northern Ireland, the Republic of Ireland, Finland, (Cuba);
K: Sweden, Norway, Israel.

The results of the factor analysis on the first two components leads to a contrast between countries with a high proportion of cerebrovascular diseases (upper right corner of Figure 14.2) and those having a high proportion of cardiovascular diseases (upper left corner of Figure 14.2). When we use the first and third component, the contrast is between those with a high proportion of deaths due to alcoholism (upper right corner of Figure 14.3) and those with a high proportion of pneumonias (bottom of Figure 14.3). Each new axis makes it possible to distinguish other diseases which are associated with more particular countries clearly; however, the multiplicity of

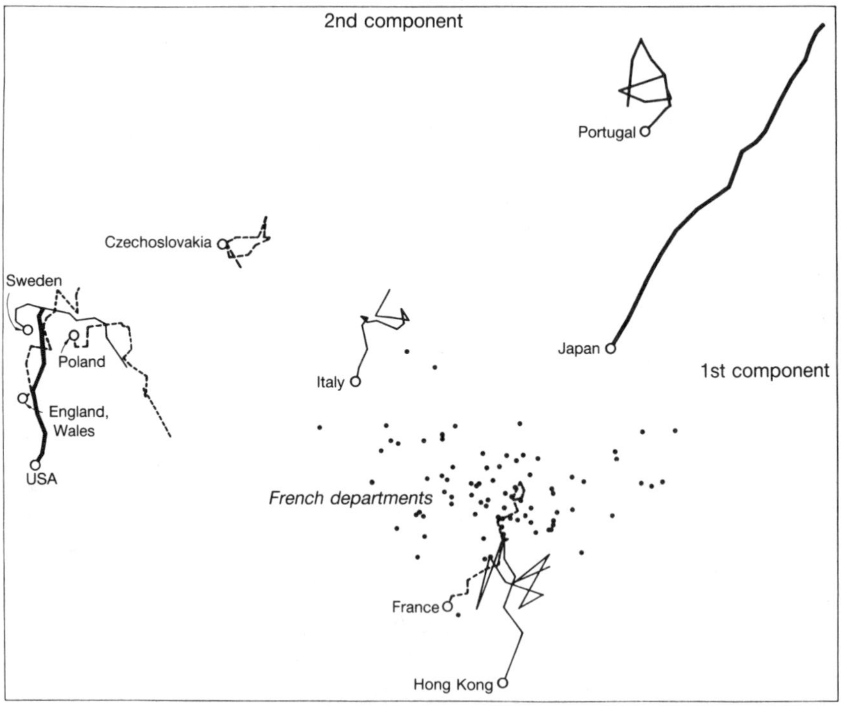

Fig. 14.4. Trajectories of representative countries from each group on the factorial plane (1,2) during the period 1968–1984, and position of the 95 French departments

planes complicates the accounting and synthesis a great deal. We should note that the factor analysis, whose objective is to summarize the initial information contributed by the 16 causes, can not lead to further reductions of the information, for the relations between causes and countries are very complex. In fact, 88% of the variance is explained by a total of six components. The graphic presentation of just two factorial planes in no way allows a correct translation of the initial information and only the final classification is valid. The graphs presented here are only an illustration of this classification.

The results of this factor analysis can be used to visualize the relative location of other countries or the same country at different periods of time in order to assess the evolution of the cause-specific distribution independently from the total level of mortality. It is this technique that we have used to visualize the trajectory of developed countries over the last 15 years or so.

3. Longitudinal Analysis

All the data from developed countries have been projected on to various factorial planes but, given that the projection of 640 or so points on to the

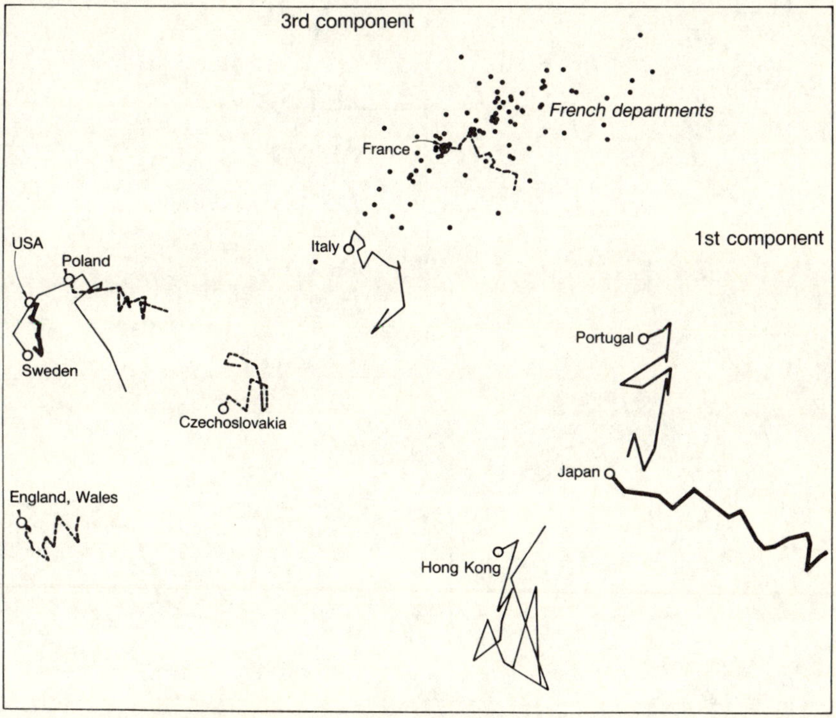

Fig. 14.5. Trajectories of representative countries from each group on the factorial plane (1,3) during the period 1968–1984, and position of the 95 French departments

same graph is illegible, we have chosen to display only one representative for each of the groups defined. Figures 14.4 and 14.5 show the trajectories of the countries on planes (1, 2) and (1, 3). Apart from Japan, where mortality decline has been particularly rapid, we observe that over 16 years the cause-of-death profiles are remarkably stable. The case of Japan is interesting as

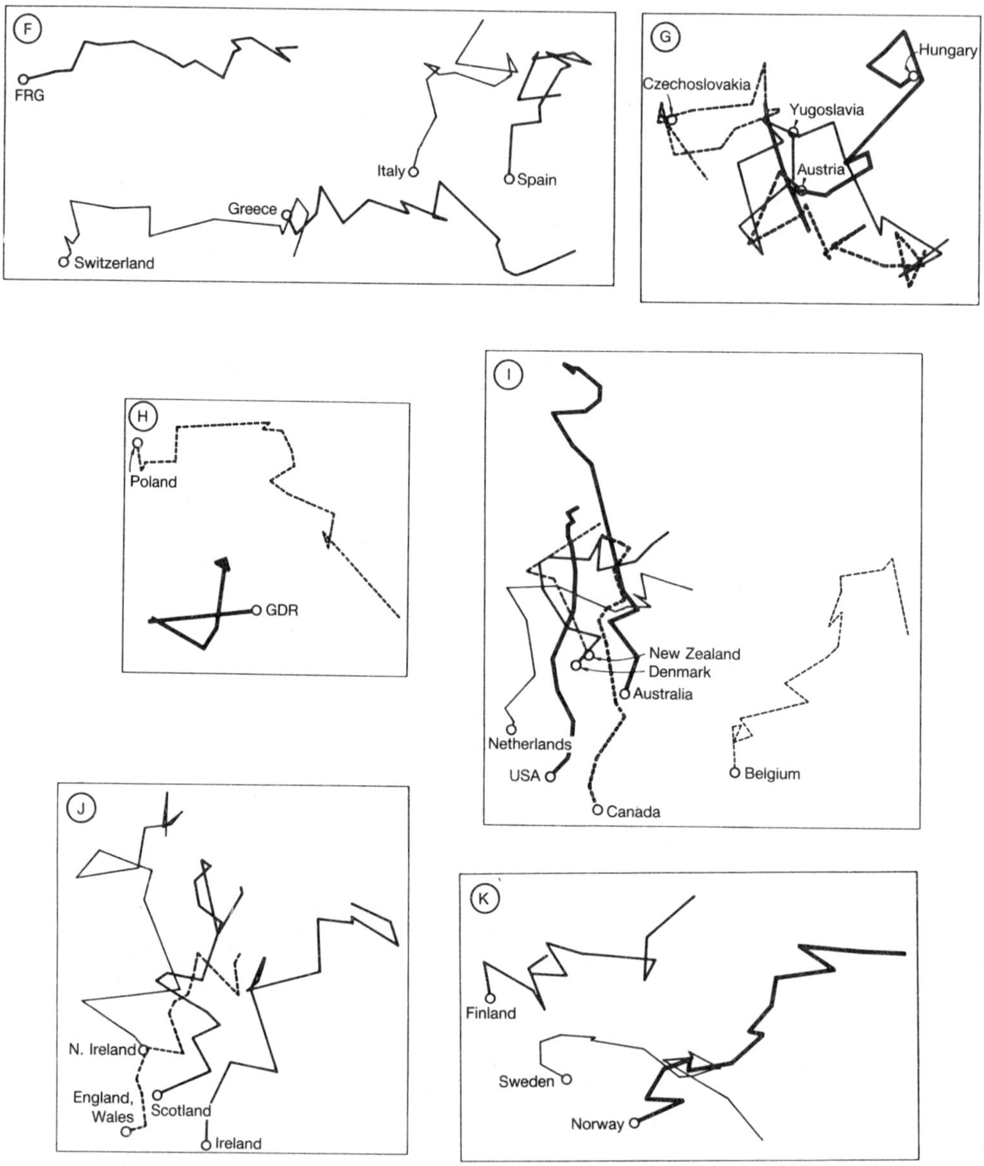

Fig. 14.6. Trajectories of various countries of a same group on the plane (1,2) during the period 1968–1984

its trajectory reveals a very clear decline in cerebrovascular diseases, which accompanies the general improvement in life expectancy between 1968 and 1984 (with gains of about four years for males and six years for females). The sloping trajectories like those of France or of Italy also reveal a decline in cerebrovascular disease.

In order to better analyse the development of countries within a group, we have drawn specific contours for each group (see Figure 14.6). In group I containing the overseas Anglo-Saxon countries (with Denmark, Belgium, and the Netherlands) one observes that the trajectories descend revealing a decline in cardiovascular and cerebrovascular disease and an increase in lung cancers. Group J (British Isles and Finland) reveals absolutely identical trends. Group K (other Nordic countries and Israel) is quite different as they show a very marked increased in the proportion of cardiovascular disease. Here it is possible to detect the recent turn-about of Sweden. In group F (the Mediterranean countries) Greece differs from Italy and Spain due to a higher proportion of cardiovascular disease. Trends are identical for Spain and Italy but different for Greece. The FRG and Switzerland, which were regrouped with these Mediterranean countries, tend to deviate from the group because of a steady increase in cardiovascular disease and lung cancers. The FRG is remarkably similar to the GDR. In group G (containing the Central European countries of Hungary, Czechoslovakia, Austria, Yugoslavia, and Romania) the most important and disturbing characteristic is the increase in the male death rates at given ages. Table 14.1 shows that, apart from Austria, male death rates increased at all ages between 1968 and 1984. For Austria, the death rate at ages 45–54

Table 14.1. *Death rates per 1000 in age groups 45–54 and 65–74 for Central European countries where male mortality stagnated or increased between 1968 and 1984* (per 1000)

Country	Year	Males		Females	
		45–54	65–74	45–54	65–74
Austria	1969	77	582	44	321
	1984	76	437	33	231
Bulgaria	1968	59	411	36	311
	1983	80	509	37	325
Czechoslovakia	1968	77	542	41	318
	1983	103	606	42	342
Hungary	1969	72	520	43	339
	1984	133	589	53	335
Poland	1969	77	507	42	302
	1984	108	514	41	284
Romania	1969	70	496	46	360
	1983	93	495	45	334
Yugoslavia	1968	70	470	44	340
	1982	86	500	40	314

Source: WHO, data bank.

has stagnated. The increase is less in the last age group and for females the trend is not systematic at all. The graphical representation reveals that the mortality increase in Central European countries is not due to any particular cause but is the outcome of a simultaneous increase of all causes. What we are witnessing in these countries is a general degradation of health conditions. Thus, for example, in the case of Hungary the trajectory is ascending according to component 3 (see Figure 14.7), which means that deaths by cirrhosis are increasing more than deaths by other causes. But the major role played by alcoholism in Hungary is not necessarily present in other countries. For example, it is not characteristic of the mortality increase in Czechoslovakia. And, since the case of Austria is not very favourable either, it is difficult to associate the Central European course of mortality with the political context. A complete explanation would require recourse to typical behaviours and practices in Central Europe. Table 14.1 also shows that mortality increases in Bulgaria and in Poland are attributable to cardiovascular diseases and lung cancers. These countries, together with the GDR, are part of group H. In the GDR mortality is decreasing rather than increasing.

In summary, this method of graphical representation makes it possible to detect major cause-specific mortality trends. It does not replace, however, classical quantitative work, which demands a return to basic data. We should also note that the method is unsuitable for a small country or for a country where the statistics are not very reliable for, by and large, the trajectories will be unclear and uninterpretable.

With this kind of graphical representation it is possible to look at the spread obtained when a country is divided into geographical sub-regions. Data are only shown for France and its 95 departments for the period

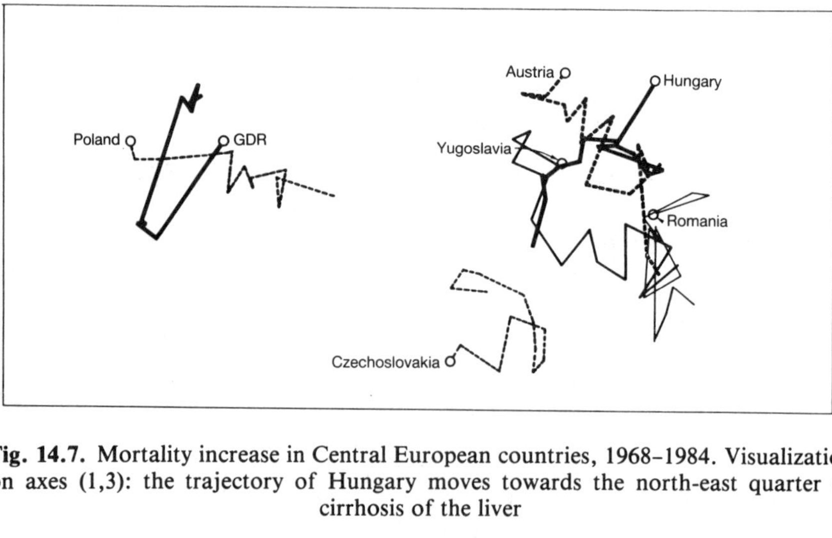

Fig. 14.7. Mortality increase in Central European countries, 1968–1984. Visualization on axes (1,3): the trajectory of Hungary moves towards the north-east quarter of cirrhosis of the liver

1974–6 (Figures 14.4 and 14.5). It may be noted that the scatter is relatively concentrated to only a small zone around the point 'France'. French departments which are geographically close to Italy (like Corsica) are close to the point 'Italy'. With this kind of statistical analysis and graphical representation we will be able to discover the structure of mortality by cause of death.

4. Death Risk in the Age Group 15–24

Although the two age groups we have chosen for analysis, 45–54 and 65–74, cover the majority of deaths, mortality in the age group 15–24 deserves special attention due to the increases in mortality experienced by some countries during the past decade. Figure 14.8 shows the position of all developed countries with regard to causes of death at this age and Figure 14.9 illustrates some remarkable national trajectories. Thus in Norway suicides have taken on an ever-growing proportion, Belgium is still particularly vulnerable to road accidents, and Italy, and especially Spain, depart from the characteristic profiles of developing countries (low proportion of suicides and road accidents) to join the other developed countries. Finally, Czechoslovakia draws away from the group of developed countries, while Bulgaria's position stagnates.

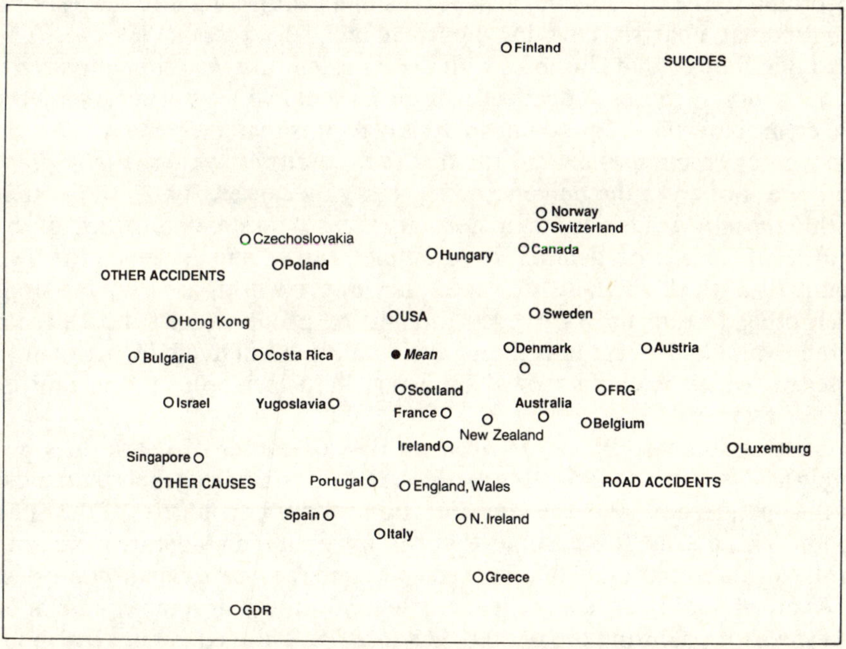

Fig. 14.8. Static position of countries according to causes of death at 15–24 years, men

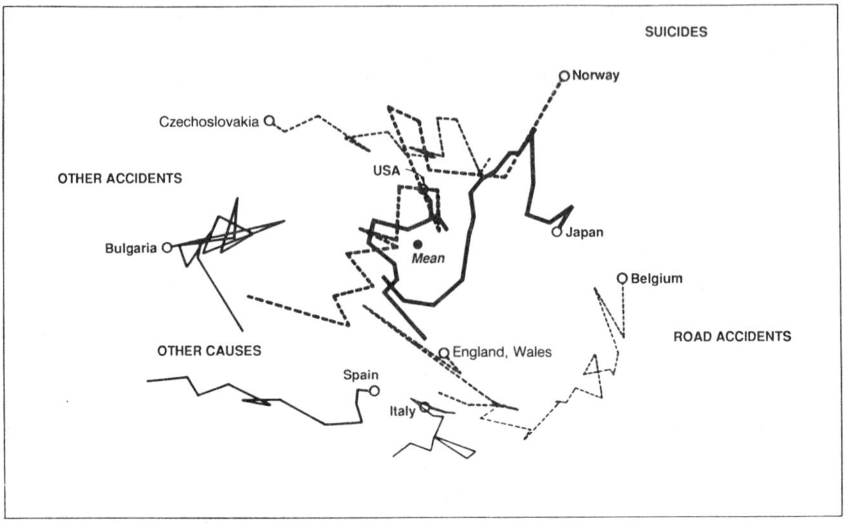

Fig. 14.9. Trajectories of certain countries within a causal interval at 15–24 years, men

5. Summary and Conclusion

Our objective in this chapter has been to validate a typology of countries according to their profile of causes of death by subjecting it to the test of a longitudinal analysis covering approximately 15 years (1968–84). The typology showed that countries with the same 'culture' had similar patterns of mortality by causes of death even if the *level* of overall mortality was very different. Now the longitudinal analysis confirms that over at least 15 years the typology remains the same, i.e. that each country stays at the same place and does not cross the universe of mortality by causes. It also shows that within the worldwide decline in mortality there is no standardization of the profile of causes of death. For example, Japan and Sweden, the two countries with the highest life expectancy in the world, are very far from each other in that universe: even though the proportion of deaths from cerebrovascular disease is declining in Japan, it is still much higher than in Sweden, which is characterized by a high and increasing proportion of cardiovascular disease.

Lack of standardization between profiles of causes of death does not imply a lack of trends within the profiles. But our finding again is that, most of the time, trends are in the same direction for all the countries of the same group. This means, for example, that cardiovascular diseases are increasing in all Nordic countries, while decreasing in overseas Anglo-Saxon countries.

Another finding concerns the increase in mortality in some countries of Central Europe during the period 1968 to 1978. This general increase does not have its origin in the increase of particular causes such as alcohol-related causes, but was the result of an increase in all causes: the health system in general has deteriorated over the period.

This study illustrates also the *dependency* of the causes of death: progress in particular fields of medicine have repercussions in all other fields. On the other hand, when health services are disorganized, mortality from any cause is increased.

At present, model life tables concern the universe of mortality *by age*, and not yet *by cause of death*, but we hope that these results will be helpful for the construction of model life tables by cause.

References

Benzecri, J.P. (1973), *L'Analyse des données, l'analyse des correspondances*, ii, 2nd edn. (Paris: Dunod).

Brouard, N., and Lopez, A. (1985), 'Causes of Death Patterns in Low Mortality Countries: A Classification Analysis', in *International Population Conference*, Florence, pp. 385–406 (Liège: IUSSP, Ordina Éditions).

Coale, A.J., and Demeny, P. (1966), *Regional Model Life Tables and Stable Populations* (Princeton, NJ: Princeton University Press).

Duchène, J. (1980), *Un éssai de modélisation de la répartition des décès selon l'âge et la cause dans les pays industrialisés*, Département de Démographie, Université de Louvain la Neuve (Louvain la Neuve: Cabay).

Lebart, L., Morineau, A., and Fénelon, J.P. (1980), *Traitement des données statistiques: Méthodes et programmes* (Paris: Dunod).

Lederman, S., and Bréas, J. (1959), 'Les dimensions de la mortalité', *Population*, vol. 4.

Appendix 14.1

Table 14.A1. *ICD codes of chosen causes according to the Eighth and Ninth Revisions*

Causes	Eighth Revision rubrics of List A	Ninth Revision rubrics
All causes	A00	B00
Cancer		
Mouth, pharnyx, oesophagus	A45, A46, A50	08, 090, 100
Stomach	A47	091
Lung	A51	101
Breast	A54	113
Prostate	A57	124
Heart diseases		
Ischaemic	A83	27
Other	A84, A86, A87, A88	28, 30
Cerebrovascular diseases	A85	29
Bronchitis, asthma	A93	32
Pneumonia	A91, A92	321
Cirrhosis of the liver	A102	347
Ill-defined	AE136–7	46

Table 14.A1. *Continued*

Causes	Eighth Revision rubrics of List A	Ninth Revision rubrics
Other accidents	AE139–46, 148–50	E470, E472, E479
Road accidents	AE138	E471
Suicides	AE147	E54
Other causes	The rest	The rest

Appendix 14.2 Results of the Correspondence Factor Analysis

Consider the table for the age group 45–54 which provides the death rate for country $i(i=1, n)$ and for cause $k(k=1, e)$ between ages 45 and 54 as well as the table for ages 65–74 years. Place them side by side as shown in Figure 14.A1. Let m_{ij} be the complete matrix with j columns ($j \in [1, 2c]$). Then, $m_{i.} = \Sigma_j m_{ij}$ is the total mortality rate within the age group 45–70 for country i. Similarly, $m.j = \Sigma_j m_{ij}$ is a measure of the average death rate by cause.

$$\begin{array}{cc} & \begin{array}{cccccc} & 45\text{–}54 & & & 65\text{–}74 & \\ & k & & & k & \end{array} \\ \text{Country} \quad i & \left[\begin{array}{cccccc} \vdots & \vdots & \vdots & \vdots & \vdots & \vdots \\ \cdots & m_{ik}^{50} & \cdots & \cdots & m_{ik}^{70} & \cdots \\ \vdots & \vdots & \vdots & \vdots & \vdots & \vdots \end{array}\right] \end{array}$$

Fig. 14.A1. Matrix of death rates by age, country, and cause of death

We calculate the casual profile for the two age groups as the set of values $\frac{m_{ij}}{m_{i.}}$ and define a distance χ^2 between the profiles of two countries i and i' as

$$d^2(i, i') = \sum_j \frac{1}{m.j}(m_{ij} - m_{i'j})^2.$$

The weight m_j attenuates the contribution of frequent causes of death and re-establishes the relative distribution of deaths of both age groups.

This distance $\chi^2(i, i^1)$ can also be written as:

$$d^2(i, i') = \sum_j \left[\frac{m_{ij}}{m_{i.\sqrt{(m.j)}}} - \frac{m_{i'j}}{m_{i'.\sqrt{(m.j)}}}\right]^2,$$

and appears as the Euclidean distance between two points i and i' of coordinates x_{ij} and $x_{i'j}$ where x_{ij} is defined as:

$$I_{ij} = \frac{m_{ij}}{m_{i.\sqrt{(m.j)}}}.$$

Moreover, if we give weight m_i to each country where m_i equals the total death rates in both age groups, the correspondence factor analysis is a principal component analysis of table x_{ij} with weights m_{ij}. We thus seek the best possible representation of this balanced form of n countries in a two-dimensional space. The centre of gravity has as jth coordinate $\bar{x}_j = \sqrt{(m.j)}$. If $\vec{u}$ is a unitary vector of $\mathbf{R}^{2c}$, then

$$\psi_i = (\vec{x} - \bar{x}\ \vec{u}).\vec{u} = \sum_j \left(\frac{m_{ij} - m_{i\cdot}.m_{.j}}{m_{i.}\sqrt{(m.j)}} \right)$$

is the coordinate of a country point on axis u. The vector $\vec{u}$ will be the major inertia vector provided it minimizes the quantity: $\sum_i m_{i.}\psi_i{}^2$ which, after simplification, corresponds to the diagonalization of the matrix $X'X$ where X is the common matrix:

$$X_{ij} = [m_{ij} - m_{i.}m_{.j}/\sqrt{(m_{i.}m_{.j})}].$$

It is easy to show that this symmetrical matrix $X'X$ has the same eigen values as the matrix $X^{*\prime}X^{*}$. In simple terms:

$$X^{*}_{ij} = m_{ij}/\sqrt{(m_{i.}m_{.j})}.$$

The vector $\bar{x}\ \vec{u}$ is thus an eigen vector associated with the eigen value 1. Apart from this singular eigen value, there is also the eigen value of the order of α with eigen vectors $\vec{u}_\alpha$; the coordinate of a country i on the factorial axis of the order of α is then:

$$\psi_{\alpha i} = \sum_{j=1}^{2c} \left(\frac{m_{ij}}{m_{i.}\sqrt{m_{.j}}} \right) u_{\alpha j} \tag{1}$$

Dual Representation

We can demonstrate (Benzecri 1973) that the matrix $X^{*}X^{*\prime}$ has the same non-void eigen values as the matrix $X^{*\prime}X^{*}$. If we take v_α for the new eigen vector associated with the same eigen value $\lambda\alpha$, we get the following double relations:

$$\begin{cases} v_\alpha = \dfrac{1}{\sqrt{(\lambda_\alpha)}} x^{*} u_\alpha \\ u_\alpha = \dfrac{1}{\sqrt{(\lambda_\alpha)}} x^{*\prime} v_\alpha. \end{cases} \tag{2}$$

If instead of considering n country points in a $2c$th dimensional interval we consider $2c$ cause points in an nth dimensional interval with weight m_j, we obtain a relation which is symmetric to the relation $1/u\alpha i$, which gives the coordinates $\phi_{\alpha j}$ of the jth cause on the αth eigen vector connected with the eigen value $_\alpha$; this coordinate is:

$$\phi_{\alpha j} = \sum_{i=1}^{n} \left(\frac{m_{ij}}{m.j\sqrt{(m_{i.})}} \right) v_{\alpha i} \tag{3}$$

Relations 1, 2, and 3 lead to a fundamental barycentric correspondence relation between cause and country:

$$\phi_{\alpha j} = \frac{1}{\sqrt{(\lambda_\alpha)}} \sum_{i=1}^{n} (m_{ij}/m.j)\psi_{\alpha i}$$

$$\psi_{\alpha j} = \frac{1}{\sqrt{(\lambda_\alpha)}} \sum_{j=1}^{2c} (m_{ij}/m_{i.})\phi_{\alpha i.} \tag{4}$$

Graphic Representation and Results

When cause points and country points are superimposed on the same graph, the barycentric correspondence is obtained. As in a principal component analysis, the percentage of inertia explained by the first components is associated with the eigen values. Table 14.A2 gives the eigen values as well as the percentages of inertia of each axis.

Table 14.A2. *Percentage of inertia of the first axes*

Axis	Eigen value	%	Accumulated (%)
1	0.03903	36.26	36.26
2	0.01926	17.90	54.16
3	0.01515	14.08	68.24
4	0.00967	8.99	77.23
5	0.00771	7.16	84.39
6	0.00423	3.93	88.32
7	0.00393	3.66	91.98
8	0.00196	1.82	96.75

We are restricted here to the representation of the first three axes (accounting for 68% of the variance). Starting with the causal groups of Table 14.A1, we have excluded all ill-defined causes and regrouped accidents on the one hand and all heart diseases on the other. We therefore had ten causes in all for each group or a total of 20 variables. For each of these variables, Table 14.A3 gives the weights $m._j$ and the coordinates on the first six axes ($\phi_{\alpha j}$).

Once the form of the causes is placed, the countries are positioned according to the barycentric relation 4. Table 14.A4 gives the coordinates on the three first factors of the 38 countries which have been used to build the form. In a previous paper, we selected the most recent data from countries having a life expectancy of over 70 years (Brouard and Lopez 1985). Here we have repeated this selection (thus diagonalizing the same matrix) and used these results to project either other countries or the same countries at different dates.

Take the example of a French department for which we know the male death rates according to the ten causes retained (Table 14.A1) and for the two age groups 45–54 and 65–74 years. Let $[j\epsilon(1, 2\times 10)]$ be the death rates, and $t = \sum_t t_j$ the total death rates in both age groups (all causes included). The coordinate ψ_{α}, of this department on the axis α is given by the equation:

Table 14.A3. *Coordinates of cause points on the first factors*

Cause	Weight *mj*	Coordinates of the factors					
		1	2	3	4	5	6
45–54							
Lung	0.012	−0.09	−0.02	0.00	−0.21	0.01	0.01
Mouth	0.005	0.45	−0.39	0.13	−0.27	−0.18	0.17
Stomach	0.004	0.34	0.20	−0.07	0.17	0.21	−0.02
Cardiovascular	0.043	−0.21	0.05	−0.04	0.00	−0.09	0.02
Cerebrovascular	0.000	0.27	0.18	−0.16	0.01	−0.03	0.05
Bronchitis	0.002	0.04	0.13	0.00	−0.03	0.06	0.25
Pneumonia	0.002	0.31	−0.15	−0.35	0.15	−0.20	0.27
Cirrhosis	0.008	0.41	0.06	0.35	−0.06	−0.17	0.04
Other accidents	0.021	0.09	0.06	0.15	0.16	−0.05	0.08
Other causes	0.038	0.13	−0.09	0.06	0.04	−0.05	0.05
65–74							
Lung	0.072	−0.17	−0.12	−0.08	−0.22	0.12	−0.02
Mouth	0.017	0.27	−0.34	0.04	−0.08	−0.10	0.09
Stomach	0.028	0.29	0.09	0.00	0.23	0.29	−0.05
Cardiovascular	0.318	−0.20	0.05	0.02	0.04	−0.04	−0.01
Cerebrovascular	0.102	0.27	0.26	−0.11	−0.10	−0.04	−0.05
Bronchitis	0.034	−0.09	0.15	0.06	−0.07	0.21	0.23
Pneumonia	0.024	0.15	−0.20	−0.58	0.16	−0.08	0.13
Cirrhosis	0.016	0.46	0.05	0.37	−0.12	−0.15	0.09
Other causes	0.031	0.16	0.00	0.13	0.12	0.09	0.02
Other accidents	0.215	0.08	−0.15	0.03	0.02	0.01	−0.05

$$\psi_\alpha = \frac{1}{\sqrt{(\lambda_\alpha)}} \sum_{j=1}^{20} (t_j/t)\phi_{\alpha j}$$

For the first axis, we obtain (see Tables 14.A2 and 14.A3):

$$\psi_\alpha = \frac{1}{t\sqrt{0.03903}} (-0.05 \text{ (lung 45–54)} + 0.45 \text{ (mouth 45–54)} + 0.34 \text{ (stomach)} + 0.27 \text{ (brain)} + \ldots + 0.12 \text{ (lung 65–74)} + \ldots + 0.08 \text{ (all other causes 65–74)}). \quad (6)$$

By using this process it is also possible to visualize other causes or sub-causes. This is what we have done in Figure 14.2 by breaking up all cardiovascular diseases into 'ischaemic heart diseases' and 'other heart diseases'. For more detail on the method see Lebart *et al.* 1980.

Table 14.A4. *Coordinates of the 38 countries on the three first factors*

Country	Year	CIM	Weight	Axis 1	Axis 2	Axis 3
Mean	80		0.026	0.00	0.00	0.00
Australia	80	9	0.025	−0.16	−0.01	0.04
Austria	80	9	0.028	0.07	0.09	0.11
Belgium	78	8	0.028	−0.06	−0.07	0.01
Bulgaria	81	9	0.028	0.20	0.33	−0.22
Canada	78	8	0.024	−0.18	−0.04	0.06
Costa Rica	80	9	0.021	0.23	−0.10	0.09
Cuba	78	8	0.015	0.00	−0.10	−0.17
Czechoslovakia	82	9	0.036	0.01	0.15	−0.05
Denmark	81	8	0.024	−0.22	0.01	0.05
England, Wales	81	9	0.027	−0.21	−0.02	−0.19
Finland	79	8	0.031	−0.23	0.11	−0.04
France	81	9	0.023	0.28	−0.28	0.25
FRG	82	9	0.027	−0.03	0.04	0.09
GDR	76	8	0.031	−0.13	−0.05	0.13
Greece	82	9	0.019	0.09	−0.08	0.03
Hong Kong	82	9	0.022	0.35	−0.33	−0.21
Hungary	82	9	0.037	0.10	0.16	0.09
Ireland	82	9	0.028	−0.18	−0.05	−0.13
Israel	80	9	0.021	−0.07	0.02	0.00
Italy	80	9	0.025	0.16	−0.01	0.15
Japan	82	9	0.019	0.48	0.08	−0.16
N. Ireland	81	9	0.031	−0.25	0.05	−0.15
New Zealand	81	9	0.026	−0.21	0.01	0.02
Norway	82	8	0.021	−0.16	0.02	0.02
Panama	80	9	0.016	0.20	−0.08	0.04
Poland	80	9	0.033	−0.15	0.04	0.08
Portugal	79	8	0.026	0.46	0.31	−0.03
Puerto Rico	82	9	0.022	0.17	−0.22	0.20
Romania	82	9	0.029	0.15	0.16	0.06
Scotland	83	9	0.031	−0.20	−0.02	−0.14
Singapore	81	9	0.032	0.21	−0.30	−0.32
Spain	79	8	0.022	0.22	0.01	0.05
Sweden	82	8	0.021	−0.23	0.06	0.03
Switzerland	81	8	0.022	−0.04	−0.08	0.11
The Netherlands	82	9	0.023	−0.24	−0.08	−0.01
Uruguay	78	8	0.027	0.07	−0.06	0.05
USA	80	9	0.025	−0.20	−0.10	0.07
Yugoslavia	81	9	0.027	0.07	0.10	0.08

15 Cause of Death and Expectation of Life: Some International Comparisons[1]

JOHN H. POLLARD

Macquarie University Sydney, Australia

1. Introduction

Formulae were recently derived which allow the analysis of the contributions of various causes of death at different ages to the change in expectation of life at birth of a population over a given time period. These formulae can also be used to analyse the sex differential in life expectancy and differentials in life expectancy between populations (Pollard 1982).

In this chapter, the changes in expectation of life of selected populations over the 1970s are analysed. Comparisons between some of these populations are also made. Among the populations studied, Japan is an interesting example in that it represents a large population with one of the highest life expectancies in the world. Hungary is also unusual in that it is one of the few experiencing deteriorating mortality for most major causes of death, and, for males, a reduction in expectation of life at birth.

Although the formulae employed in this chapter relate solely to the expectation of life at birth, the methods and formulae are readily adapted to expectations of life at other ages and indeed temporary expectations, which have been proposed as useful mortality indicators (Arriaga 1982).

2. The Relationship between Mortality and Expectation of Life

An improvement of ϕ in the force of mortality in the small age range $(x, x+\Delta x)$ causes the expectation of life at birth in the population under consideration to increase by an amount

$$ {}_{x}p_{0}\,\mathring{e}_{x}\,\phi\Delta x, \tag{1}$$

assuming that there are no changes in mortality at other ages. This formula is well known and leads to the following approximate formula for the gain

[1] I should like to thank the following colleagues, who kindly sent me cause-of-death data in respect of their own national population; in many cases the collating of the data in the form requested involved considerable work, which is much appreciated: Professor B. Benjamin, Dr. A.M. Bolander, Dr. J. Duchene, Dr. N. Keilman, Dr. A. Klinger, Dr. M. Minowa, Dr. I. Shigematsu, Dr. E. Valkovics, Professor G. Wunsch.

in expectation of life at birth in a population between time 1 and time 2.

$$\mathring{e}_x^2 - \mathring{e}_x^1 \simeq \int_0^\infty (\mu_x^1 - \mu_x^2)\,{}_xp_0^1\,\mathring{e}_x^1\,dx. \tag{2}$$

In this, and other formulae in this section of the chapter, a superscript 1 will indicate that the function concerned is evaluated at time 1 and superscript 2 will indicate that the function is evaluated at time 2; μ_x represents the force of mortality (or instantaneous death rate) at age x, ${}_xp_0$ the probability of survival x years from age 0, and $\mathring{e}_x$ the complete expectation of life at age x.

Formula (2) is reasonably accurate, provided the improvements in mortality are modest. It always *underestimates* the gain in expectation of life when mortality improvements are positive. The reason the formula is only approximate and underestimates the actual gain in expectation of life when positive improvements in mortality take place at all or most ages is that it ignores interaction effects between mortality improvements at the different ages.

An exact formula explaining the gain in expectation of life of a population in terms of mortality improvements at the individual ages was recently derived (Pollard 1982). In its simplest form, the formula may be written

$$\mathring{e}_0^2 - \mathring{e}_0^1 = \int_0^\infty (\mu_x^1 - \mu_x^2)\,w_x\,dx, \tag{3}$$

with

$$w_x = \tfrac{1}{2}({}_xp_0^2\,\mathring{e}_x^1 + {}_xp_0^1\,\mathring{e}_x^2). \tag{4}$$

These formulae will be used later in the chapter to analyse the effects of mortality changes in selected populations on the expectations of life of those populations and differentials among the populations. The weight function w_x declines with age in an almost linear fashion from the average life expectancy at birth of the population at the two points of time to zero.

A variant of (3) allows the calculation of the main effects of mortality improvements at the various ages (formula (2)) and the interaction effects of various orders. If we define

$$M_x = \int_0^x \mu_t\,dt = -\ln {}_xp_0, \tag{5}$$

the differential in expectation of life at birth may be written

$$\mathring{e}_0^2 - \mathring{e}_0^1 = \int_0^\infty (\mu_x^1 - \mu_x^2) \exp(M_x^1 - M_x^2)_x p_0 \, \mathring{e}_x^1 dx. \tag{6}$$

When the exponential term in (6) is expanded in terms of the powers of $M_x^1 - M_x^2$, the main effects term (2) and various interaction terms are obtained. The jth-order interaction term, for example, takes the form

$$\frac{1}{j!} \int_0^\infty (M_x^1 - M_x^2)^j (\mu_x^1 - \mu_x^2)_x p_0^1 \mathring{e}_x^1 dx. \tag{7}$$

An explanation of this result, not given in the original paper, is set out in Appendix 15.1.

Interaction terms tend to be difficult to interpret, and they are certainly not easy to explain to the layman. Fortunately, even in quite extreme cases, the interaction effects of mortality improvements at different ages are relatively minor. For all practical purposes therefore, the interaction effects may be merged with the main effects and the relatively simple formula (3) employed (Pollard 1982). An interesting interpretation of interaction when mortality changes are all improvements are outlined later in this chapter.

For numerical work, the integral in (3) may be approximated by

$$\mathring{e}_0^2 - \mathring{e}_0^1 \simeq ({}_1Q_0^1 - {}_1Q_0^2)\, w_0 + ({}_4Q_1^1 - {}_4Q_1^2) w_2 + ({}_5Q_5^1 - {}_5Q_5^2) w_{7.5} + ({}_5Q_{10}^1 - {}_5Q_{10}^2) w_{12.5} + \cdots, \tag{8}$$

with

$${}_nQ_x = -\ln(l_{x+n}/l_x) \tag{9}$$

or by

$$\mathring{e}_0^2 - \mathring{e}_0^1 \simeq ({}_1m_0^1 - {}_1m_0^2) w_0 + 4({}_4m_1^1 - {}_4m_1^2) w_2 + 5({}_5m_5^1 - {}_5m_5^2) w_{7.5} + \cdots, \tag{10}$$

where ${}_nm_x$ is the central mortality rate at age x. In the age ranges (0–1) and (1–5) deaths are concentrated at the younger ages, and for this reason, weights at ages 0 and 2 have been chosen, rather than at mid-range (0.5 and 3 respectively).

3. Causes of Death

According to the usual mode of mortality by cause, the force of mortality at age x is the sum of the forces of mortality for the various causes. In other words,

$$\mu_x = \sum_i \mu_x^{(i)}, \tag{11}$$

where $\mu_x^{(i)}$ is the force of mortality by cause i. Substituted in (3), formula (11) allows the partition of $\mathring{e}_o^2 - \mathring{e}_o^1$ in (10) by cause, as well as age.

The central mortality rate for cause i, ${}_nm_x^{(i)}$ will usually provide an accurate estimate of

$$\frac{1}{n}\int_0^n \mu_{x+t}^{(i)}\,dt\,,$$

provided n is not too large. Furthermore, central mortality rates by age and cause are readily available for most developed populations. The following formula can therefore be used to analyse change in expectation of life at birth according to mortality trends by age and cause:

$$\begin{aligned} \mathring{e}_0^2 - \mathring{e}_0^1 \simeq\; & \sum_i ({}_1m_0^{(i)1} - {}_1m_0^{(i)2})w_0 \\ & + 4\sum_i ({}_4m_1^{(i)1} - {}_4m_1^{(i)2})w_2 \\ & + 5\sum_i ({}_5m_5^{(i)1} - {}_5m_5^{(i)2})w_{7.5} \\ & + 5\sum_i ({}_5m_{10}^{(i)1} - {}_5m_{10}^{(i)2})w_{12.5} + \cdots. \end{aligned} \qquad (12)$$

In applying this formula, it is important to ensure that the values on the left-hand side and right-hand side are calculated on a consistent basis. It would be wrong, for example, to use expectations of life based on a graduated mortality table derived from data over the three-year period 1970–2 with ungraduated central mortality rates by cause for 1971. If rates have been graduated, the graduated expectations of life and graduated central mortality rates by cause must be consistent. If a single year of deaths has been used to determine the central mortality rates, the expectations of life must be based on the same single year of deaths. To proceed otherwise would be to impair the precision of the formula, which is designed to analyse the accurate *difference* between expectations.

Mortalities for six different countries and at different epochs are studied in this chapter. The published life tables for these populations are based on different periods of exposure and have been graduated in different ways. All the central mortality rates by cause used are ungraduated and based on a single year's deaths. To be consistent with these rates, therefore, survivor-

Table 15.1. *Expectation of life at birth in selected countries, 1970 and 1980*

Country	Males			Females		
	1970	1980	Change	1970	1980	Change
Australia[a]	67.97	71.42	3.45	74.54	78.45	3.91
Belgium[b]	67.87	69.91	2.04	74.22	76.44	2.22
England and Wales[a]	69.42	71.25	1.83	75.57	77.12	1.55
Hungary	66.42	65.54	−0.88	72.18	72.80	0.62
Japan	69.53	73.41	3.88	74.82	78.99	4.17
Netherlands	70.08	72.49	2.41	75.88	79.17	3.29

[a] 1971 and 1981.
[b] 1970–71 and 1979.

ship proportions and expectations of life have been recalculated directly from the ungraduated central mortality rates by cause.

Table 15.1 shows the expectations of life at birth of males and females in Australia, Belgium, England and Wales, Hungary, Japan, and the Netherlands. In each case, the expectation of life was calculated directly from the ungraduated central mortality rates.

4. Comparisons between Populations

Formula (12) was derived by considering changes in mortality of a given population over time. There is no reason, however, why it should be restricted to this use. It can in fact be used to make comparisons between the mortality and life expectancies of any two populations. The superscripts 1 and 2 in the formulae will then refer to populations 1 and 2.

Where the superscript 1 refers to the male component of the population and the superscript 2 to the female component, formula (12) can be used to analyse the sex differential in life expectation at birth. The sex differentials in mortality and life expectancy, of course, have always been of interest to demographers, biologists, actuaries, and public health authorities, and many reasons have been suggested for these differentials and their size (Lopez and Ruzicka 1983; Waldron 1983).

5. Mortality Change in Selected Countries

Tables 15.2 to 15.7 summarize the effects of mortality change in the 1970s by age and group of causes on the expectation of life at birth for males and females separately in Australia, Belgium, England and Wales, Hungary, Japan, and the Netherlands. The tables also provide the contributions to the sex differential in each of these countries at the end of the decade.

For details, the reader must refer to the relevant table. In the text, therefore, we shall restrict our comments to the more important features which emerge. The ICD Codes of the 22 groups of causes used in this paper are set out in Appendix 15.2.

5.1. Australia (see Table 15.2)

Australian males and females both enjoyed substantial improvements in life expectancy over the decade (3.45 and 3.91 years respectively), which were remarkable after the previous decade of mortality stagnation. For both sexes, the major contributing cause to the improvement was ischaemic heart disease (IHD), and, for males, reduced IHD mortality in the age range 50–69 alone made the single most important contribution to the sizeable improvement in life expectancy (0.7 years). Reduced mortality from all diseases of the circulatory system in fact contributed 54% of the male and 58% of the female increase in expectation of life at birth.

Mortality from neoplasms had a slight negative effect on the trend in male

Table 15.2. *Contributions of the various causes of death to the change in expectation of life at birth, 1971–1981, and to the 1981 sex differential, Australia* (100ths of a year)

Code	Cause	Males 1971–1981 (age group)								Females 1971–1981 (age group)								Males, females 1981 (age group)							
		0	1–4	5–14	15–29	30–49	50–69	70+	All	0	1–4	5–14	15–29	30–49	50–69	70+	All	0	1–4	5–14	15–29	30–49	50–69	70+	All
1	Infective, parasitic	7	3	0	0	1	0	0	12	5	2	1	0	0	0	0	9	−1	0	0	0	0	0	0	1
2	Neoplasm, excl. 3, 4, 5	0	0	1	1	0	−2	−4	−3	0	1	1	2	2	4	−3	8	0	1	1	3	4	18	30	55
3	Neoplasms, digestive	0	0	0	0	0	0	2	1	0	0	0	0	0	4	4	9	0	0	0	0	2	20	15	37
4	Neoplasms, respiratory	0	0	0	0	0	0	−4	−5	0	0	0	0	0	−4	−2	−6	0	0	0	0	5	43	30	77
5	Neoplasms of breast	0	0	0	0	0	0	0	0	0	0	0	0	0	1	0	3	0	0	0	0	−11	−21	−9	−42
6	Endocrine disease	1	0	0	0	1	2	1	7	1	1	0	0	1	4	5	12	0	0	0	0	0	3	2	4
7	Blood, mental, sense	0	1	0	−2	2	3	0	3	0	−1	0	0	4	2	−1	5	1	−1	0	4	1	4	4	13
8	Circulatory, excl. 9, 10	0	0	0	0	4	6	13	22	0	0	0	1	5	10	31	46	0	0	0	0	4	19	16	40
9	Ischaemic heart disease	0	0	0	0	19	70	33	122	0	0	0	0	6	41	55	101	0	0	0	0	25	133	86	244
10	Cerebrovascular disease	0	0	0	0	3	17	21	40	0	0	0	0	8	24	47	80	0	0	0	0	1	12	6	18
11	Respiratory disease	16	2	1	0	3	5	5	32	13	2	0	1	2	4	7	30	0	0	0	0	0	4	8	14
12	Bronchitis, emphysema, asthma	0	0	0	0	0	7	3	11	0	0	0	0	1	−4	−3	−5	0	0	0	0	1	17	38	54
13	Digestive, excl. 14	1	0	0	0	0	1	−1	2	1	0	0	0	2	1	−2	2	0	0	0	0	2	4	3	10
14	Cirrhosis of liver	0	0	0	0	0	−3	0	−5	0	0	0	0	0	−1	0	−1	0	0	0	0	5	10	1	16
15	Genito-urinary system	0	0	0	0	1	3	2	8	0	0	0	0	5	6	0	13	0	0	0	0	−1	−1	3	2
16	Pregnancy, childbirth	0	0	0	0	0	0	0	0	0	0	0	0	2	0	0	2	0	0	0	−1	0	0	0	−1
17	Skin, tissue disease	0	0	0	0	0	0	0	1	0	0	0	0	0	0	0	3	0	0	0	0	0	0	0	−1
18	Congenital anomalies	48	2	0	0	0	0	0	51	46	1	1	0	0	0	0	50	12	0	0	0	0	0	0	13
19	Ill-defined conditions	−13	−1	0	0	0	0	0	−10	−9	0	0	0	0	0	2	−5	5	0	0	0	0	0	0	8
20	External, excl. 21, 22	3	2	2	3	5	3	1	22	3	2	0	2	0	2	5	15	1	3	2	14	14	6	2	42
21	Motor vehicle accidents	0	1	2	14	6	4	1	29	0	1	1	3	1	3	0	10	0	0	3	36	13	4	3	60
22	Suicide, self inflicted	0	0	0	0	2	1	0	3	0	0	0	2	5	2	0	11	0	0	0	12	10	5	1	29
	TOTAL ALL CAUSES	63	10	7	22	47	121	74	342	60	10	6	17	50	102	149	394	20	3	8	70	76	276	240	693

Table 15.3. *Contributions of the various causes of death to the change in expectation of life at birth, 1970/71–1979, and to the 1979 sex differential, Belgium* (100ths of a year)

Code	Cause	Males 1970/71–1979 (age group)								Females 1970/71–1979 (age group)								Males, females 1979 (age group)							
		0	1–4	5–14	15–29	30–49	50–69	70+	All	0	1–4	5–14	15–29	30–49	50–69	70+	All	0	1–4	5–14	15–29	30–49	50–69	70+	All
1	Infective, parasitic	5	2	1	0	2	4	1	13	5	1	0	0	1	1	0	10	0	−1	0	0	0	4	0	5
2	Neoplasm, excl. 3, 4, 5	0	2	2	2	0	−3	−4	0	0	1	1	−1	4	5	−1	10	0	−1	1	0	1	20	27	47
3	Neoplasms, digestive	0	0	0	0	1	2	3	6	0	−1	0	0	2	3	4	9	0	−1	0	0	3	19	13	34
4	Neoplasms, respiratory	0	0	0	0	−1	−5	−9	−16	0	0	0	0	−1	−1	0	−3	0	0	0	0	10	71	41	124
5	Neoplasms of breast	0	0	0	0	0	0	0	0	0	0	0	0	−1	−3	0	−3	0	0	0	0	−15	−24	−7	−48
6	Endocrine disease	0	1	0	0	0	1	3	5	0	−1	0	0	0	8	10	18	0	−1	0	0	0	−1	−5	−7
7	Blood, mental, sense	1	1	−1	0	−1	0	−2	−3	1	1	0	0	0	2	−2	3	0	0	0	3	4	6	2	14
8	Circulatory, excl. 9, 10	1	0	0	−1	−1	3	−2	0	1	−1	0	0	4	11	4	15	0	0	0	0	7	35	26	68
9	Ischaemic heart disease	0	0	0	0	7	22	7	39	0	0	0	0	1	15	13	30	0	0	0	1	20	88	36	145
10	Cerebrovascular disease	0	0	0	0	1	13	18	31	0	−1	0	0	0	14	28	43	0	0	0	0	0	10	8	18
11	Respiratory disease	6	2	0	0	3	9	5	24	4	0	0	0	2	3	10	19	0	−1	0	0	1	18	17	34
12	Bronchitis, emphysema, asthma	1	0	0	0	0	4	0	4	0	0	0	0	−1	0	1	0	0	0	0	0	0	16	19	36
13	Digestive, excl. 14	1	0	0	0	0	1	0	4	1	0	0	0	0	3	0	3	0	0	0	0	1	5	2	8
14	Cirrhosis of liver	0	0	0	0	−3	−1	0	−3	0	0	0	0	−1	0	0	−2	0	0	0	0	3	6	1	13
15	Genito-urinary system	0	0	0	0	0	2	0	3	0	0	0	0	2	1	−3	0	0	0	0	0	0	0	4	5
16	Pregnancy, childbirth	0	0	0	0	0	0	0	0	0	0	0	0	1	0	0	1	0	0	0	0	0	0	0	−1
17	Skin, tissue disease	0	0	0	0	0	0	0	2	0	0	0	0	0	0	0	0	0	0	0	0	0	0	0	−1
18	Congenital anomalies	50	0	1	1	0	0	0	52	33	1	0	0	0	0	0	35	14	1	0	−2	0	0	0	14
19	Ill-defined conditions	2	2	0	2	−1	4	9	18	1	0	0	1	0	3	19	25	3	−1	0	3	5	14	8	31
20	External, excl. 21, 22	5	1	3	1	1	0	3	14	4	−1	0	−2	0	−1	3	2	1	−1	2	8	12	6	−1	26
21	Motor vehicle accidents	0	2	5	1	8	5	0	19	0	0	0	1	3	3	1	9	0	1	0	34	16	7	1	58
22	Suicide, self inflicted	0	0	−1	−5	−7	0	0	−14	0	0	0	−4	−4	−1	0	−9	0	0	1	7	11	5	3	27
	TOTAL ALL CAUSES	72	11	8	4	12	63	31	199	49	0	2	−3	16	66	85	216	18	−5	2	53	81	304	195	649

Table 15.4. *Contributions of the various causes of death to the change in expectation of life at birth, 1971–1981, and to the 1981 sex differential, England and Wales* (100ths of a year)

Code	Cause	Males 1971–1981 (age group)								Females 1971–1981 (age group)								Males, females 1981 (age group)							
		0	1–4	5–14	15–29	30–49	50–69	70+	All	0	1–4	5–14	15–29	30–49	50–69	70+	All	0	1–4	5–14	15–29	30–49	50–69	70+	All
1	Infective, parasitic	4	0	0	0	0	0	0	7	4	1	0	0	0	0	0	7	1	0	0	0	0	0	0	3
2	Neoplasm, excl. 3, 4, 5	0	1	2	1	1	−4	−3	−3	0	0	1	0	3	−2	−4	−1	0	0	0	2	−3	4	20	23
3	Neoplasms, digestive	0	0	0	0	2	4	2	7	0	0	0	0	0	5	3	9	0	0	0	0	2	22	14	40
4	Neoplasms, respiratory	0	0	0	0	3	10	−2	12	0	0	0	0	1	−6	−4	−9	0	0	0	0	4	48	36	89
5	Neoplasms of breast	0	0	0	0	0	0	0	0	0	0	0	0	0	−2	0	−3	0	0	0	0	−17	−31	−10	−58
6	Endocrine disease	1	0	0	0	0	0	0	2	0	0	0	0	0	1	2	5	0	0	0	0	0	0	0	2
7	Blood, mental, sense	1	0	0	0	0	0	0	0	0	−1	0	0	0	1	−2	−1	0	0	0	3	0	1	2	6
8	Circulatory, excl. 9, 10	0	0	0	1	5	10	11	26	0	0	0	0	6	11	23	40	0	0	0	0	3	14	13	30
9	Ischaemic heart disease	0	0	0	0	8	7	2	17	0	0	0	0	0	0	6	7	0	0	0	1	37	148	74	259
10	Cerebrovascular disease	0	0	0	0	1	11	13	26	0	0	0	0	3	13	25	41	0	0	0	0	0	9	8	17
11	Respiratory disease	14	1	0	0	2	4	−5	19	13	2	0	1	1	3	−12	8	3	1	0	0	0	7	22	34
12	Bronchitis, emphysema, asthma	0	0	0	0	1	15	7	23	0	0	0	0	1	0	1	4	0	0	0	0	0	14	33	50
13	Digestive, excl. 14	2	0	0	0	1	0	0	5	2	0	0	0	1	0	−2	2	0	0	0	0	1	4	2	7
14	Cirrhosis of liver	0	0	0	0	0	0	0	−2	0	0	0	0	−1	−1	0	−3	0	0	0	0	0	0	0	1
15	Genito-urinary system	0	0	0	0	1	3	0	5	0	0	0	0	3	1	−2	3	0	0	0	0	0	1	4	5
16	Pregnancy, childbirth	0	0	0	0	0	0	0	0	0	0	0	0	0	0	0	1	0	0	0	0	0	0	0	−1
17	Skin, tissue disease	0	0	0	0	0	0	0	0	0	0	0	0	0	0	0	1	0	0	0	0	0	0	0	−2
18	Congenital anomalies	33	1	1	0	0	0	0	35	27	1	0	0	0	0	0	30	16	0	0	0	0	0	0	16
19	Ill-defined conditions	−11	0	0	0	0	0	1	−10	−8	0	0	0	0	0	3	−5	4	0	0	0	0	0	0	4
20	External, excl. 21, 22	2	1	1	−6	−1	0	0	−1	2	1	0	−1	0	0	3	4	0	1	2	14	9	3	0	28
21	Motor vehicle accidents	0	2	3	6	4	4	0	17	0	1	2	2	1	2	1	10	0	0	2	14	4	0	0	22
22	Suicide, self inflicted	0	0	0	−2	−1	0	0	−4	0	0	0	0	0	0	0	1	0	0	0	5	6	2	0	13
	TOTAL ALL CAUSES	46	6	8	3	27	64	26	180	41	5	4	4	24	25	45	148	25	2	6	37	48	251	221	589

Table 15.5. *Contributions of the various causes of death to the change in expectation of life at birth, 1970–1980, and to the 1980 sex differential, Hungary* (100ths of a year)

Code	Cause	Males 1970–1980 (age group)								Females 1970–1980 (age group)								Males, females 1980 (age group)							
		0	1–4	5–14	15–29	30–49	50–69	70+	All	0	1–4	5–14	15–29	30–49	50–69	70+	All	0	1–4	5–14	15–29	30–49	50–69	70+	All
1	Infective, parasitic	12	1	0	1	4	7	3	27	10	2	0	0	1	3	2	18	0	0	0	0	2	8	2	13
2	Neoplasm, excl. 3, 4, 5	0	0	0	−1	−4	−8	−3	−17	0	1	0	1	2	−2	−1	0	0	1	1	3	−2	10	14	29
3	Neoplasms, digestive	0	0	0	0	−3	−3	0	−4	0	0	0	0	−1	1	3	4	0	0	0	1	6	27	11	43
4	Neoplasms, respiratory	0	0	0	0	−7	−12	−5	−22	0	0	0	0	−1	−3	0	−4	0	0	0	0	11	38	16	66
5	Neoplasms of breast	0	0	0	0	0	0	0	0	0	0	0	0	−3	−4	−1	−7	0	0	0	0	−12	−16	−4	−32
6	Endocrine disease	0	0	0	0	0	−3	−1	−6	−1	0	0	0	0	−4	−4	−10	0	0	0	0	0	−3	−4	−5
7	Blood, mental, sense	3	1	1	1	−1	−1	0	5	3	2	0	1	0	1	0	8	3	0	1	1	3	4	0	14
8	Circulatory, excl. 9, 10	0	1	0	1	−3	1	2	3	0	1	0	1	3	4	8	16	0	0	0	0	10	25	14	50
9	Ischaemic heart disease	0	0	0	−2	−22	−26	9	−41	0	0	0	0	−3	−3	18	10	0	0	0	3	45	93	28	167
10	Cerebrovascular disease	0	0	0	0	−8	−22	−8	−38	0	0	0	0	−5	−9	−4	−18	0	0	0	0	8	32	13	53
11	Respiratory disease	16	3	0	0	−1	2	3	22	14	2	0	0	0	1	4	20	5	0	−1	0	5	6	2	19
12	Bronchitis, emphysema, asthma	0	0	0	0	−3	−14	−16	−33	0	0	0	0	−2	−5	−10	−16	0	0	0	0	4	21	21	45
13	Digestive, excl. 14	0	1	1	−1	−3	−2	−1	−6	0	0	0	0	0	0	0	0	1	0	0	2	6	7	2	17
14	Cirrhosis of liver	0	0	0	−1	−11	−12	−1	−27	0	0	0	0	−6	−4	0	−11	0	0	0	1	9	18	3	32
15	Genito-urinary system	0	0	0	0	0	0	2	4	0	0	0	0	7	1	0	8	0	0	0	0	0	1	5	7
16	Pregnancy, childbirth	0	0	0	0	0	0	0	0	0	0	0	1	2	0	0	2	0	0	0	0	0	0	0	−2
17	Skin, tissue disease	1	0	0	0	0	0	0	2	0	0	0	0	0	0	0	2	0	0	0	0	0	0	0	−1
18	Congenital anomalies	67	1	0	−1	0	0	0	67	53	1	−1	0	0	0	0	51	31	0	1	0	0	0	0	32
19	Ill-defined conditions	−1	0	0	0	0	0	0	0	−1	0	0	0	0	0	2	1	0	0	0	0	0	0	0	0
20	External, excl. 21, 22	−2	3	1	3	−5	−6	−1	−6	−2	3	1	0	−2	0	0	−3	0	2	4	15	23	16	2	61
21	Motor vehicle accidents	0	0	1	1	1	0	0	4	0	1	1	−2	0	0	0	0	0	0	2	15	11	6	1	38
22	Suicide, self inflicted	0	0	−1	−3	−11	−3	−2	−20	0	0	0	0	−6	−2	0	−9	0	0	1	19	34	15	3	75
	TOTAL ALL CAUSES	94	12	3	−1	−75	−100	−19	−87	76	12	3	−1	−16	−27	15	63	38	5	9	62	163	309	136	721

Table 15.6. *Contributions of the various causes of death to the change in expectation of life at birth, 1970–1980, and to the 1980 sex differential, Japan* (100ths of a year)

Code	Cause	Males 1970–1980 (age group)								Females 1970–1980 (age group)								Males, females 1980 (age group)							
		0	1–4	5–14	15–29	30–49	50–69	70+	All	0	1–4	5–14	15–29	30–49	50–69	70+	All	0	1–4	5–14	15–29	30–49	50–69	70+	All
1	Infective, parasitic	5	2	0	1	8	14	7	37	4	2	1	1	6	8	7	28	1	0	0	0	1	7	5	14
2	Neoplasm, excl. 3, 4, 5	0	1	0	0	4	11	2	19	0	1	0	0	7	13	4	28	0	1	1	1	−1	2	8	12
3	Neoplasms, digestive	0	0	0	0	−2	−3	−5	−8	0	0	0	0	2	6	−4	5	0	0	0	0	7	55	36	98
4	Neoplasms, respiratory	0	0	0	0	0	−4	−6	−11	0	0	0	0	0	−1	−3	−6	0	0	0	0	1	16	16	34
5	Neoplasms of breast	0	0	0	0	0	0	0	0	0	0	0	0	0	−1	0	−3	0	0	0	0	−5	−6	−1	−13
6	Endocrine disease	0	0	0	0	0	2	1	3	0	0	0	0	0	4	0	5	0	0	0	0	0	0	0	3
7	Blood, mental, sense	1	0	0	0	2	1	0	8	1	1	1	2	3	3	0	10	0	0	0	2	1	1	0	7
8	Circulatory, excl. 9, 10	0	0	0	0	−2	3	5	6	0	0	0	2	6	10	10	26	0	0	0	3	8	17	10	48
9	Ischaemic heart disease	0	0	0	0	2	8	−4	6	0	0	0	0	0	8	2	11	0	0	0	0	4	21	16	41
10	Cerebrovascular disease	0	0	0	0	8	65	50	125	0	0	0	0	3	49	62	115	0	0	0	0	11	35	38	85
11	Respiratory disease	11	2	1	0	3	7	−4	21	10	2	2	1	4	6	8	33	0	0	0	0	1	10	34	46
12	Bronchitis, emphysema, asthma	0	0	0	0	0	4	4	11	1	0	0	0	0	2	4	9	0	0	0	0	0	3	12	16
13	Digestive, excl. 14	3	0	0	2	4	8	5	24	2	0	0	1	4	8	9	24	0	0	0	0	3	7	5	16
14	Cirrhosis of liver	0	0	0	0	2	−1	1	3	0	0	0	0	0	0	0	3	0	0	0	0	6	14	4	26
15	Genito-urinary system	0	0	0	3	4	2	0	9	0	0	0	3	4	3	−1	8	0	0	0	0	0	1	4	5
16	Pregnancy, childbirth	0	0	0	0	0	0	0	0	0	0	0	2	2	0	0	4	0	0	0	0	0	0	0	−2
17	Skin, tissue disease	0	0	0	0	0	0	0	2	0	0	0	0	0	3	5	9	0	0	0	0	0	0	1	0
18	Congenital anomalies	27	1	0	0	0	0	0	28	20	0	0	0	0	0	0	21	10	0	0	0	0	0	0	10
19	Ill-defined conditions	1	0	0	3	7	8	23	41	1	1	0	0	3	6	47	58	0	0	0	0	0	0	−2	0
20	External, excl. 21, 22	1	5	3	10	9	4	0	34	1	2	1	1	0	0	0	10	1	3	3	7	12	10	4	41
21	Motor vehicle accidents	0	2	3	12	11	9	2	39	0	2	1	2	3	4	2	13	0	1	1	15	7	4	1	31
22	Suicide, self inflicted	0	0	0	−4	−6	−1	0	−12	0	0	0	3	0	0	0	3	0	0	0	10	12	6	0	29
	TOTAL ALL CAUSES	50	14	13	31	57	137	84	385	39	10	9	22	47	130	156	415	13	5	7	38	72	208	203	549

Table 15.7. *Contributions of the various causes of death to the change in expectation of life at birth, 1970–1980, and to the 1980 sex differential, The Netherlands* (100ths of a year)

Code	Cause	Males 1970–1980 (age group)								Females 1970–1980 (age group)								Males, females 1980 (age group)							
		0	1–4	5–14	15–29	30–49	50–69	70+	All	0	1–4	5–14	15–29	30–49	50–69	70+	All	0	1–4	5–14	15–29	30–49	50–69	70+	All
1	Infective, parasitic	3	7	0	0	0	0	0	11	2	7	0	0	0	0	0	12	0	0	0	0	0	0	0	4
2	Neoplasm, excl. 3, 4, 5																								
3	Neoplasms, digestive																								
4	Neoplasms, respiratory																								
5	Neoplasms of breast	1	10	0	3	4	2	−20	−1	0	8	1	2	12	14	10	47	0	1	1	1	−3	87	99	184
6	Endocrine disease	−1	1	0	0	0	0	0	2	0	1	0	0	0	1	7	11	1	0	0	0	0	0	−2	−1
7	Blood, mental, sense	1	8	1	0	1	3	1	15	1	4	1	0	1	1	6	14	1	0	0	0	2	3	3	9
8	Circulatory, excl. 9, 10																								
9	Ischaemic heart disease																								
10	Cerebrovascular disease	1	2	0	0	7	31	22	64	1	2	0	0	1	30	82	117	0	0	0	1	28	156	98	284
11	Respiratory disease																								
12	Bronchitis, emphysema, asthma	0	8	1	0	2	16	4	31	2	6	0	0	2	6	15	32	2	0	0	0	0	15	37	56
13	Digestive, excl. 14																								
14	Cirrhosis of liver	2	1	0	0	−1	1	1	5	0	2	0	0	0	1	2	7	0	0	0	0	3	7	4	13
15	Genito-urinary system	0	0	0	0	0	3	3	9	0	1	0	0	1	3	−1	4	0	0	0	0	0	0	5	7
16	Pregnancy, childbirth	0	0	0	0	0	0	0	0	0	0	0	0	0	0	0	1	0	0	0	0	0	0	0	−1
17	Skin, tissue disease	0	0	0	0	0	0	0	−1	0	0	0	0	0	0	−1	−1	0	0	0	0	0	0	−1	−2
18	Congenital anomalies	32	11	2	1	0	0	0	46	26	12	2	0	0	0	0	41	9	−1	1	0	0	0	0	10
19	Ill-defined conditions	−7	1	0	−2	−4	−3	−1	−17	−5	1	0	−1	−1	−2	2	−9	3	0	0	4	7	15	9	40
20	External, excl. 21, 22																								
21	Motor vehicle accidents																								
22	Suicide, self inflicted	2	40	9	8	6	9	5	80	1	25	2	4	3	3	13	50	0	2	4	30	19	9	3	66
	TOTAL ALL CAUSES	33	89	13	12	19	62	16	244	28	69	6	9	19	62	131	325	17	1	7	39	54	293	256	669

life expectancy and a small positive effect on the trend in female life expectancy. Disturbing in relation to the females are the increasing trends in mortality from both neoplasms of the respiratory system and bronchitis, emphysema, and asthma. Contrary perhaps to popular belief, declining mortality from motor vehicle accidents (and external causes generally) produced an appreciable increase in life expectancy for both sexes.

The right-hand section of Table 15.2 reveals that IHD is responsible for 35% of the sex differential in expectation of life at birth. Cancers explained 18% and external causes a further 19%. Excess male mortality in the age range 50–69 produced 40% of the total differential.

5.2. Belgium (see Table 15.3)

While Belgian expectations of life at birth were at the same general level as those in Australia at the beginning of the decade, the Belgian improvements over the period were only about half those recorded in Australia. Diseases of the circulatory system and IHD in particular made substantial contributions to the improvements, but at a much lower level than in Australia. Once again, deteriorating male mortality from neoplasms reduced life expectancy, while reduced female mortality from these causes resulted in a slight improvement in expectation of life. A disturbing feature is the appreciable decline in life expectancy for both sexes due to increased levels of suicide.

The proportions of the Belgian sex differential ascribable to IHD, neoplasms, and external causes were 22%, 24%, and 17% respectively. Diseases of the circulatory system generally contributed 36% compared with 44% in Australia. Excess male mortality in the age range 50–69 produced 47% of the differential.

5.3. England and Wales (see Table 15.4)

At the beginning of the decade, expectations of life in England and Wales were generally higher than those in Australia and Belgium. Improvements over the period were below those recorded in Belgium (well below for females). As a result, life expectancy at birth in England and Wales at the end of the decade was about the same as in Australia for males, but rather less for females.

Decreased IHD mortality produced only marginal improvements in expectation of life for both sexes, and cerebrovascular-disease mortality improvements had much the same effect as in Belgium. Unlike the situation in Australia and Belgium, mortality from neoplasms moved in such a manner as to increase male and decrease female life expectancy.

Causes of death making major contributions to the relatively small sex differential in expectation of life at birth (5.87 years) included IHD (44%), diseases of the circulatory system generally (52%), external causes (11%), and neoplasms (16%).

5.4. Hungary (see Table 15.5)

Among modern developed populations, Hungary is unusual in that it is one of the few experiencing deteriorating mortality for most major causes of death, and, for males, a reduction in expectation of life at birth. Poland (Okolski 1985) and the USSR are other examples.

In the case of Hungarian males, improvements in mortality from infective and parasitic diseases, respiratory diseases (excluding bronchitis, emphysema, and asthma), and neonatal causes and congenital anomalies added 1.16 years of life. Neoplasms of the respiratory system, bronchitis, emphysema, and asthma on the other hand reduced the expectation of life by 0.55 years, and diseases of the circulatory system (codes 8, 9, 10) led to a decline of 0.76 years. Increases in mortality from cirrhosis of the liver and suicide reduced the expectation of life at birth by 0.47 years.

Suicide is, of course, an area in which coding is often a problem and practices may vary from country to country. If, as is likely to be true, coding practices *within* individual countries have remained substantially unchanged over the decade, the above observations on suicide certainly stand.

Coding practices do, however, change within countries from time to time and may produce spurious results. The 'ill-defined conditions' trend in Australia (Table 15.2), for example, is essentially the result of a change in the coding practice for so-called 'cot deaths'.

Hungarian male mortality improved generally at ages under 30, and led to an increase in life expectancy of 1.09 years. At the older ages, however, mortality deteriorated considerably, reducing life expectancy by 1.95 years. Hungarian females enjoyed a modest increase in expectation of life at birth of 0.62 years between 1970 and 1980 (Table 15.5). We see that the following causes made important contributions to this improvement: infective and parasitic diseases, diseases of the circulatory system (excluding bronchitis, emphysema, and asthma), and neonatal causes and congenital anomalies. Deteriorating mortality from cerebrovascular disease, bronchitis, emphysema, and asthma, and cirrhosis of the liver, however, considerably reduced the overall improvement.

As with the males, mortality at ages under 30 generally improved, producing an increase in life expectancy of 0.90 years. There was also an improvement in overall mortality at ages over 70. Deteriorating mortality at middle ages reduced the overall improvement in life expectancy by 0.28 years.

According to Table 15.1, the 1980 sex differential was 7.26 years, which is large by international standards. Table 15.5 indicates that 37% of the differential (2.7 years) is due to heavier male circulatory-system mortality (codes 8, 9, 10). External-cause mortality explained 24% of the differential, and respiratory diseases (codes 4, 11, and 12) a further 18%. Cirrhosis of the liver contributed only 4% of the differential.

5.5. Japan (see Table 15.6)

Life expectancy for Japanese males improved by 3.8 years between 1970 and 1980 (Table 15.1). Table 15.6 indicates that all causes except neoplasms of the digestive and respiratory system and suicide contributed to this improvement. The major contributor was cerebrovascular disease (32%).

All causes except neoplasms of the respiratory system and breast made positive contributions to the improvement of 4.17 in the Japanese female expectation of life between 1970 and 1980 (Table 15.6). The largest contributor was cerebrovascular disease (28%).

Compared with many other developed populations, the sex differential in Japan is relatively narrow (5.58 years in 1980). Diseases which made the major contributions to the differentials in Japan and the other countries in this study were rather different. In Japan (Table 15.6), the six most important in order of magnitude were neoplasms of the digestive system (18%), cerebrovascular disease (15%), circulatory disease excluding IHD and cerebrovascular disease (9%), respiratory disease excluding bronchitis, emphysema, and asthma (8%), external causes excluding motor vehicle deaths and suicide (7%), and ischaemic heart disease (7%). In Hungary on the other hand (Table 15.5) the major contributors were ischaemic heart disease (23%), suicide (10%), neoplasms of the respiratory system (9%), external causes excluding motor vehicle deaths and suicide (8%), cerebrovascular disease (7%), and circulatory system excluding IHD and cerebrovascular disease (7%).

5.6. The Netherlands (see Table 15.7)

Expectations of life in the Netherlands are among the highest in the world, being slightly higher in the case of females than the Japanese and slightly lower in the case of males (Table 15.1). The improvements in Dutch life expectancies in the 1970s in fact were lower than the very high Japanese and Australian increases, but higher than any of the other changes recorded in this study.

Improved circulatory-system mortality contributed 24% of the male increase in life expectancy and 36% of the female increase. Neoplasms reduced male life expectancy negligibly, but made an important contribution to the female increase. Remarkably, large reductions in mortality due to external causes contributed 33% of the male increase in life expectancy and 15% of the female increase.

The most important causes contributing to the sex differential of 6.68 years are circulatory-system diseases (42%), neoplasms (28%), and external causes (10%). Excess male mortality in the age range 50–69 produced 44% of the differential.

5.7. General Comments

In the past, the age groups which have made the major contributions to improvements in expectation of life at birth are the younger ones (particularly under age 5). Many populations with high expectations of life are now experiencing a 'mortality revolution' in that the age group making the major contribution to increased life expectancy is now the over-50 group. This change is clear in the case of Australia, England and Wales, Japan, and the Netherlands. It has yet to emerge among Hungarian and Belgian males.

A feature common to all the 'European' populations in this study (including Australia) is the negative effect of increased mortality from cirrhosis of the liver. Apart from the Hungarian case, the negative trends observed are small. Whether these are genuine developments or represent changed societal attitudes to cirrhosis of the liver reflecting on cause-of-death coding practices is not known.

6. International Comparisons

We have seen that formula (12) allows the calculation of the contributions of the various age groups and causes of death to the sex differential in expectation of life at birth. The formula also allows the analysis of the differential between any two populations, and in Tables 15.8 and 15.9 the analyses are between English and Welsh and Hungarian males and females. English and Welsh mortality is often used for international comparative purposes, and the other two populations were chosen because of their special features which we mentioned earlier.

In the England and Wales/Hungary comparison, we see that 31% or 1.76 years of life of the large male sex differential and 57% or 2.42 years of life of the female sex differential of 4.28 years can be ascribed to mortality from diseases of the circulatory system. Suicide contributes almost a year of life to the male differential and a third of a year of life to the female differential. External causes as a whole in fact contribute more than 1.5 years to the male differential. Mortality due to congenital anomalies added most of a year of life to the differential for both sexes. Cirrhosis of the liver is also an important contributor for both sexes. More than 60% of the male differential is due to excess Hungarian mortality in the age range 30–69.

The Japanese/Hungarian differentials are large: 7.85 years for males and 6.12 years for females. In the case of males, 2.26 years can be attributed to excess Hungarian IHD mortality (1.17 years to age group 50–69 alone), and 1.29 years to excess Hungarian external-cause mortality. The only cause which, for males, produced an important negative contribution to the differential was neoplasms of the digestive system, a cause for which the Japanese are well known to have high mortality. For females, excess Hungarian mortality from diseases of the circulatory system contributed

Table 15.8. *Contributions of the various causes of death to the differential in expectation of life at birth between England and Wales, 1981, and Hungary, 1980* (100ths of a year)

Code	Cause	Males 1981 (1980) (age group)								Females 1981 (1980) (age group)							
		0	1–4	5–14	15–29	30–49	50–69	70+	All	0	1–4	5–14	15–29	30–49	50–69	70+	All
1	Infective, parasitic	0	0	0	0	2	8	3	13	1	−1	0	0	0	2	2	7
2	Neoplasm, excl. 3, 4, 5	0	1	1	2	5	8	2	20	0	0	0	0	4	2	1	8
3	Neoplasms, digestive	0	0	0	1	6	20	9	36	0	0	0	0	5	17	14	37
4	Neoplasms, respiratory	0	0	0	0	6	−6	−6	−6	0	0	0	0	−1	−10	−3	−14
5	Neoplasms of breast	0	0	0	0	0	0	0	0	0	0	0	0	−5	−12	−2	−18
6	Endocrine disease	1	0	0	0	0	4	1	5	1	0	0	0	0	11	7	17
7	Blood, mental, sense	5	0	2	1	3	1	−4	10	3	0	0	0	0	−2	−7	−2
8	Circulatory, excl. 9, 10	0	0	0	2	18	40	42	100	0	0	0	2	14	39	82	136
9	Ischaemic heart disease	0	0	0	3	18	−21	−18	−20	0	0	0	0	5	0	−3	2
10	Cerebrovascular disease	0	0	0	1	13	48	35	96	0	0	0	0	6	36	60	104
11	Respiratory disease	6	1	0	0	5	−4	−25	−18	4	1	1	1	0	−7	−40	−41
12	Bronchitis, emphysema, asthma	0	0	0	0	3	9	7	20	0	0	0	0	2	1	16	17
13	Digestive, excl. 14	1	0	0	2	8	9	3	22	1	0	0	0	3	7	5	16
14	Cirrhosis of liver	0	0	0	1	15	23	3	43	0	0	0	0	7	9	2	20
15	Genito-urinary system	0	0	0	1	2	4	3	11	0	0	0	1	3	5	2	10
16	Pregnancy, childbirth	0	0	0	0	0	0	0	0	0	0	0	0	0	0	0	1
17	Skin, tissue disease	0	0	0	0	0	0	0	−1	0	0	0	0	0	0	−3	−4
18	Congenital anomalies	88	1	1	1	0	0	0	91	79	1	1	0	0	0	0	82
19	Ill-defined conditions	−12	0	0	0	0	0	0	−13	−9	0	0	0	0	0	−1	−11
20	External, excl. 21, 22	2	2	1	2	17	15	6	46	3	0	0	0	1	3	13	21
21	Motor vehicle accidents	0	0	1	4	9	6	1	22	0	0	1	1	0	1	0	6
22	Suicide, self inflicted	0	0	2	19	40	21	5	87	0	0	1	6	14	11	5	35
	TOTAL ALL CAUSES	90	4	8	40	171	181	69	565	82	1	3	14	66	112	149	428

Table 15.9. *Contributions of the various causes of death to the differential in expectation of life at birth between Japan and Hungary, 1980* (100ths of a year)

Code	Cause	Males (age group) 0	1–4	5–14	15–29	30–49	50–69	70+	All	Females (age group) 0	1–4	5–14	15–29	30–49	50–69	70+	All
1	Infective, parasitic	0	0	0	0	2	2	0	4	1	0	0	0	0	0	0	1
2	Neoplasm, excl. 3, 4, 5	0	0	1	4	11	25	16	57	0	0	0	1	13	24	12	51
3	Neoplasms, digestive	0	0	0	0	−6	−13	−5	−22	0	0	0	−1	−6	0	6	0
4	Neoplasms, respiratory	0	0	0	0	10	32	9	50	0	0	0	0	1	4	0	5
5	Neoplasms of breast	0	0	0	0	0	0	0	0	0	0	0	0	8	16	7	33
6	Endocrine disease	1	0	0	0	0	3	0	5	1	0	0	0	0	10	6	18
7	Blood, mental, sense	6	0	2	2	4	3	0	16	4	0	0	2	3	1	−3	8
8	Circulatory, excl. 9, 10	−1	0	0	−1	12	42	41	92	−1	0	0	2	11	41	76	127
9	Ischaemic heart disease	0	0	0	3	50	117	56	226	0	0	0	0	11	55	74	141
10	Cerebrovascular disease	0	0	0	0	3	22	7	32	0	0	0	0	5	21	24	51
11	Respiratory disease	11	1	0	0	5	−2	−21	−5	7	1	1	1	0	−2	−17	−7
12	Bronchitis, emphysema, asthma	0	0	0	0	6	25	24	55	0	0	0	0	2	11	20	33
13	Digestive, excl. 14	2	0	0	2	6	8	1	19	1	0	0	0	3	9	3	17
14	Cirrhosis of liver	0	0	0	1	11	13	1	25	0	0	0	0	8	6	0	15
15	Genito-urinary system	0	0	0	1	1	4	4	10	0	0	0	1	2	4	1	7
16	Pregnancy, childbirth	0	0	0	0	0	0	0	0	0	0	0	1	0	0	0	0
17	Skin, tissue disease	0	0	0	0	0	0	0	−1	0	0	0	0	0	0	0	0
18	Congenital anomalies	103	1	1	1	0	0	0	107	89	0	1	2	0	0	0	93
19	Ill-defined conditions	0	0	0	0	0	0	−13	−14	1	0	0	0	0	0	−32	−33
20	External, excl. 21, 22	0	−2	0	10	18	13	7	45	1	−2	0	3	5	6	15	28
21	Motor vehicle accidents	0	−1	2	4	7	3	0	17	0	−1	1	2	0	0	0	5
22	Suicide, self inflicted	0	0	2	10	33	17	4	67	0	0	0	0	11	7	1	21
	TOTAL ALL CAUSES	122	−1	7	41	172	315	131	785	104	−1	5	15	84	213	193	612

3.19 years of the 6.12-year differential. In respect of both sexes, the age group making the largest contribution to the differential was 50–69 years.

Belgian and Dutch mortalities have both been included in this study. In passing, it may be noted that Dutch mortality is sometimes used with a time lag of about ten years to project Belgian mortality (Wattelar 1984). For this reason, the analysis in Table 15.10 of the male and female life expectancy differentials between Belgium in 1979 and the Netherlands in 1970 is of interest. The table suggests that the use of Dutch mortality with a time lag of about nine years for males and slightly less for females is not unreasonable. Even the age and cause patterns appear to be tolerably similar.

7. Detection and Correction of Errors

Approximations (8) and (10) are very accurate and can be used to detect and correct errors in life expectancy calculations (Pollard 1982). We saw in Section 4 that, in addition to facilitating the analysis of changes over time in expectation of life of a given population, formulae (8) and (10) permit comparisons of mortality and expectation of life between different populations. In such a comparison, errors were revealed in the published expectations of life of Hungarian males and females. The error is a consistent feature in all recent life tables, with the expectation of life at all ages exaggerated by 0.5, and is possibly the result of a programming error. The correct formula for $\mathring{e}_x$ is of course

$$\mathring{e}_x = \frac{1}{l_x} \sum_{r=x}^{\infty} L_x . \tag{13}$$

It would appear that l_x has been inadvertently substituted for L_x.

8. Interactions Revisited

In a recent paper, Vaupel and Yashin (1985) have interpreted changes in mortality which are all improvements in an interesting manner—in terms of lives saved. The authors show that, in a population now experiencing the reduced force of mortality μ_x^2, having experienced the heavier mortality μ_x^1 in the past, life table survivors 1_x^2 under the new mortality regime can be partitioned into those who have been saved 0 times as a result of the mortality improvement ($1_0(x)$), those who have been saved once ($1_1(x)$), those who have been saved twice ($1_2(x)$), and so on. In a simple manner, they show that (in our notation)

$$1_r(x) = 1_x^1 \frac{(M_x^1 - M_x^2)^r}{r!} \quad (r = 0, 1, 2, \ldots), \tag{14}$$

and of course

Table 15.10. *Contributions of the various causes of death to the differential in expectation of life at birth between the Netherlands, 1970, and Belgium, 1979* (100ths of a year)

Code	Cause	Males (age group)								Females (age group)							
		0	1–4	5–14	15–29	30–49	50–69	70+	All	0	1–4	5–14	15–29	30–49	50–69	70+	All
1	Infective, parasitic	−1	−7	0	0	0	2	0	−7	0	−6	0	0	0	0	0	−8
2	Neoplasm, excl. 3, 4, 5																
3	Neoplasms, digestive																
4	Neoplasms, respiratory																
5	Neoplasms of breast	−1	−10	0	−4	2	13	17	16	0	−6	−1	−1	−7	−8	−5	−29
6	Endocrine disease	0	−1	0	0	0	3	3	5	−1	−1	0	0	0	3	8	10
7	Blood, mental, sense	0	−9	0	1	3	3	3	−1	0	−5	0	0	0	2	4	2
8	Circulatory, excl. 9, 10																
9	Ischaemic heart disease																
10	Cerebrovascular disease	−1	−1	0	2	−3	−18	−1	−23	−1	0	1	2	1	−9	−29	−35
11	Respiratory disease																
12	Bronchitis, emphysema, asthma	−1	−8	0	0	0	5	2	1	−2	−5	0	0	1	−1	−11	−18
13	Digestive, excl. 14																
14	Cirrhosis of liver	−2	−2	0	1	6	9	3	12	−1	−2	0	0	2	4	−2	5
15	Genito-urinary system	0	0	0	0	0	−1	−2	−4	0	−1	0	0	0	1	0	0
16	Pregnancy, childbirth	0	0	0	0	0	0	0	0	0	0	0	0	0	0	0	−1
17	Skin, tissue disease	0	0	0	0	0	0	0	−1	0	0	0	0	0	0	0	−2
18	Congenital anomalies	−7	−10	−2	−1	0	0	0	−21	−3	−12	−2	0	0	0	0	−20
19	Ill-defined conditions	9	−2	0	−1	3	3	13	26	8	−1	0	0	1	4	21	33
20	External, excl. 21, 22																
21	Motor vehicle accidents																
22	Suicide, self inflicted	0	−37	−8	23	22	9	1	10	0	−21	1	8	10	9	−3	6
	TOTAL ALL CAUSES	−4	−86	−11	17	33	27	39	14	2	−61	0	8	7	2	−15	−56

$$1_x^2 = \sum_{r=0}^{\infty} 1_r(x). \tag{15}$$

For a newly born child, the expected lifetime $\mathring{e}_o^2(r)$ spent in the 'saved r times' category under the improved mortality conditions is obtained by integrating (14) with respect to x over the range 0 to ∞.

We note that

$$(M_x^1 - M_x^2)^j(\mu_x^1 - \mu_x^2)$$

is the derivative of $(M_x^1 - M_x^2)^{j+1}/(j+1)$ and that the derivative of ${}_xp_o^1\mathring{e}_x^1$ is $-l_x^1$. If (7) is integrated by parts using these results, the integral turns out to be the integral of (14) with respect to x, with $r=j+1$. In other words, the jth-order interaction term (7) is equal to the expected lifetime spent in the 'saved $j+1$ times' category when all the mortality changes are improvements.

9. Concluding Remarks

The complete expectation of life at birth is often used as a convenient summary measure of the mortality of a population (Benjamin and Pollard 1980). It does, of course, suffer from the usual disadvantages of single-figure indices (Keyfitz and Golini 1975). At the same time it does have a number of advantages, not least of which is its ease of interpretation. Even the experienced observer has little feeling for the difference between, say, q_x values of 0.00307 and 0.00921. On the other hand, both layman and expert have some feeling for the expectation of life, and differences between such expectations.

Changes in expectations of life of a population over time can be analysed using (12) to determine the contributions of the various ages and causes of death to the change in life expectancy. The technique was applied to males and females over the 1970s in Australia, Belgium, England and Wales, Hungary, Japan, and The Netherlands.

We also demonstrated how the formula can be used to make direct comparisons between different populations, and in particular comparisons between the sexes, i.e. analyse the sex differential in life expectancy. In the case of Hungary, for example, the differential is large, some 7.26 years in 1980. We also found that, compared with other European populations, the proportion of the differential attributable to ischaemic heart disease is relatively low: 23% or 1.67 years of life.

Of some interest, but peripheral to the other analyses in this paper, is the comparison between expectation of life in the Netherlands in 1970 and in Belgium in 1979, since Dutch mortality is sometimes used with a lag of about a decade for the projection of Belgian mortality. The comparison suggests that a time lag of about nine years for males and slightly less for females is appropriate and that with such a lag even the mortality rates by age and cause are tolerably similar.

The chapter concluded with a theoretical analysis connecting the methods underlying the present analysis and some recent work on lives saved by mortality improvements by Vaupel and Yashin.

References

Arriaga, E. (1982), 'Contribution to the Discussion', in WHO/ESCAP, *Mortality in South and East Asia: A Review of Changing Trends and Patterns, Manila* (Geneva: WHO; Bangkok: ESCAP).

Benjamin, B., and Pollard, J.H. (1980), *The Analysis of Mortality and Other Actuarial Statistics* (London: Heinemann).

Keyfitz, N., and Golini, A. (1975), 'Mortality Comparisons: The Male–Female Ratio', *Genus*, 31, 1–34.

Lopez, A.D., and Ruzicka, L.T. (eds). (1983), *Sex Differentials in Mortality*, (Canberra: Australian National University, Printing Service for the Department of Demography).

Okolski, M. (1985), 'The Case of Poland', in J. Vallin and A.D. Lopez (eds.), *Health Policy, Social Policy and Mortality Prospects* (Liège: Ordina Éditions).

Pollard, J.H. (1982), 'The Expectation of Life and Its Relationship to Mortality', *Journal of the Institute of Actuaries*, 109, 225–40.

Vaupel, J.W., and Yashin, A.I. (1985), *Repeated Resuscitation: How Lifesaving Alters Life Tables*, Working Paper no. WP-85-85 (Laxenburg, Austria: International Institute for Applied Systems Analysis).

Waldron, I. (1983), 'The Role of Genetic and Biological Factors in Sex Differences in Mortality', in Lopez and Ruzicka (1983).

Wattelar, C. (1984), Personal communication on the mortality projection techniques used in the updating of regional population projections for Belgium.

Appendix 15.1. The Interaction Terms

The expectation of life at birth may be written as

$$\mathring{e}_x = \int_0^\infty \exp(-\int_0^u \mu_t dt) du.$$

Let us now reduce the force of mortality between ages x and $x + dx$ from μ_x to $\mu_x - \theta(x)$. If the integral for the revised expectation of life (which we shall denote by $\mathring{e}_0(x)$) is considered in three parts $(0, x)$, $(x, x + dx)$ and $(x + dx, \infty)$, we obtain the following formula after some analysis:

$$\mathring{e}_0(x) = \mathring{e}_0 + [\phi(x) dx]\,_x p_0\, \mathring{e}_x. \tag{A1}$$

Applying formula (A1) recursively for subsequent improvements in mortality between y and $y + dy$ $(y > x)$ and between z and $z + dz$ $(z > y > x)$, we obtain

$$\mathring{e}_0(x, y) = \mathring{e}_0 + [\phi(x)dx]_x p_0 \mathring{e}_x + [\phi(y)dy]_y p_0 \mathring{e}_y + [\phi(x)dx][\phi(y)dy]_y p_0 \mathring{e}_y, \quad \text{(A2)}$$

and

$$\begin{aligned}\mathring{e}_0(x, y, x) = \mathring{e}_0 &+ [\phi(x)dx]_x p_0 \mathring{e}_x + [\phi(y)dy]_y p_0 \mathring{e}_y + [\phi(z)dz]_z p_0 \mathring{e}_z \\ &+ [\phi(x)dx][\phi(y)dy]_y p_0 \mathring{e}_y + [\phi(x)dx][\phi(z)dz]_z p_0 \mathring{e}_z \\ &+ [\phi(y)dy][\phi(z)dz]_z p_0 \mathring{e}_z \\ &+ [\phi(x)dx][\phi(y)dy][\phi(z)dz]_z p_0 \mathring{e}_z. \end{aligned} \quad \text{(A3)}$$

We see that the jth -order interaction term involving the mortality improvement $\phi(u)$ over $(u, u+du)$ and improvements at all ages younger than u is the sum of all possible products of the form

$$\phi(x_1)dx_1\phi(x_2)dx_2 \ldots \phi(x_j)dx_j\phi(u)du_u p_0 \mathring{e}_u, \quad \text{(A4)}$$

with $x_l < x_2 < \ldots < x_j < u$.

To see that this term emerges from (10), consider for example

$$\begin{aligned}\frac{1}{3!}(M_u^1 - M_u^2)^3 &= \frac{1}{3!}\left[\int_0^u \phi(t)\,dt\right]^3 \\ &= \lim_{n\to\infty}\left[\frac{1}{n}\sum_{r=0}^{n-1}\phi\left\{\frac{ru}{n}\right\}\right]^3 \Big/ 3! \\ &= \lim_{n\to\infty}\left\{\sum\sum\sum_{r_1<r_2<r_3}\left[\phi\left(\frac{r_1 u}{n}\right)\frac{1}{n}\right]\left[\phi\left(\frac{r_2 u}{n}\right)\frac{1}{n}\right]\left[\phi\left(\frac{r_3 u}{n}\right)\frac{1}{n}\right]\right. \\ &\quad + 1/2 \sum\sum_{r_1 \neq r_1}\left[\phi\left(\frac{r_1 u}{n}\right)\frac{1}{n}\right]^2\left[\phi\left(\frac{r_2 u}{n}\right)\frac{1}{n}\right]. \\ &\quad \left.\sum_{r_1}\left[\phi\left(\frac{r_1 u}{n}\right)\frac{1}{n}\right]^3\right\}. \end{aligned} \quad \text{(A5)}$$

The triple sum involves $\binom{n}{3}$ terms and is therefore of order 1. The double and single sums on the other hand involve respectively $\binom{n}{2}$ and $\binom{n}{1}$ terms and are therefore of order n^{-1} and n^{-2} respectively. It follows that

$$\frac{1}{3!}(M_u^1 - M_u^2)^3 = \lim_{n\to\infty}\left\{\sum\sum\sum_{r_1<r_2<r_3}\left[\phi\left(\frac{r_1 u}{n}\right)\frac{1}{n}\right]\left[\phi\left(\frac{ru}{n}\right)\frac{1}{n}\right]\left[\phi\left(\frac{r_3 u}{n}\right)\frac{1}{n}\right]\right\}. \quad \text{(A6)}$$

The triple sum in (A6) when multiplied by $\phi(u))_u p_0 \mathring{e}_u du$, is of course the third-order interaction term involving mortality improvement $\phi(u)$ over *(u, u+du)* and improvements at all ages younger than u (formula (A4)).

Appendix 15.2

Table 15.A1. *Groups of causes of death used in the text, according to the ICD numbers (Ninth Revision)*

Code	Cause	ICD code (Eighth Revision)
1	Infective and parasitic diseases	000–139
2	Neoplasm, excl. 3, 4, 5	140–239, excl. 150–159, 162, and 174
3	Malignant neoplasm of digestive organs and peritoneum	150–159
4	Malignant neoplasm of trachea, bronchus, and lung	162
5	Malignant neoplasm of breast	174
6	Endocrine, nutritional, and metabolic disease	240–279
7	Disease of blood, blood-forming organs	280–289
	+ mental disorders	290–319
	+ diseases of nervous system and sense organs	320–389
8	Diseases of circulatory system, excl. 9, 10	390–459, excl. 410–414, 430–438
9	Ischaemic heart disease	410–414
10	Cerebrovascular disease	430–438
11	Disease of respiratory system, excl. 12	460–519, excl. 490–493*
12	Bronchitis, emphysema, asthma	490–493*
13	Disease of digestive system, excl. 14	520–579, excl. 571
14	Cirrhosis of liver	571
15	Disease of genito-urinary system	580–629
16	Complications of pregnancy	630–679
17	Disease of skin and subcutaneous tissue	680–709
	+ diseases of musculoskeletal system and connective tissue	710–738
18	Congenital anomalies	740–759
	+ certain causes of perinatal mortality	760–779
19	Symptoms and ill-defined conditions	780–799
20	External causes, excl. 21, 22	800–999, excl. 810–819, 950–959
21	Motor vehicle traffic accidents	810–819
22	Suicide and self-inflicted injury	950–959

* 490–496 in Ninth Revision.

Part III

Approaches for Studying Differentials and Causes

16 The Causes of Death in France, 1925–1978: Reclassification according to the Eighth Revision of the International Classification of Diseases

JACQUES VALLIN AND FRANCE MESLÉ,

Institut National d'Études Démographiques (INED), Paris, France

Each successive revision of the International Classification of Diseases (ICD) has introduced a discontinuity in the cause-of-death statistics which makes it very difficult to assess the changes over a long period of time. This difficulty is aggravated, particularly in the case of France, by ill-defined or unspecified causes whose relative importance is variable over time.

Elsewhere, we proposed a method for reclassifying deaths according to two criteria, aetiology and anatomy, to reconstruct statistical series which were consistent with, but completely different from any that might be drawn directly from the ICD (Meslé and Vallin 1984). Our results were limited to the period 1958–74, which is covered in France by the Seventh and Eighth Revisions of the ICD. This work has now been completed, and we have an aetiological–anatomical reclassification of all deaths in France between 1925 and 1978 (Vallin and Meslé 1988). In proceeding with this study, we undertook an intermediate operation, which proved considerably harder than what we had anticipated and which has produced findings regarding the International Classification that are of interest independently of our aetiological–anatomical classification. This intermediate stage was the reclassification of all deaths in France between 1925 and 1978 according to the Eighth Revision of the ICD. As a result we obtained a complete and consistent set of data for the whole period, which has allowed us to carry out the dual classification (aetiology–anatomy) we intended it for, but also provides a broad range of possibilities for other analyses.

This chapter describes the difficulties we encountered in the course of this intermediate stage and procedures we adopted for dealing with them. In Section 1 we describe the problems in establishing consistency between successive revisions of the ICD; in Section 2 we outline the reconstruction of the complete statistical series based on the Eighth Revision of the ICD; and in Section 3 we discuss the treatment of ill-defined or unspecified causes of death.

1. Problems Posed by the Successive Revisions of the ICD

The two synthetic tables published by the Institut National de la Statistique et des Études Économiques (INSEE) illustrate the extent of the problems posed by the successive revisions of the ICD. The first table classifies the deaths in 1950–67 by sex and age group into 100 groups of items in the Seventh Revision A list (INSEE 1970). However, although this period is effectively covered by two Revisions (the Sixth, adopted in 1948 and applied in France from 1950 to 1957, and the Seventh, adopted in 1955 and applied in France from 1958 to 1967), the differences between them are so slight that the A list is not affected. The second table covers a much broader period (1906–64) to which apply the first seven revisions of the ICD, those adopted in 1900, 1909, 1920, 1929, 1938, 1948, and 1955 (INSEE 1966). The table is necessarily less detailed, providing only the total number of deaths (all ages) for 30 groups of causes, nine of which are only used for part of the period. Moreover, the periods 1922–4 and 1937–9 are excluded for lack of cause-of-death data. This table illustrates the types of defect affecting the statistical series that are available for studying changes over time in cause of death. We have extended it to cover the period 1965–82 by grouping the corresponding Eighth and Ninth Revision items into the 30 groups used by INSEE. Out of these 30 groups of causes, 15 are designed for well-known infectious diseases and two for ill-defined causes; the other 13 are either for specific causes such as alcoholism, nephritis, suicide, etc. or broad groups covering various forms of neoplasm, heart disease, and accidents. In spite of this, strong statistical discontinuity is often introduced in the cause-of-death data by the adoption of a new ICD revision. Excluding the missing years, eight groups of causes are only marginally affected by the transition from one revision to the next: typhoid and paratyphoid fevers, scarlatina, whooping cough, smallpox, influenza, cirrhosis of the liver, suicide, and accidents. In spite of the peak observed during epidemics, the death rate by whooping cough has decreased steadily since 1906 and disappeared almost completely by the end of the 1960s (see Figure 16.1). No obvious discontinuity is observed at any of the transition years (indicated by a vertical dotted line). However, these eight 'problem-free' groups only represent a small proportion of the total mortality: 7% in 1906 and 12% in 1982 (11.5% for cirrhosis of the liver, suicide, and accidents alone).

All the other groups of causes are affected to a considerable degree by either gaps (for instance, tetanus and leukaemia are not isolated until 1944), statistical discontinuity due to classification changes (for instance, cerebrovascular lesions in 1910, 1930, 1950, and 1968), or perturbations related to changes in the proportion of ill-defined causes (tuberculosis before World War II). In the case of cerebrovascular lesions, the discontinuity observed in 1930 and 1950 appears all the more clearly as the deaths for these two years were classified according to both the old and the new revisions (see Figure 16.2), but it is just as evident in 1910 and 1968.

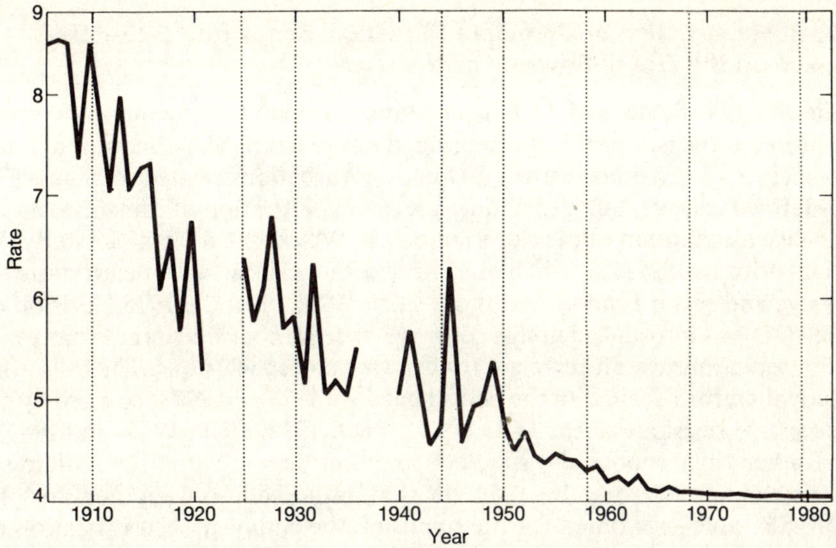

Fig. 16.1. Annual variations in the crude death rate from whooping cough, 1906–1982, according to the INSEE data (1906–1964) extended to 1982

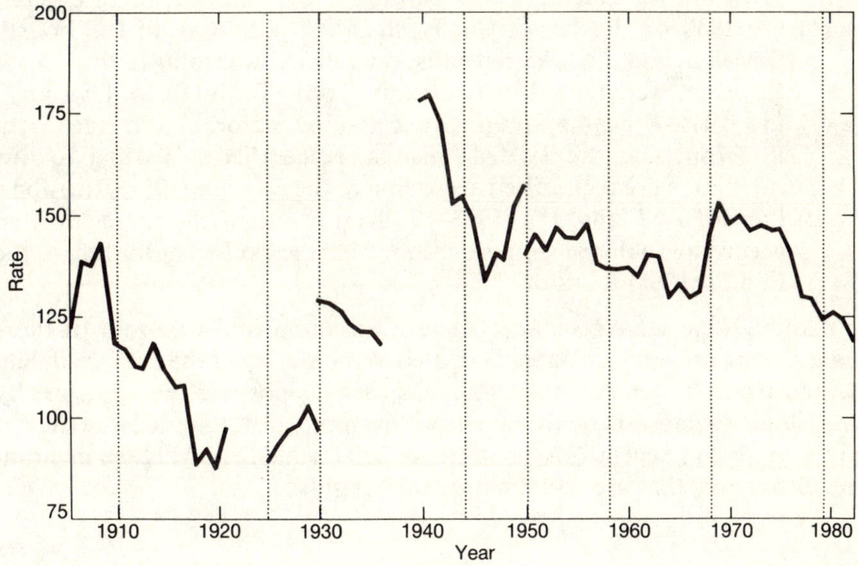

Fig. 16.2. Annual variations in the crude death rate from haemorrhage and other cerebrovascular diseases, 1906–1982, according to the INSEE data (1906–1964) extended to 1982

2. Reconstruction of Complete Statistical Series for 1925–1978 Based on the Eighth Revision of the ICD

The analysis of causes of death over a long period of time requires not only a solution to the two problems mentioned above (statistical discontinuity due to successive revisions of the ICD and perturbations related to changes in ill-defined causes), but also a more precise classification of categories and a sex–age distribution of the causes of death. We excluded from our study the years prior to 1925, for which there are no national cause-of-death statistics by age and sex in France, and those since 1979, when the 1975 Revision of the ICD was introduced in this country, since we considered that four years of observations was insufficient for drawing any conclusions. The following study therefore deals with the period between 1925 and 1978, covered by six successive revisions of the ICD (1920, 1929, 1938, 1948, 1955, and 1965).

Rather than choose the smallest common denominator for building a statistical series from these six revisions—the solution adopted for the INSEE table—we opted for the opposite: the complete reclassification of all deaths, year by year, into the Detailed List categories of the Eighth Revision, which in France covers 1968–78. Our method consists of gradually reclassifying all the deaths for a period covered by one revision into the categories in the following revision. The following are the main steps:

(*a*) the 1925–9 deaths, originally classified according to the Third (1920) Revision, were reclassified according to the Fourth (1929) Revision;
(*b*) the 1930–43 deaths, originally classified according to the Fourth Revision, and the 1925–9 deaths, reclassified according to the Fourth Revision, were reclassified according to the Fifth (1938) Revision;
(*c*) the 1944–9 deaths, originally classified according to the Fifth Revision, and the 1925–43 deaths, reclassified according to this revision, were reclassified according to the Seventh (1955) Revision;[1]
(*d*) the 1950–67 and the 1925–49 deaths, classified or reclassified according to the Seventh Revision, were reclassified according to the Eighth (1965) Revision.

Although the same general procedure was adopted for all four of these stages, there are some differences related to the quality of the data available. To illustrate the general procedure as clearly as possible, we will start by describing the most difficult transition, from the Fifth Revision (200 items) to the Sixth and Seventh Revisions (over 2000 items). We will then mention the differences that are specific to the other stages.

[1] As the differences between the Sixth and Seventh Revisions are very slight, we first used a simplified procedure for reclassifying with the Seventh Revision all 1950–67 deaths originally classified according to the Sixth Revision. The resulting series of deaths in 1950–67, classified according to the Seventh Revision, was then used for a direct transition to the Fifth Revision.

2.1. Transition from the Fifth to the Seventh Revisions

As we have already mentioned, the Seventh Revision introduced considerably fewer changes than the other revisions since it only led to subdivision of some items of the Sixth Revision and, in particular, to a splitting of three-digit items into four-digit items (Vallin *et al.* 1987*a*). As we only had data based on three-digit items for the Sixth Revision, the most straightforward solution was to reclassify these data directly (apart from a few adjustments)[2] into the four-digit items of the Seventh Revision. All the Sixth Revision three-digit items involved were therefore split up into four-digit items and the deaths in 1950–7[3] redistributed according to the proportions observed between 1958 and 1962 in the corresponding Seventh Revision items. This gave us a continuous series of deaths for 1950–67, classified according to the 2124 items in the Seventh ICD Revision. We then had to consider how to extend this series to the preceding years, classified according to the 200 items in the Fifth Revision.

The transition from the Fifth to the Seventh Revisions, like all four stages mentioned above, involved a threefold operation: firstly, we constructed a dual *correspondence table* linking the items of the two Revisions; secondly, we defined a certain number of *fundamental associations* of items which had the same medical contents at both revisions and ensured a sound statistical continuity; thirdly, we elaborated a *transition table* indicating how to redistribute the deaths so that those originally classified according to one revision could be reclassified in line with the other. In what follows we describe the nature of each of these operations.

2.1.1. The correspondence tables Unlike the rather straightforward transition from the Sixth to the Seventh Revisions, the transition from the Fifth to the Sixth–Seventh Revisions is accompanied by drastic changes, much more complicated than a mere subdivision of items used in earlier revisions. Complex inter-item exchange (exchange of parts of items) operates in both directions and the only way to untangle the situation is to start by building a dual correspondence table. The first table lists, for each Fifth Revision item,[4] all the Seventh Revision items which have in common with it one or more illnesses or causes of death. Conversely, the second table lists, for each Seventh Revision item, all the Fifth Revision items which have in common with it one or more illnesses or causes of death. These two correspondence tables were constructed by systematically comparing the

[2] In particular, taking into account the slight changes observed for a certain number of three-digit items.

[3] A preliminary stage was necessary for 1950 and 1951, when data by age group were only available according to the Sixth Revision Intermediate List (150 items). For these two years, the Detailed List classification only provided all deaths and deaths under age 1.

[4] In fact, at this stage we used not only the 200 items adopted for French death statistics, but also their alphabetical sub-items.

Table 16.1. *Extract of the Fifth–Seventh Revision correspondence table*

Fifth Revision (1938)		Seventh Revision (1955)	
Code	Name	Code	Name
93*a*	Acute myocarditis	431	Acute myocarditis not specified as rheumatic
93*b*	Chronic myocarditis specified as rheumatic	415	Other myocarditis specified as rheumatic
93*c*	Cardiac degeneration and infarction, myocardial sclerosis or chronic myocarditis not specified as rheumatic	420.0	Cardiac arteriosclerosis
		420.1	Heart disease specified as including coronary disease
		422.0	Fatty myocardial degeneration
		422.1	Myocardial degeneration with arteriosclerosis
		422.2	Other myocardial degeneration
		433.0	Heart block
		433.1	Other dysrythmias
		440	Benign essential hypertension with heart disease
		441	Malignant essential hypertension with heart disease
		443	Other or unspecified hypertension with heart disease
93*d*	Myocarditis not specified as acute or chronic	422.2	Other myocardial degeneration
		440	Benign essential hypertension with heart disease
		441	Malignant essential hypertension with heart disease
		443	Other or unspecified hypertension with heart disease

analytical tables and alphabetical indexes for the two ICD revisions. Extracts of the complete tables are presented in Tables 16.1 and 16.2.

The correspondence thus established between a 1955 (Seventh Revision) item and a 1938 (Fifth Revision) item, may, of course, involve only part of its contents. For instance, in Table 16.1, we see that for 1955 only part of the contents of item 420.1 (cardiac aneurysm and cardiac infarction) corresponds to 1938 item 93*c*; the remainder (arteriosclerosis, arteritis, etc. of the coronary arteries, cardiac embolism, myocardial ischaemia, etc.) corresponds to other 1938 items (in this case, item 94*a*, as seen in Table 16.2).

In a second stage, we must therefore use these correspondence tables to define the highest possible number of fundamental associations of items with the same overall medical and statistical contents at each of the two revisions.

2.1.2. The fundamental associations of items These fundamental associations of items were constructed in a stepwise manner. The second corres-

Table 16.2. *Extract of the Seventh–Fifth Revision correspondence table*

Seventh Revision (1955)		Fifth Revision (1938)	
Code	Name	Code	Name
420.0	Cardiac arteriosclerosis	93*c*	Cardiac degeneration and infarction, myocardial sclerosis, or chronic myocarditis not specified as rheumatic
420.1	Heart disease specified as including coronary disease	93*c* 94*a*	As above Coronary artery disease
420.2	Angina pectoris without mention of coronary disease	94*b*	Angina pectoris without mention of coronary artery disease

pondence table (1955–38) gives, for the first item in the 1955 list, the 1938 item(s) in which the different illnesses enumerated in this 1955 item are to be found. We then turn to the first correspondence table (1938–55) to see whether this/these 1938 item(s) also include illnesses found in other 1955 items. If this is not the case, the fundamental association is complete. If it is the case, we go back to the 1955–38 correspondence table to see whether these other 1955 items include, in turn, affections belonging to other 1938 items and so on, until the association is constituted of two sets of items, the total contents of which are exactly the same at both revisions. We then proceed once more down the 1955 list to the first item *not* already involved in this first association and, following the same procedure, we construct a second association. These associations do not, however, take into account all of the inter-item exchange listed in the two correspondence tables. Each case was re-examined in detail and, whenever the contents of the exchange seemed negligible in relation to the items involved, we ignored it so as to simplify the structure of the associations. Indeed, the new definitions adopted in 1948 and 1955 introduced so much more detail that, if we had attempted to take everything into account, we might only have been able to isolate a very small number of associations with the same contents at the 1938 and 1955 revisions. This first stage gave us 244 fundamental associations of items. At this level, no statistical verification was possible since the 1938 subdivisions have never been used in France for cause-of-death statistics. These associations were thus based solely on the medical definitions provided by the two analytical lists and alphabetical indexes. In a second stage, these first-level associations were combined into final associations so that all the first-level associations which isolated subdivisions of a same 1938 item became parts of a same final association. In other words, these final associations, linking three-digit 1938 items and four-digit 1955 items, became comparable with French cause-of-death statistics. The 1938 item subdivisions and the first-level associations were used merely to

Table 16.3. *Extract of the fundamental associations of items with the same total medical contents at both the Fifth and Seventh ICD Revisions*

Seventh Revision (1955)		Deaths 1950	Deaths 1949	Fifth Revision (1938)		
Code	Name			Code	P/T[a]	Name[b](1)
Association no. 2						
003.1	Exudative pleurisy without mention of cause	62	1186	110*b*	P	Other forms of pleurisy and pleurisy NOS[c]
519	Other pleurisy	644	—	110*b*	P	
.0	Without mention of effusion or tuberculosis	(579)				
.1	With effusion, with mention of a bacterial cause other than tuberculosis	(0)				
.2	With effusion, other, except tuberculosis	(65)				
(518)	Purulent pleurisy			(110*b*)	P	(Gangrenous pleurisy)
2.1	*003.1, 519*	*706*	—			*110b*
518	Purulent pleurisy	314	—	110*a*	T	Empyema
				(110*b*)	P	(Gangrenous pleurisy)
2.2	*518*	*314*	—			*110a*
2	**003.1, 518, 519**	**1020**	**1186**			**110**
Association no. 3						
010	Tuberculosis of meninges and central nervous system	2596	2926	14	T	Tuberculosis of meninges and central nervous system
				(87*e*)	P	(Acute hydrocephalus)
3	**010**	**2596**	**2926**			**14**
Association no. 4						
011	Tuberculosis of intestine, peritoneum, and mesentery lymph glands	336	350	15	T	Tuberculosis of intestine and peritoneum
4	**011**	**336**	**350**			**15**

Association no. 87

641	Other infections of the genito-urinary system during pregnancy	3	18	145	P	Other diseases and complications of pregnancy
645.1	Ectopic pregnancy with mention of infection	0	43	142*a*	P	Ectopic pregnancy with mention of infection
				(145)	P	(Septicaemia)
646	Anaemia during pregnancy	1	—	145	P	
647	Malposition of foetus in uterus	3	—	145	P	
648.0	Threatened labour	0	—	145	P	
.1	Hydatidiform mole	0	—	145	P	
.3	Other conditions due to pregnancy	27	—	142*a*	P	
			—	145	P	
				(100*a*)	P	(Varix in pregnancy)
				(100*b*)	P	(Phlegmasia alba dolens)
649	Other conditions complicating pregnancy	6	—	145	P	
(625)	Other diseases of ovary and fallopian tube			(142*a*)	P	(Rupture of tube, 15–49 years)
(640)	Pyelitis and pyelonephritis complicating pregnancy			(145)	P	(Pyelitis during pregnancy)
(645.0)	Ectopic pregnancy without mention of infection			(145)	P	(Pyelitis during pregnancy)
87.1	*641, 645.1, 646, 647, 648.0, .1, .3, 649*	*40*	—			*142a, 145*
645.0	Ectopic pregnancy without mention of infection	25	—	142*b*	P	Ectopic pregnancy with haemorrhage but not infection
				142*c*	P	Other cases of ectopic pregnancy
				(145)	P	(Lithopedion)
(625.)	Other diseases of ovary and fallopian tube			(142*b*)	P	(Rupture of tube)
				(142*c*)	P	(Rupture of tube)
87.2	*645.0*	*25*	—			*142b, c*
87	**641, 645–647, 648.0, .1, .3, 649**	**65**	**61**			**142, 145**

[a] P = the 1938 item is partially included in the 1955 item; T = totally included.
[b] Name of the item or condition belonging to the item which is considered negligible for the association (the latter in parenthesis).
[c] NOS = not otherwise specified.

give precision to the final associations. From now on the expression 'association of items' will refer to the final assoications remaining after this second stage (an extract is presented in Table 16.3).

For a given fundamental association, the comparison of the number of deaths in 1949, classified according to the three-digit 1938 items, and the number of deaths in 1950, classified according to the four-digit 1955 items (the two central columns in Table 16.3), can only give an indication of its validity. No strict statistical verification is possible since, on the one hand, the total number of deaths[5] decreased from 565 561 in 1949 to 530 294 in 1950 and, on the other, the cause structure of mortality may change from year to year. However, a graphic representation of each association showing the annual number of deaths between 1944 and 1967 enabled us to detect any flagrant discontinuity at the transition between 1949 and 1950. We then re-examined the medical contents of the associations which revealed irregularities in the statistical series. Certain inter-item exchanges listed in the correspondence tables but ignored as negligible when we defined the associations were reconsidered when they were seen to be statistically more important than we had assumed. In other cases, a certain symmetry was apparent in the discontinuity observed for two associations. This suggested the existence of a relationship between the two which was independent of changes in definition reflected by the two correspondence tables and due to changes in cause-of-death coding practices introduced when the new ICD revision came into effect in France. In all cases, the associations involved were recombined to provide a better balance, in terms of both medical definition and statistical contents. We thus finally obtained 121 well-balanced associations which can be grouped and characterized as follows:

(*a*) 34 link a single four-digit 1955 item to a single three-digit 1938 item;
(*b*) 73 are the result of a single 1938 item being split up into several 1955 items;
(*c*) 14 reflect complex inter-item exchange between several 1938 and several 1955 items.

Unlike transitions between other revisions, there are no cases of several items in the old revision being merged into a single item in the new revision. In fact, it was natural for most of the 200 items in 1938 to be split up into several items, as they become 2124 four-digit items in 1955. The 34 associations in the first group reflect the unchanged part of the International Classification, while the 73 associations in the second group show its increasing precision. These first two groups therefore constitute the part of the ICD which has not been deeply modified at the detailed-items level, even if its organization into chapters has sometimes changed a great deal. The 14 complex associations in the third group, on the other hand, reflect radical transformations in the ICD.

[5] These numbers do not include babies who died before the registration of their birth (in France, such cases are counted as stillbirths; they are called 'false stillbirths').

Table 16.4. *Distribution of deaths in 1950 according to the type of fundamental association of items*

Associations		Deaths 1950	
Type	No.	No.	%
Simple one-to-one transfer	34	8842	1.7
Splitting up of items	73	71587	13.5
Complex inter-item exchange,	14	449865	84.8
including association no. *13*		*317249*	*59.8*
47		*62945*	*17.5*
49		*43797*	*8.3*
83		*18153*	*3.4*

The distribution of the deaths in 1950 among these three types of fundamental association of items is interesting in itself (see Table 16.4). Out of a total of 530 000 deaths, 450 000 (or 85%) belong to the third group, the complex-associations, whereas less than 10 000 (under 2%) belong to the straightforward first group. Moreover, most of the deaths in the complex associations group belong to only four associations (see Table 16.4), one of which accounts for over half of all deaths in 1950. This reflects the scope of the changes that the ICD underwent following World War II. It also illustrates the particular difficulties we faced in attempting to reconstitute continuous series of deaths classified according to the 1955 ICD. However, like the three series presented as examples in Figure 16.3, none of the statistical series corresponding to our 121 associations shows any drastic discontinuity at the transition between 1949 and 1950 (dotted vertical line). This therefore provides solid grounds for proceeding to the next stage.

2.1.3. The transition table and the reclassification of the 1944–1949 deaths In order to reclassify the 1944–9 deaths according to the Seventh Revision, a dual classification of deaths according to the 1955 and 1938 Detailed Lists was first necessary for a transition period. This required the construction of a *transition table* enabling us to reclassify according to the 1938 ICD the deaths originally classified according to the 1955 ICD. As the 1955 Detailed List (2124 items) is much longer than the 1938 List (200 items, since we cannot use the subdivisions), it was much easier to divide the 1955 items among the corresponding 1938 items than the other way round. To construct the transition table (see extract in Table 16.5), we started with the fundamental associations of items described above and the deaths in 1950, but classified according to the 1938 Abbreviated List.[6] The distributions defined in the transition table are the result of our attempts to

[6] In 1950, INSEE published two tables, one in which the deaths were classified according to the Sixth ICD Detailed List, the other to the Fifth ICD Abbreviated List.

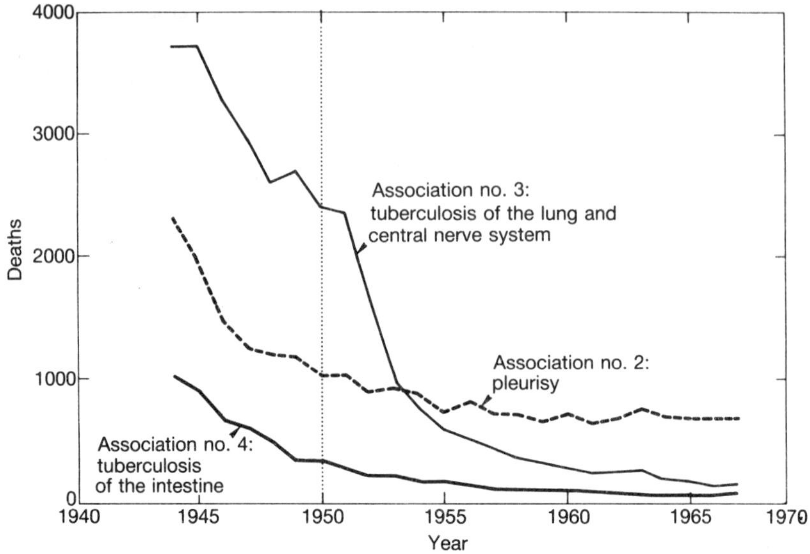

Fig. 16.3. Annual variations in the number of deaths in three of the fundamental associations of items in Table 16.3, 1944–1967

obtain a statistical balance between the two classifications within each 1938 Abbreviated List item. This balance was achieved by comparing the number of deaths published by INSEE for a given 1938 Abbreviated List item with the number of deaths obtained by adding together the deaths in all the items or parts of items in the 1955 Detailed List which correspond to the 1938 Detailed List items that make up this Abbreviated List item. Thus, for each item in the 1938 Detailed List, the transition table shows the part of each of the items in the 1955 Detailed List corresponding to it. The estimation of this part was based on two elements which in some cases proved to be in conflict: (*a*) the 121 associations of Detailed List items, and (*b*) the deaths in 1950 published according to the 1938 Abbreviated List categories. The estimation procedure was more or less laborious, depending on the complexity of inter-item exchange. For the 34 simple associations resulting from cases of one-to-one correspondence between a 1955 and a 1938 item, there is a straightforward transfer of all the contents of the 1955 item into the 1938 one. For instance, all the deaths in 1955 item 040 (typhoid fever) go into 1938 item 1 (same name). The 73 associations where a 1938 item is simply split up into several 1955 items are just as straightforward. For instance, all the deaths in 1955 items 012.0 (tuberculosis of the spine) and 013.0 (tuberculosis of the spine, sequelae) go into 1938 item 16 (tuberculosis of the spine). The other cases require a more complicated procedure. This is illustrated by our 87th association which groups 1955 items 641, 645 to 647, 648.0, .1, .3, and 649, and 1938 items 142 and 145 (see Table 16.3), although the distribution was facilitated by the fact that 1938 items 142 and 145 belong to the same 1938 Abbreviated List item (number 36, other complications of pregnancy, childbirth, and the puerperium). The operation was therefore conducted within the framework of a single 1938 Abbreviated List item. Moreover, as all the deaths in 1955 item 645 come under 1938 item 142 and all the deaths in 1955 items 641, 646, 647, 648.0, .1, and 649 come under 1938 item 145, we only had to redistribute the deaths in 1955 item 648.3. By comparing the number of deaths in 1938 items 142 and 145 with the number of deaths in all the above-mentioned 1955 items added together, we could directly redistribute the deaths in 1955 item 648.3 between these two 1938 items so that the totals balanced (see Table 16.5).

When the associations include 1938 items belonging to different Abbreviated List categories, the procedure is even more difficult. The deaths classified according to the 1955 items must be distributed so as to obtain the best possible balance between the number of deaths in each of the Abbreviated List items involved and the cumulated total in the corresponding 1955 items. Thus, step by step, we define the ratios enabling us to redistribute the contents of the 1955 detailed items among the 1938 detailed items. In most cases, this redistribution of 1950 deaths according to the 1938 Detailed List was made independently of age, as we assumed that the latter had generally little influence. In some cases, however, this assumption proved unacceptable and we were forced to differentiate between certain age groups (three at most). The redistribution of the deaths in each 1955 item among the 1938 items defined by the transition table was then applied to the combined 1950 and 1951 deaths in order to obtain a *dual classification of these deaths* according to the 1938 Detailed List. The dual classification enabled us to calculate for these two years the proportional distribution by sex and age group of the deaths in each 1938 item corresponding to the 1955 items. When applied to the deaths by sex, age group, and 1938 item between 1944 and 1949, this proportional distribution finally enabled us to reclassify the deaths by sex and age group for each of these years according to the 1955 Detailed List items. Added to the 1950–67 data available, these estimates give a complete series of deaths by sex, age group, and 1955 Detailed List item for the entire 1944–67 period.

2.2. Problems Specific to Other Transitions between Different ICD Revisions

The example developed above illustrates the general procedure we adopted for reclassifying deaths originally classified according to one ICD revision so that they were in line with the following revision. Within this general procedure, however, there are variations due to the particular context of each transition. The transition we have just described was peculiar in that it concerned three successive ICD revisions, going directly from the Fifth to the Seventh Revisions. Some of the other transitions required additional

Table 16.5. *Extract from the transition table used for reclassifying deaths classified according to the Seventh ICD Revision into the Fifth ICD Revision*

Association		Seventh Revision Detailed List item				Fifth Revision Detailed List item			
No.	P/T	Code	Deaths 1950	Deaths taken		Deaths		Code	Name of item
				%	No.	1950[a]	1949		
									Abbreviated List item 1: Typhoid and paratyphoid fevers
15	T	040	241	100	241	237	562	1	Typhoid fever
16	T	041	104	100	104	106	252	2	Paratyphoid fevers
TOTAL					345	343			
									Abbreviated List item 7: All other forms of tuberculosis
3	T	010	2596	100	2596	2614	2926	14	Tuberculosis of meninges and central nervous system
4	T	011	336	100	336	313	350	15	Tuberculosis of intestines and peritoneum
5	T	012.0, 013.0	202	100	202	188	210	16	Tuberculosis of spine
6	T	012.1, .2, .3 013.1, .2, .3	132	100	132	197	221	17	Tuberculosis of bones and joints (excluding spine)
7	T	014	13	100	13	13	15	18	Tuberculosis of skin and subcutaneous cellular tissue
8	T	015	27	100	27	19	21	19	Tuberculosis of lymph glands
9	T	016	326	100	326	305	341	20	Tuberculosis of genito-urinary system
10	T	017, 018	37	100	37	44	49	21	Tuberculosis of other sites
11	T	019	701	100	701	677	757	22	Disseminated tuberculosis
TOTAL					4370	4370			

									Abbreviated List item 36: Other complications of pregnancy, childbirth, and the puerperium
89	T	650, 652	46	100	46	58	58	141	Abortion non-specified as septic
87	P	645.0, .1	25	100	25	43	43	142	Ectopic pregnancy
		648.3	27	55	15				
Subtotal					40	43			
88	T	643, 644	31	100	31	26	26	143	Haemorrhage in pregnancy
		648.2	0	100	0				
Subtotal					31	26			
86	T	640, 642.1 to .5	126	100	126	104	103	144	Toxaemia in pregnancy
87	P	641, 646, 647	7	100	7	18	18	145	Other complications and disorders of pregnancy
		648.0, .1	0	100	0				
		648.3	27	45	12				
		649	6	100	6				
Subtotal					25	18			
92	T	670, 672	103	100	103	111	103	146	Haemorrhage in childbirth or the puerperium
		688.0	0	100	0				
Subtotal					103	111			
95	T	685, 686	23	100	23	32	32	148	Toxaemia of the puerperium
91	P	660	0	50	0	201	200	149	Other complications of childbirth
		673, 677	18	100	18				
		678	232	80	186				
Subtotal					204	201			
91	P	660	0	50	0	64	63	150	Other and unspecified disorders of childbirth and the puerperium
		678	232	20	46				
		688.1, .3, 689	15	100	15				
Subtotal					61	64			
TOTAL					659	657			

[a] The 1950 deaths were only classified according to the Fifth Revision Abbreviated List. We redistributed the deaths in each Abbreviated List item among the Fifth Revision Detailed List items in proportion to the 1949 data given in the next column.

work. Thus, the transition from the Third to the Fourth Revisions required cause-of-death statistics for Paris to make up for gaps in the national statistics. Unlike this transition, the transition from the Fourth to the Fifth Revisions presented no particular difficulties, despite the fact that it occurred during the war years, in 1943–4. Finally, the transition from the Seventh to the Eighth Revisions is undoubtedly the one to which our method is best adapted as it involves two highly detailed classifications and very complete statistical data. In return, it requires much more thorough treatment.

2.2.1. Transition from the Third to the Fourth Revisions: The problem of incomplete and insufficiently detailed data Although Detailed Lists of 205 and 200 items respectively exist for the Third and Fourth ICD Revisions, French national statistics were based on the Abbreviated Lists (38 and 43 items) until 1942 (see Vallin and Meslé 1986). The Statistique Générale de la France (SGF) used the Detailed List classification for the first time in 1943, the last year covered by the Fourth Revision. This problem was aggravated by the fact that no national statistics were published between 1937 and 1939. In contrast to these incomplete national statistics, a complete series of annual-mortality data classified according to the Detailed List is available for the city of Paris for the period 1925–31, covered by the Third Revision, and for the period 1932–43, covered by the Fourth Revision (the change from one revision to the other occurred in 1930 at the national level but two years later in Paris). This second source enabled us to fill in the gaps in the national data. We proceeded in three stages:

(*a*) reclassification according to the Fourth Revision Abbreviated List of the 1925–9 national data originally classified according to the Third Revision Abbreviated List;
(*b*) reclassification according to the Fourth Revision Detailed List of the deaths in Paris in 1925–31 originally classified according to the Third Revision Detailed List;
(*c*) reconstitution of continuous series of national death statistics for 1925–43, according to the Fourth Revision Detailed List.

The first two stages are simply specific applications of the general method described earlier: the construction of fundamental associations of items with the same medical contents, the construction of a transition table permitting the dual classification of deaths for a transition period, and, finally, reclassification among the Fourth Revision items of the deaths originally classified according to the Third Revision. These first two stages gave us three sets of data based on the Fourth Revision of the ICD: (*a*) national deaths in 1925–36 and 1940–3 according to the Abbreviated List, (*b*) national deaths in 1943 according to the Detailed List, published by the SGF, and (*c*) deaths in Paris between 1925 and 1943 according to the Detailed List.

A double operation was required in the third stage: first, we decomposed the Abbreviated List items already available for 1925–36 and 1940–2; secondly, we reconstructed the data missing for the period 1937–9. These two operations are peculiar to the procedure applied to the 1925–43 period, and as such require some explanation.

Decomposition. The national classification of deaths according to the Detailed List published for 1943 provides, for that year, the proportion of the Abbreviated List category to which it belongs, represented by the Detailed List item. If these proportions were invariable over the entire period studied, the decomposition would be straightforward. However, as the cause structure of mortality changed over this period, it is likely that the distribution of deaths within certain Abbreviated List categories has also changed. Furthermore, since 1943 was a war year, it is not representative of other periods: certain causes of death, in particular accidents or violence, were exceptionally high that year while others were exceptionally under-represented. The Paris data do not share this drawback as they provide annual proportions for each Detailed List item within the corresponding Abbreviated List category. However, they present another difficulty: the distribution of deaths within each Abbreviated List category is probably not the same in Paris as in the rest of France. The comparison of the Paris and national data for 1943 enabled us to assess the major divergences. When the differences within an Abbreviated List item were slight or involved very small absolute numbers, we applied the distribution observed for Paris to the national data. In other cases we took into consideration both the trend revealed by the Paris series and the difference between the Paris and national data in 1943. The national proportions finally adopted for each Detailed List item were obtained by applying weighting factors to the proportions observed for Paris: within an Abbreviated List category, certain Detailed List items were thus increased and others decreased by a certain percentage so that the total remained the same. We illustrate with the example of Abbreviated List item 40 (suicide). The proportions observed in 1943 for Detailed List items 164 (gas poisoning) and 169 (falling from a high place) are respectively 48.7% and 3.9% in Paris and 10.5% and 1.4% in France (see Table 16.6). Conversely, items 165 (hanging), 166 (drowning), and 171 (other means) respectively account for 23.2%, 10.3%, and 1.3% of this category in Paris and for 52.7%, 14.8%, and 10.8% in France as a whole. These differences are significant, gas distributed by pipeline and high-rise buildings being typically urban, whereas hanging and drowning are more characteristic of rural areas. The proportions provided by the Paris data were therefore weighted so as to obtain roughly the pathological structure found in the 1943 national data.

The weighting factors thus obtained were applied to the distributions by sex and age group observed in Paris in 1925–36 and 1940–2. To alternate the fluctuations due to the low number of deaths in Paris, we used three-year moving averages and seven large age groups (0 years, 1–4 years, 5–24 years,

Table 16.6. *Distribution within Abbreviated List item 40 (suicide) of the corresponding Detailed List items, comparison between the Paris and national data, 1943*

Detailed List item		Deaths				Corrected distribution adopted for France 1925–1942
		Paris		France		
Code	Name	No.	% (p)	No.	% (f)	(f')
163	Poisoning by solid or liquid substances	20	6.5	172	4.0	$f'=p$
164	Gas poisoning	151	48.7	455	10.5	$f'=0.2p$ $x=0.8p$
165	Hanging or strangulation	72	23.2	2285	52.7	$f'=p+0.8x$
166	Submersion	32	10.3	639	14.8	$f'=p+y$ (see 169 for y)
167	Firearms	12	3.9	155	3.6	$f'=p$
168	Cutting and piercing instruments	5	1.6	57	1.3	$f'=p$
169	Jumping from a high place	12	3.9	59	1.4	$f'=0.36p$ $y=0.64p$
170	Being crushed	2	0.6	37	0.9	$f'=p$
171	Other forms of suicide	4	1.3	469	10.8	$f'=p+0.2x$
TOTAL		310	100.0	4328	100.0	

25–49 years, 50–64 years, 65–74 years, and 75 years and above). These mean distributions by broad age group were applied to the national deaths in 1925–36 and 1940–2 classified by sex, age, and Fourth Revision Abbreviated List. For each narrow age group, we used the distribution computed for the broad age group to which it belonged.

Reconstruction. In view of the length of the period we set out to study (1925–78), a gap of three years would not have been too important. But as their reconstruction seemed feasible we attempted to complete the series. Although the detailed results should be considered with caution, in most cases the series obtained appear very consistent. The only national mortality data available for 1937–9 are the sex–age distribution of deaths from all causes combined. For Paris we have available for these three years the distribution of deaths according to the Fourth Revision Detailed List. However, the application of the cause structure of mortality observed in Paris to such limited national data is even more unthinkable than in the previous case, when we were considering distributing deaths within Abbreviated List categories. On the other hand, we now have a Detailed List distribution of national deaths for neighbouring years. Since the years 1940 and after are of little interest because of the war, we attempted an extrapolation based on the years prior to 1937.

Combining the deaths in 1934, 1935, and 1936, we computed the age- and sex-specific ratios D_i/d_i of the national deaths from a cause $i(D_i)$ to the deaths in Paris from the same cause $i(d_i)$. These ratios were then applied to the three-year moving averages of the deaths in Paris (δ_i) centred on 1937, 1938, and 1939. Finally, we applied to the deaths thus computed a weighting factor, the ratio between the national deaths from all causes ($D.$) and the sum of the deaths computed for each cause. This gave us the national deaths (D_i) for each cause *i* and each of the three years 1937, 1938, and 1939. For instance, we have for each sex and age group the equation:

$$D_i(1937) = D.(1937) \times \frac{r_i}{\sum_i r_i},$$

where:

$$\mathrm{r}_i = \frac{D_i(1934\text{–}6)}{d_i(1934\text{–}6)} \times \delta_i(1936\text{–}8).$$

We thus obtained the national death structure by sex, age, and Fourth Revision Detailed List for each of the years between 1925 and 1943.

2.2.2. Transition from the Fourth to the Fifth Revisions: No major problems As a Detailed List classification of national deaths was available for both Revisions, our general procedure presents no particular difficulties (see Vallin and Meslé 1987*a*). It is noteworthy that the Third, Fourth, and Fifth Revisions, although introducing a number of changes, do not radically modify the structure of the ICD: the nomenclature comprises roughly 200 items in each case. Without the important gaps in the national death statistics prior to 1943, our method would have been relatively easy to apply to the whole of this period. Therefore, these two reclassifications, from the Third to the Fourth and from the Fourth to the Fifth Revisions, provide us with a complete, homogeneous series of deaths for 1925–49, classified according to the Fifth Revision of the ICD.

2.2.3. Transition from the Seventh to the Eighth Revisions: 889 fundamental associations of items Contrary to these pre-war revisions, the transition from the Fifth to the Sixth–Seventh Revisions introduced a complete transformation in the structure of the ICD, which grew from 200 to over 2000 items. This is why we chose it, in Section 2.1, to illustrate our procedure. This stage enabled us to reclassify the deaths for the entire 1925–67 period according to the 1955 ICD (Seventh Revision) (see Vallin and Meslé 1987*b*; Vallin *et al.* 1987*a*). Our final stage was the reclassification of all these deaths according to the Eighth ICD Revision, in order to extend our series to include 1968–78. As both the Seventh and the Eighth Revisions comprise several thousand items, our constitution of fundamental associations of items could in this case operate in full. Despite the complexity of certain transfers between Seventh and Eighth Revision items, the information available was so detailed that it became possible to construct associations of items at a much finer level than for the earlier

Table 16.7. *No. of fundamental associations of items constituted between successive ICD Revisions*

Revisions	Scope	No. of associations
Third–Fourth	France, Abbreviated List	17
	Paris, Detailed List	150
Fourth–Fifth	France, Detailed List	137
Fifth–Sixth–Seventh	France, Detailed List, including alphabetical sub-items	244
	200 whole items	121
Seventh–Eighth	France, Detailed List	889

transitions: 889 associations in all, a much more operative number, were obtained (see Table 16.7).

Over half of these associations (491) reflect the complete transfer of a Seventh Revision item to an Eighth Revision one; 155 are the result of a Seventh Revision item being split up into several Eighth Revision ones; and 68 are the result of several Seventh Revision items being merged into one Eighth Revision one. Only 20% of the associations (175 out of 889) are concerned with complex inter-item exchange involving two-way transfer of parts of items. Although these are more numerous than for the other transitions and account for the vast majority (75%) of the deaths observed in the transition years (1967–8), no single complex association covers a majority of items or deaths, as was the case with the earlier transitions. This advantage compensates for the absence of dual classification. For the earlier transitions, the modest dual classification based on the Abbreviated Lists, which was available for the transition years, served as a guideline for verifying our associations of items. In the present case, no such dual classification existed, but the extreme precision of our associations made it possible to construct the transition table enabling us to reclassify according to the Eighth Revision Detailed List the 1925–67 deaths classified or reclassified according to the Seventh Revision Detailed List.

2.3. Complete, Detailed Statistical Series, 1925–1978

These different stages finally provided a complete series of 54 years of cause-of-death statistics (1925 to 1978) based on the Eighth Revision. If we compare these results with the broad groups of causes used by INSEE, which we presented in the first part of this study, we can verify that our reclassification eliminates the strong discontinuities observed at each transition from one ICD revision to the next. Figures 16.4 and 16.5 show for two of these groups of causes (cerebral haemorrhage, and pneumonia and bronchopneumonia) the trends observed in the crude death rates obtained from the INSEE data and from our series reclassified according

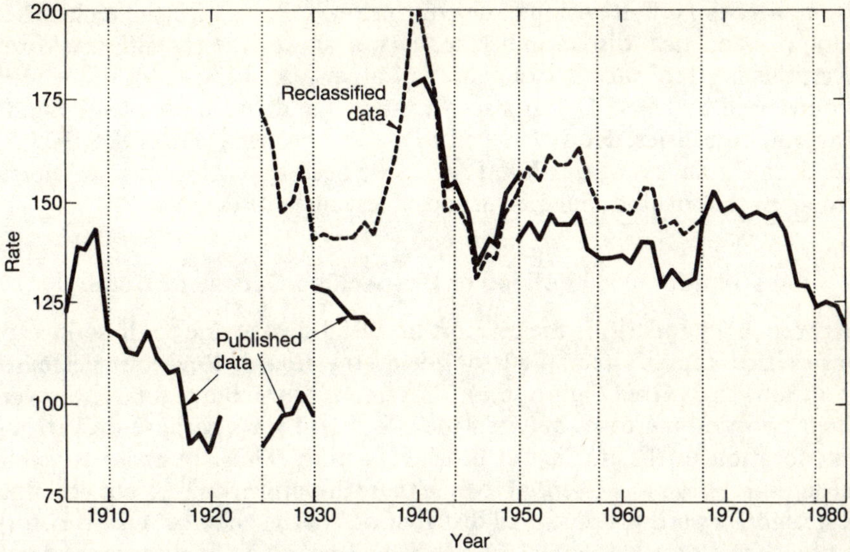

Fig. 16.4. Comparison of the crude death rates by cerebral haemorrhage from the INSEE data and the series reclassified according to the Eighth Revision of the ICD (1925–1978)

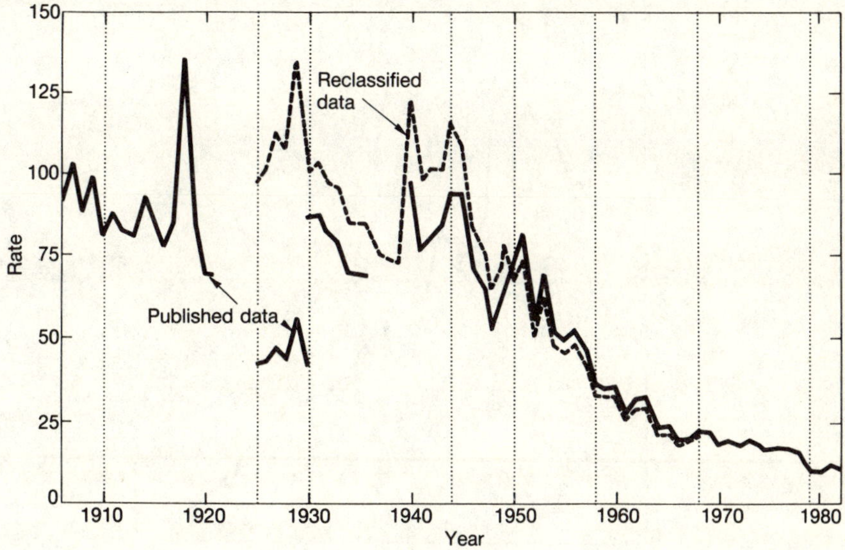

Fig. 16.5. Comparison of the crude death rates by pneumonia and bronchopneumonia from the INSEE data and the series reclassified according to the Eighth Revision of the ICD (1925–1978)

to the Eighth ICD Revision. The conclusion is clear: with the reclassified data, no statistical discontinuity appears at any of the transitions. Moreover, this is established not only in the framework of the very rudimentary groups used by INSEE, but also with the most detailed list of the Eighth Revision categories. However, before we can properly assess the 1925–78 trends in the causes of death in France, we need to resolve another thorny problem, that of ill-defined or unspecified causes of death.

3. Redistribution of Ill-defined or Unspecified Causes of Death

Between 1925 and 1978, the proportion of all deaths due to ill-defined or unspecified causes (items 780 to 799 of the Eighth Revision) decreases from 38% to 6% (see Figure 16.6). This trend is far from regular, however: the proportion increases between the two world wars because of a strong deterioration in the quality of death reporting. Thus, in order to avoid misinterpretation it is essential to take these ill-defined causes into account. A straightforward proportional distribution was considered unsatisfactory because the assumption on which it is based (that all defined causes of death are equally susceptible to inclusion in this category) is questionable. Jacques Breil (1947) then Sully Ledermann (1955) have suggested solutions which take into account the possible correlations between variations in the proportion of ill-defined causes and in the proportions of certain defined

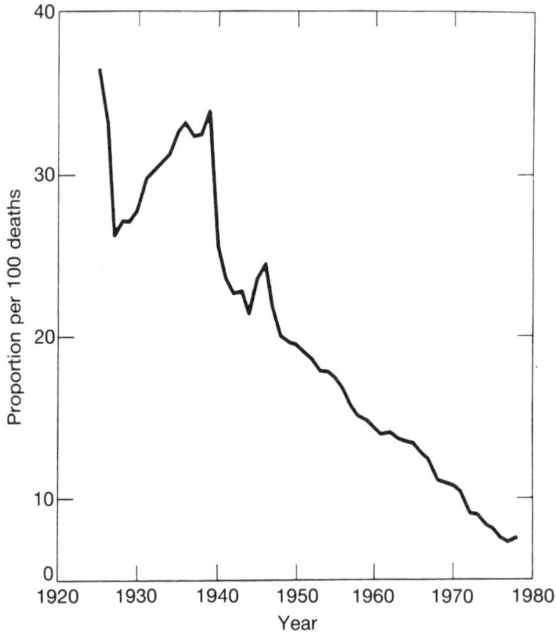

Fig. 16.6. Proportions of all deaths due to ill-defined or unspecified causes, 1925–1978

causes. We adopted an idea conceived by Sully Ledermann and applied it to our data. Ledermann elaborated a method based on the fluctuations in the proportions of all deaths that were observed in 1948–9 across French departments with available data on cause of death. His method could only be applied to our cause-of-death series of 1968–78, for which Detailed List data are available by department. Even in this case, we would be faced with the problem of small numbers of deaths by department for a majority of causes. Moreover, it would be difficult to extrapolate the findings to earlier years, as the proportion of ill-defined causes has changed markedly over time. For the inter-war period during which the problem is most acute, only very rudimentary departmental data are available. However, the annual fluctuations in the proportions of ill-defined causes and certain defined causes reveal a striking symmetry, particularly between the two world wars, when these fluctuations are quite strong. The symmetry between ill-defined causes and certain causes of infant mortality remains very pronounced until the mid-1960s. This prompted us to adapt Ledermann's method by applying it to annual variations (instead of geographical ones) in the proportions of deaths due to the different causes and by paying particular attention to the 1925–43 period[7] (deaths under age 1 are considered separately).

3.1. Redistribution of Deaths due to Ill-defined or Unspecified Causes, 1925–1943

The modified Ledermann's procedure enabled us to use our detailed cause-of-death series even for the earliest period. However, it was necessary to group causes to facilitate the systematic study of correlations. We determined our groups of causes as follows: in a first stage, we defined 117 groups within the Detailed List according to two criteria: (*a*) similar medical contents for each item within the group, and (*b*) similar annual fluctuations in the number of deaths due to each item. Out of these 117 groups, 110 involved defined causes and seven ill-defined or unspecified causes. A systematic study of the correlations observed among these 117 groups in their annual variations between 1925 and 1943 revealed the existence of very strong statistical correlations. This led us to combine these groups into 32 groups of defined and three of ill-defined causes. The proportion of all deaths due to the first of these three groups of ill-defined causes (coma, symptoms involving cardiovascular system, sudden death, unknown or unspecified cause) plummeted between 1925 and 1927, increased rapidly until the end of the 1930s, and thereafter plummeted once more before decreasing very regularly after World War II (see Figure 16.7). In contrast, the second group of ill-defined causes (in reality, a single cause, uraemia) reveals a strong increase in 1925–9, followed by a very slight decrease after 1945. Finally, the third group of ill-defined causes (covering senility and

[7] 1943 is the last year covered by the Fourth Revision of the ICD.

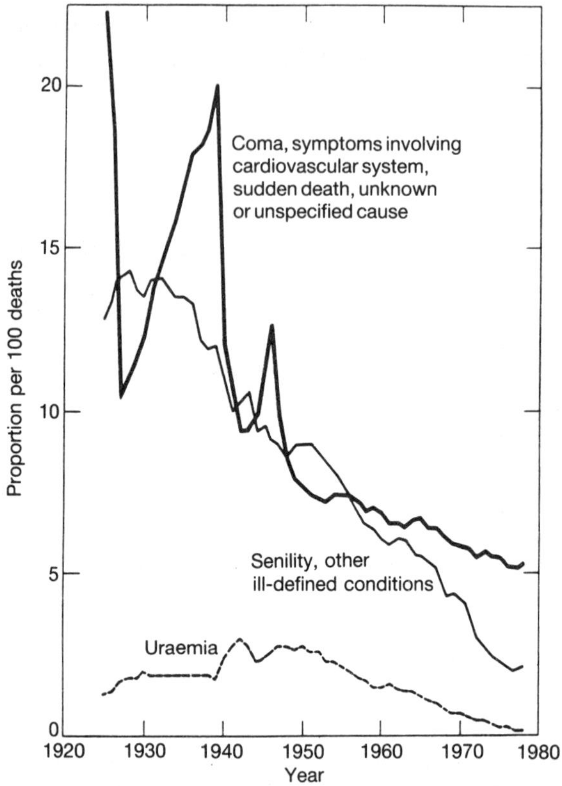

Fig. 16.7. Proportions of all deaths due to three groups of ill-defined or unspecified causes, 1925–1978

other ill-defined conditions) decreases regularly throughout the period. Thus the three groups of ill-defined causes show very different trends during the period we are most concerned with here, 1925–43. We therefore decided to consider each of these three groups separately. To compute the redistribution coefficients we distinguished between sexes and between seven broad age groups which may have been differently affected by changes in the quality of cause-of-death reporting: 0 years, 1–4 years, 5–29 years, 30–44 years, 45–59 years, 60–74 years, and 75 years and above.

The procedure adopted for each group of ill-defined causes was directly inspired by Ledermann's technique. Ledermann showed that the fraction of the deaths in the ill-defined causes category which can be attributed to a given defined cause can be estimated by the slope of the regression line of the proportion of all deaths due to this cause over the proportion of all deaths due to ill-defined causes. Comparing the variations by departments in the proportions of all deaths due to tuberculosis and to ill-defined causes for men aged 45–64 in 1948–9, Ledermann obtained a regression line with a slope of −0.36. He concluded that for this age group, 36% of the male deaths in France due to ill-defined causes could be attributed to tuberculosis. Our method followed this same line of reasoning, but was based on annual fluctuations instead of variations by department. We attempted to apply it to each of our 32 groups of defined causes, for each sex, age group, and group of ill-defined causes. Sometimes we obtained results which were not as straightforward as those in Ledermann's example. This was particularly true for the age groups where the number of deaths was relatively low (1–45 years). In these cases we obtained implausible positive slopes, correlations that were too weak to be considered significant, and sums of the negative slopes attaining values less than −1. In order to apply the method systematically we were therefore forced to adopt certain rules for eliminating any peculiar results, even if this is to the detriment of a perfect mathematical rigour. We studied, for each sex, age group, and each of the three groups of ill-defined causes, the correlation between the annual variations observed in 1925–43 in the proportion of all deaths due to each group of defined causes and in the proportion due to the group of ill-defined causes considered. Taking only the groups of causes for which this correlation was negative and sufficiently strong (correlation coefficient under −0.7), we computed the slopes of the corresponding regression lines. This gave us the proportion of deaths in the group of ill-defined causes considered to be attributed to each group of defined causes. Two types of situation were then observed: (*a*) the sum of the proportions thus estimated was less than one; in this case, the remainder was proportionally redistributed among all the groups of defined causes; (*b*) the sum of the proportions was more than one; in this case, the proportions were first reduced to a total of one.

Finally, the proportions of the deaths due to ill-defined causes thus attributed to each defined cause were computed for broad age groups. As strong variations were observed from one age group to the next, we decided to perform a linear interpolation before applying the proportions to five-year age groups.

3.2. Redistribution of Deaths due to Ill-defined or Unspecified Causes, 1944–1978

The pattern of deaths due to ill-defined causes changes completely after World War II. It becomes very homogeneous, with very little difference between the three groups of ill-defined causes distinguished before. The annual variations are slight and unimportant when compared with the general trend. Thus, the method we adopted for the earlier period is probably ill-suited for the later period. It would probably reveal strong negative correlations between the decrease observed for ill-defined causes and the steady increase in some well-defined causes but without implying that the former alone accounts for the latter. On the other hand, changing our method of redistribution for this second period would have meant creating a statistical discontinuity for the transition years. We consequently decided to base our redistributions after 1943 on a dual assumption: (*a*) immediately after World War II, when the proportion of deaths due to ill-defined causes is still high, there is little reason to assume that the factors to which they can be attributed are very different from the first period; (*b*) in the 1970s, when this proportion is low, there is little reason not to use a straightforward proportional distribution. Starting in 1944, we therefore decreased by 5% per year the regression coefficients estimated for the first period, distributing the remaining deaths from ill-defined causes proportionally among the defined causes. Thus the part of the deaths from ill-defined causes redistributed proportionally increases each year, gradually taking over from the method of redistribution we used for the pre-war period. In all cases, the redistribution coefficients calculated for a given broad age group were applied to each of the component five-year age groups. We subsequently calculated for each group of causes a standardized mortality rate based on the population in 1978. Figure 16.8 illustrates the evolution of the standardized male mortality rate by pulmonary cardiopathy before and after the redistribution of the deaths due to ill-defined causes. Ledermann's method led us to attribute a large part of the first group of ill-defined causes (over 10% after age 60) to this cause before 1944. A simple proportional redistribution of the deaths due to ill-defined causes, also presented in Figure 16.8, gives a much less plausible pattern.

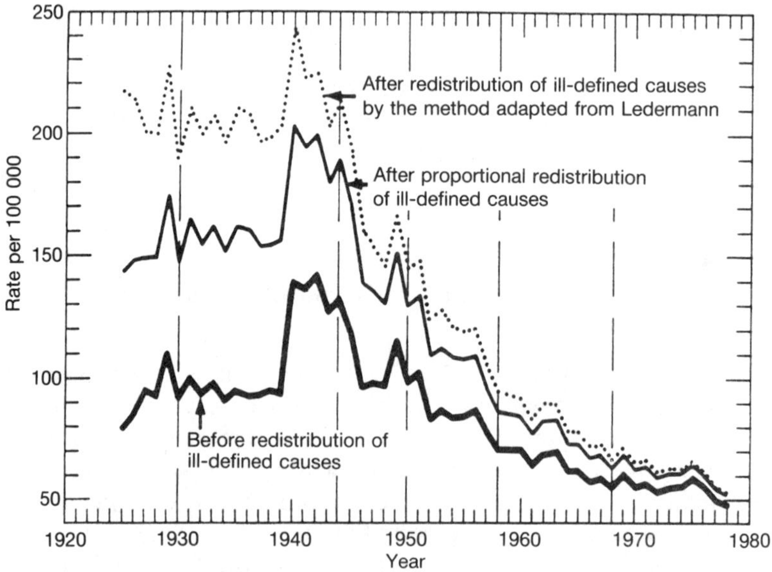

Fig. 16.8. Standardized mortality rate by pulmonary cardiopathy before and after the redistribution of deaths due to ill-defined causes, males, 1925–1978

3.3. The Case of Infant Mortality

The proportion of all deaths due to certain early-childhood diseases follows a peculiar pattern which led us to modify slightly our redistribution method for this age group. In particular, the groups of causes 'congenital anomalies', 'other asphyxia of the new-born', and 'prematurity and debility', which represented only a very small proportion of all deaths under age 1 in pre-war years, sharply increased between 1945 and 1970, then considerably decreased in the 1970s. In fact, it was only after World War II that a real attempt was made to classify the rather complex causes of infant mortality according to specific items. Until then, they tended to be classified in the ill-defined categories. The apparent rise in these causes is therefore essentially related to improved diagnosis. The decrease observed since the beginning of the 1970s is, on the contrary, the sign of real headway made in this field since the end of the 1960s. This led us to examine the correlations between defined and ill-defined causes of infant mortality observed during the period 1925–43, as well as those observed during 1925–64.

As with the general procedure used for the other age groups, the correlations observed in 1925–43 provided a series of coefficients for redistributing the ill-defined deaths among certain defined causes. But, as the remainder was high at this stage, instead of redistributing it proportionally among all the groups of defined causes as we did for the other age groups, we obtained from the correlations observed in 1925–64 a new series

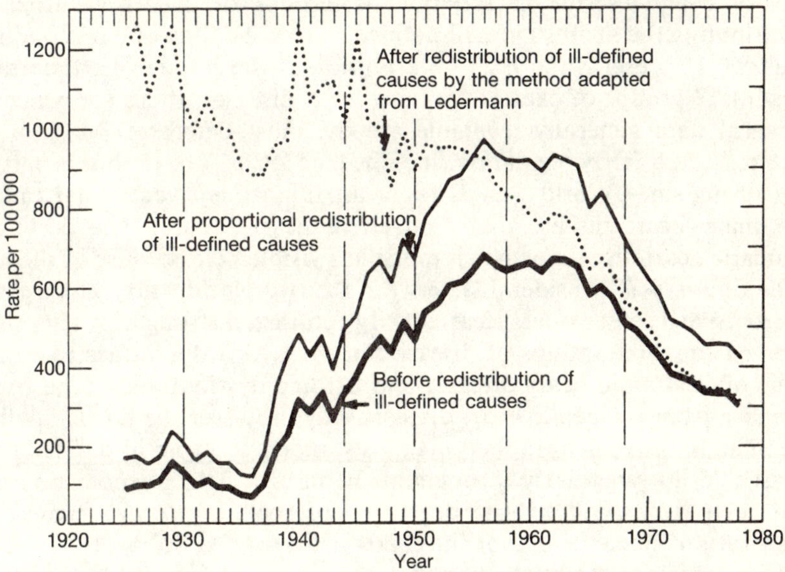

Fig. 16.9. Infant mortality rate by congenital anomalies before and after the redistribution of deaths due to ill-defined causes, males, 1925–1978

of coefficients for redistributing the ill-defined deaths among other groups of defined causes (in particular the three mentioned above, but also the group 'pneumonia, bronchopneumonia'). We then conducted a proportional redistribution of the ill-defined deaths remaining after these two successive operations. In reality, the first set of correlations enabled us above all to redistribute the contents of the first two groups of ill-defined causes, and the second set, those of the third group.

Figure 16.9 illustrates the results we obtained for 'congenital anomalies'. It displays the infant mortality rate from this cause before and after redistribution of the ill-defined causes. Once more, our results seem much more credible than those provided by a simple proportional redistribution.

4. Concluding Remarks

The results obtained by redistributing the deaths due to ill-defined causes among broad groups of defined causes can be easily broken down according to a more detailed classification. Whether we choose the 110 groups of causes constructed in Section 3.1 or even the ICD Detailed List, the principle remains the same. As the groups of causes we formed are relatively homogeneous both in terms of medical contents and the annual variations from which we computed the redistribution coefficients, it seems fairly safe to assume that the deaths from ill-defined causes attributed to a broad group of causes can be proportionally redistributed among the different causes in this broad group. It would be impossible to present all our results here, or even describe in full the 110 groups of causes we used for redistributing the deaths from ill-defined causes. To illustrate our findings, Figure 16.10 presents the 1925–78 evolution of the standardized mortality rates for 12 groups of causes involving heart diseases, those for which the statistical data generally available are the most defective. Most of the diseases show a fairly regular decline since the 1920s. The decline is particularly strong since World War II for heart failure, left ventricular failure, pulmonary heart disease, and congestive heart failure. The decline in rheumatic heart disease, which is much less visible here because of the scale of the figure, was considerable between the two world wars. The trend in hypertensive diseases is less clear-cut. By contrast, a strong mortality rise is observed for three groups of diseases: acute myocardial infarction, other forms of ischaemic heart disease, and cardiac dysrhythmias. The rise in mortality from ischaemic heart diseases may, however, be partly attributable to an improvement in their diagnosis: mortality from ill-defined heart diseases declines regularly throughout the period. This phenomenon may have been intensified by the decrease observed for 'other myocardial degeneration' since the end of the 1950s.

The painstaking reconstruction that we have carried out leads not only to a final series devoid of discontinuities and abrupt jumps but also appears to reveal features of cause-specific mortality trends that are marked by definitional shifts and changes in conventionally available data.

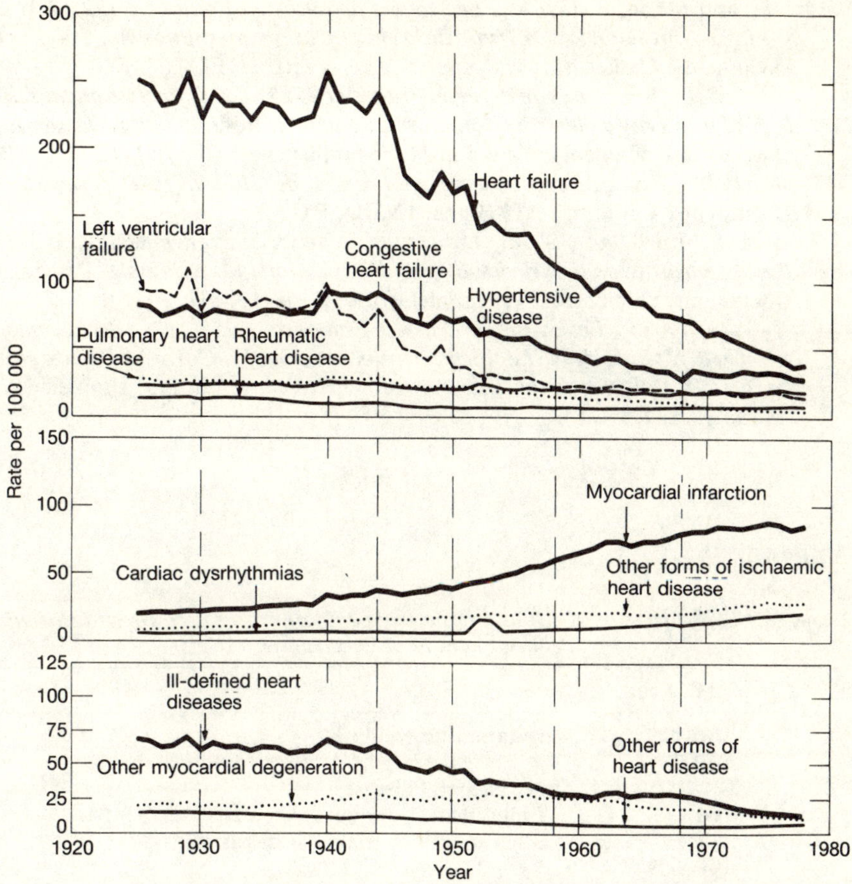

Fig. 16.10. Evolution of standardized mortality rates for 12 groups of causes related to heart diseases, 1925–1978

References

Breil, J. (1947), *Statistique du mouvement de la population, année 1943, 2e partie: Les causes de décès* (Paris: Imprimerie Nationale).

INSEE (1966), *Annuaire statistique de la France, 1966, résumé rétrospectif* (Paris: Imprimerie Nationale).

INSEE (1970), *Statistique des causes de décès, aperçu rétrospectif, années 1950 à 1967* (Paris: Imprimerie Nationale).

Ledermann, S. (1955), La Répartition des décès de cause indéterminée, *Revue de l'Institut International de Statistique*, 23(1–3).

Meslé, F., and Vallin, J. (1984), The problem of studying mortality patterns by cause over a long period of time: an example from France, in Vallin, J., Pollard, J., and Meligman, L. (eds), *Methodologies for the collection and analysis of mortality data* (Liège, IUSSP, Ordina Editions).

Vallin, J., and Meslé, F. (1986), *Les causes de décès en France de 1925 à 1943: Reclassement selon la 4e révision de la classification internationale*, Travaux et Documents, Cahier no. 115, supplement 1 (Paris: INED, PUF).

Vallin, J., and Meslé, F. (1987*a*), *Les causes de décès en France de 1925 à 1949: Reclassement selon la 5e révision de la classification internationale*, Travaux et Documents, Cahier no. 115, supplement 2 (Paris: INED, PUF).

—— —— (1987*b*), *Les causes de décès en France de 1925 à 1978: Reclassement selon la 8e révision de la classification internationale, 2: Résultats du reclassement*, Travaux et Documents, Cahier no. 115, supplement 5 (Paris: INED, PUF).

—— —— (1988), *Les causes de décès en France, de 1925 à 1978*, Travaux et Documents, Cahier no. 115 (Paris: INED, PUF).

—— —— and Nizard, A. (1987*a*), *Les causes de décès en France de 1925 à 1967: Reclassement selon la 7e révision de la classification internationale*, Travaux et Documents, Cahier no. 115, supplement 3 (Paris: INED, PUF).

—— —— —— (1987*b*), *Les causes de décès en France de 1925 à 1978: Reclassement selon la 8e révision de la classification internationale, 1: Correspondance entre les 7e et 8e révisions*, Travaux et Documents, Cahier no. 115, supplement 4 (Paris: INED, PUF).

Appendix 16.1

Table 16.A1. *Broad groups (G) and basic groups (g) of causes used for redistributing deaths due to ill-defined causes*

G	g	ICD code	Name
1	3	010–012, 019.0	**Respiratory tuberculosis**
2			Other bacterial diseases
	1	001–003	Typhoid and paratyphoid fevers
	2	000, 004–009	Other intestinal infectious diseases
	4	013–018, 019.1–019.9	Other forms of tuberculosis
	5	020–039	Other bacterial diseases
3			**Other infectious diseases**
	6	040–079	Viral diseases
	7	090–099	Syphilis and other venereal diseases
	8	080–089 100–136	Other infectious and parasitic diseases
4			**Neoplasms of digestive organs**
	10	150	Malignant neoplasm of oesophagus
	11	151	Malignant neoplasm of stomach
	12	152–154	Malignant neoplasm of intestine and rectum
	13	155–159	Malignant neoplasm of other digestive organs
	23	197.7, .8	Secondary malignant neoplasm of liver
5			**Neoplasms of respiratory organs**
	9	140–149	Malignant neoplasm of oral cavity and pharynx
	14	160, 161	Malignant neoplasm of nasal cavities and larynx
	15	162, 163	Malignant neoplasm of other respiratory organs
6			**Neoplasms of breast and uterus**
	17	174	Malignant neoplasm of breast
	18	180–182	Malignant neoplasm of uterus

Table 16.A1. *Continued*

G	g	ICD code	Name
7			**Other neoplasms**
	16	170–173	Malignant neoplasm of bone, connective tissue, and skin
	19	183–187	Other malignant neoplasms of genital organs
	20	188, 189	Malignant neoplasm of urinary system
	21	190–192	Malignant neoplasm of nervous system
	22	195.0, .1, 199	Malignant neoplasm of abdomen or unspecified site
	24	193, 194, 195.9–197.6, 197.9, 198	Malignant neoplasm of other sites and secondary malignant neoplasms
	25	200–209	Malignant neoplasm of lymphatic and haematopoietic tissue
	26	210–239	Benign neoplasms or of unspecified nature
8	27	250	**Diabetes mellitus**
9	28	268	**Nutritional deficiency**
10			**Alcoholism and cirrhosis of liver**
	30	291, 303	Alcoholism
	78	571	Cirrhosis of liver
11			**Diseases of the nervous system**
	31	290, 292–302, 304–315	Mental disorders
	32	320–324	Inflammatory diseases of the central nervous system
	33	343, 344	Cerebral paralysis
	34	330–342 345–389	Other diseases of the nervous system
12			**Ischaemic heart diseases**
	37	410	Acute myocardial infarction
	38	411–414	Other forms of ischaemic heart disease
13			**Pulmonary heart diseases**
	39	426	Pulmonary heart disease
	40	427.0	Congestive heart failure
	41	427.1	Left ventricular failure
14			**Other heart diseases**
	35	390–398	Rheumatic heart diseases
	44	428.0	Other myocardial degeneration
	45	420–425	Other forms of heart disease
15			**Hypertension, cardiac dysrhythmias, ill-defined heart diseases**
	36	400–404	Hypertensive diseases
	42	427.2, .9	Cardiac dysrhythmias
	43	428.9	Heart failure
	46	429	Ill-defined heart diseases
16			**Cerebrovascular diseases**
	47	430	Meningeal haemorrhage
	48	431	Cerebral haemorrhage
	49	432–435	Cerebral occlusion, thrombosis, and embolism
	50	436	Acute and ill-defined cerebrovascular diseases
	51	437	Generalized ischaemic cerebrovascular diseases
	52	438	Other and ill-defined cerebrovascular diseases

Table 16.A1. *Continued*

G	g	ICD code	Name
17			**Other diseases of the circulatory system**
	53	440	Arteriosclerosis
	54	441–444	Arterial aneurysm, embolism, and thrombosis
	55	445	Gangrene
	56	446–448	Other diseases of arteries and arterioles
	57	450–458	Other diseases of the circulatory system
18			**Influenza and acute infections of respiratory tract**
	58	460–466	Acute infections of respiratory tract
	59	470–474	Influenza
19			**Pneumonia bronchopneumonia**
	60	480–484	Viral or bacterial pneumonia
	61	485	Bronchopneumonia
	62	486	Pneumonia, unspecified
20			**Bronchitis**
	63	490	Bronchitis, unspecified
	64	491	Chronic bronchitis
	65	492–508	Emphysema, asthma, and other diseases of respiratory tract
21			**Other diseases of the respiratory system**
	66	510–511	Pleurisy
	67	515–517	Chronic interstitial pneumonia
	68	512–514, 518–519.0	Spontaneous pneumothorax, abscess of lung, pulmonary congestion, bronchiectasis, and atelectasis
	69	519.1	Acute oedema of lung
	70	519.2	Other diseases of lung
	71	519.9	Other diseases of the respiratory system
22			**Diseases of the digestive system**
	72	520–530	Diseases of mouth and oesophagus
	73	531–537	Diseases of stomach and duodenum
	74	540–543	Appendicitis
	75	550–560	Hernia and occlusion of intestine
	76	561–566	Gastroenteritis, colitis, and functional disorders of intestine
	77	567–569	Peritonitis and other diseases of intestine and peritoneum
	79	570, 572–577	Other disorders of liver, gall-bladder, and pancreas
23			**Diseases of kidney and ureter**
	80	580–592	Nephritis, nephrosis, renal infections, and calculus
	81	593	Other diseases of kidney and ureter
24			**Congenital anomalies and obstetrical lesions**
	87	746	Congenital anomalies of heart
	88	740–745, 747–759	Other congenital anomalies
	89	772	Obstetrical lesions
25			**Hyaline membrane syndrome, foetal distress**
	90	776.1	Hyaline membrane syndrome
	92	776.3	Foetal distress
26	93	776.0, 776.4–.9	**Other asphyxia of new-born**
27	94	777	**Prematurity and debility**

Table 16.A1. *Continued*

G	g	ICD code	Name
28			**Other conditions of new-born**
	91	776.2	Respiratory distress syndrome
	95	760–771, 773–775, 778, 779	Other conditions of foetus and new-born
29			**Other diseases**
	29	240–246 251–267 269–289	Other endocrine, nutritional, metabolic, or blood diseases
	82	594–599	Other diseases of the urinary system
	83	600–629	Diseases of genital organs
	84	630–678	Diseases of pregnancy and the puerperium
	85	680–709	Diseases of the skin
	86	710–738	Diseases of musculo-skeletal system and connective tissue
30			**Accidents**
	96	810–823	Motor vehicle accidents
	97	800–807, 825–845	Other transport accidents
	98	850–879	Accidental poisoning
	99	880–887	Accidental falls
	100	890–899	Accidents caused by fire and flames
	101	900–909	Accidents due to natural factors
	102	910	Drowning
	103	911–913	Accidental mechanical suffocation
	104	914–928	Accidents due to other defined causes
	105	929, 940–946, 948,949	Accidents, other or unspecified
	106	930–936, 947	Complications of surgical or medical procedure
	110	980–989	Injury undetermined whether accidentally or purposely inflicted
31			**Suicide**
	107	953	Suicide by hanging
	108	950–952, 954–959	Suicide by other or unspecified means
32	109	960–978, 990–999	**Homicide**
33	111	780.0, 782	Coma and symptoms involving the circulatory system
	113	784–791, 796.1–.9, 793, 799	Symptoms involving other symptoms and other ill-defined causes of morbidity
	116	795	Sudden death
	117	798	Unspecified cause
34	114	792	Uraemia
35	112	780.1–781, 783, 796.0, 797	Other symptoms involving the nervous and respiratory systems and other ill-defined conditions
	115	794	Senility

17 Mortality Differentials from a Multiple-Cause-of-Death Perspective

CHARLES B. NAM

Center for the Study of Population, Florida State University Tallahassee, Florida, USA

A basic principle of demographic measurement is that attempts should be made to refine each concept and adjust its indicators so that the true or underlying meaning can be revealed and studied. In the area of mortality analysis, this principle has been upheld to a great extent. We make every effort to decompose mortality into its age components, even detailing the beginning stages of life into months, weeks, days, and hours. We differentiate the various measures of mortality by sex, ethnicity, geographical location, and other personal and environmental characteristics, often controlling for these factors in statistical measurement. We analyse mortality patterns to separate out variations due to cohort effects from those that are associated with particular time periods. We develop mortality rates that ideally pin-point populations at risk. In each case, we strive to free mortality effects from confounding factors at the same time as trying to move from simplicity to complexity in identifying the concept and measuring it.

More and more, we have come to realize that the complexity of mortality analysis is reflected in the fact that death is best viewed not just as a momentary vital event in time but also, and more meaningfully, as the end of a life-long process during which the individual has experienced some brief and some prolonged episodes of illness and disability which have combined with the person's genetic and physiological stock to alter the risks of survival. This very process must be complex because numerous factors are being introduced at different points of time and are known to be interacting with other factors in an environment that is also undergoing change.

Emphasis on this process as a prelude to death requires us to examine morbidity in conjunction with mortality and to become more cognizant of morbid conditions contributing to the mortal state. The notion of 'cause of death' provides demographers and other analysts of mortality with a tool for uncovering the prior morbid process. The critical questions we must address are 'How adequate is our conceptualization and measurement of cause of death in mortality analysis?' and 'If inadequate, how can we improve on such conceptualization and measurement?'

The general purpose of this chapter is to answer those questions. The more specific aim is to indicate how results vary when alternative cause-of-death measures are used to portray demographic differentials in mortality. The sections of the chapter which follow include a brief examination of the history of cause-of-death conceptualization and classification, summary of the literature about multiple-cause-of-death approaches, review of alternative measures based on a multiple-cause-of-death framework, illustrations of differential mortality analysis comparing alternative cause-of-death approaches, and discussions of a future research agenda and possible strategies for implementing it.

1. Prevailing Concept of Cause of Death

Attention to categorization of diseases and causes of death has a long history. Graunt, in his London Bills of Mortality, specified certain disease states, and in the eighteenth century the works of Sauvages, Linnaeus, and Cullen on the European continent attempted to classify diseases systematically. In the nineteenth century, William Farr produced a more refined nomenclature and classification system, which was adopted for statistical use by the General Register Office of England and Wales. Farr's scheme, following several revisions at the International Statistical Congresses in the mid-1800s, became the basis for the subsequent International List of Causes of Death (WHO 1977).

The International Statistical Institute (successor to the Congress) in 1893 accepted a further revision of the classification by Bertillon, which was adopted by a number of countries and cities around the world. The original Bertillon formula included three lists: an abridged group of 44 titles, an intermediate group of 99 titles, and a detailed group of 161 titles. All followed Farr's principle of distinguishing between those diseases which are general in nature and those localized to a particular organ or anatomical site. Bertillon was the guiding force at subsequent meetings, which prepared decennial revisions of the classification to take account of new knowledge and development in health and medicine (Fales and Moriyama 1949).

The Sixth Decennial Revision Conference in 1948 went beyond the usual modifications in the classification system to (*a*) agree on rules for selecting a cause when a death occurred, (*b*) include non-fatal conditions in the classification, and (*c*) lay the groundwork for international co-operation in the field of vital and health statistics. The Ninth Revision (ICD-9), which is now in force, was established in 1975 at a conference attended by 46 countries.

The categories offered by the ICD represent choices to be made by an attending physician or other medical or health staff in assigning a cause of death. The principles of assignment are founded on assumptions of a single underlying cause of death and information on a death record that will permit logical selection of that cause.

The notion of a principal or primary cause of death has been the traditional basis of cause-of-death statistics throughout history and continues today. It has had great validity from the public health point of view in that it enables targeting of a particular disease (Moriyama 1956). Moreover, specific infectious and parasitic diseases that took heavy tolls of the population were dominant through so much of history. Further, limiting the choice to a single cause simplifies statistical compilations.

For a long while, there was little incentive for a physician to enter more than one cause of death on a certificate since only one cause was desired. As time went on, it became increasingly apparent that more than one condition may be present and there were no rules for the physician to select the principal cause. Some physicians tended to report the terminal disease or condition responsible for the death. Others identified the main disease condition under treatment (Moriyama 1956).

Beginning in about 1900, rules for deciding on the principal cause when two or more causes were jointly reported were suggested and appropriate manuals devised. Later this approach was discarded in favour of the physician's opinion of the underlying cause of death. The underlying cause follows from the ordering of causes presented on a death certificate. It represents the single condition or injury that initiated the chain of morbid events leading directly to death.

In order to arrive at the underlying cause, medical certification on the death record is divided into two parts. The first part includes several lines for the physician to state either a single condition or a pathologically or aetiologically related sequence of conditions with the most recent manifestation stated first. The second part allows for the reporting of other significant conditions contributing to the death but not directly related to the disease last named in the first part. The underlying cause is thus regarded as the last-mentioned condition in the first part (Erhardt 1958).

Given this approach, it is still the case that determination of the underlying cause is subject to great variability. Some pertinent conditions may not be recognized at all, others may be misdiagnosed, and the sequential chain may be misperceived. Physicians are more apt to report conditions that have been called to their attention as problematic. Comparisons of physician's reports with autopsies and hospital records reveal substantial discrepancies (National Center for Health Statistics 1982). The reports are obviously more accurate when broad disease classifications, rather than detailed ones, are used, and aggregate statistical interpretations may hardly be affected.

A more pressing concern is the validity of the single underlying cause-of-death notion as a representation of the significant morbid circumstances leading to death. In response to this concern, efforts to consider alternative conceptions of cause of death have been extended in recent years.

2. Multiple-Cause-of-Death Approaches

The idea of shifting from a single underlying cause of death to a multiple-cause-of-death approach is hardly new. When the early conferences on disease classification took place a century or more ago, multiple conditions were discussed. The procedures that were later developed for dealing with selecting a single cause of death from joint causes recognized the multiplicity of conditions. What was absent from these considerations was the acceptance of more than a single cause accounting for the fact and timing of death.

Nearly 70 years ago, the US Census Bureau's annual report on mortality included a cross-tabulation of every disease and form of violence covered by title headings in the ICD by each other cause, but no interpretations were made of it. Dublin and Van Buren (1924) analysed the material to show the incidence of secondary causes, the extent to which each disease is associated with every other disease, and the frequency of certain combinations. They pointed to the value of multiple-cause analysis as well as the problems involved in interpreting the data.

Manton and Stallard (1984) have argued that it is critical that the underlying-cause concept be replaced by a concept of human mortality sufficiently general to (*a*) permit the possibility that any death may have more than a single cause, (*b*) allow the cause of death to operate over a period of time, and (*c*) reflect the possibility that diseases may interact in various ways or interact with the characteristics of the host. Two multiple-cause-of-death models which are derived from these stipulations are an extended underlying-cause model, in which associated or contributory causes are seen to modify behaviour of the underlying cause, and a pattern-of-failure model, in which all medical conditions noted at the time of death are assumed to contribute to death but no effort is made to disentangle the causal sequence.

Janssen (1940) adopted the first of these approaches in his analysis of 1936 US death certificates. Dorn and Moriyama (1964) took the position that both approaches have merit and that each introduces another dimension to mortality analysis by cause. They argued for using all the diagnostic information provided by the reporting physician.

One concern about carrying out either of the aforementioned approaches is that there has been no great incentive for a physician to identify and report more causes than are needed to establish the underlying cause. Guralnick (1966) found that in the USA in 1955 only 58% of death certificates showed more than one condition and only 24% more than two conditions. By 1969, with greater attention to multiple-condition reporting, the death records of 84% of white females and 75% of non-white males included more than one cause, and a quarter of the records for whites and one-fifth of the records for non-whites showed four or more conditions! The mean number of conditions for all decedents was 2.7. Thus, it appears that multiple conditions are believed to be present and they are likely to be reported if their importance is emphasized.

The fact that multiple-cause reporting and tabulations are beginning to appear with greater frequency in modernized nations has led many to assume that multiple causes are of significance only for recent time periods, in economically advanced countries, at older ages, and for complications associated with chronic diseases. It may be the case that the generation of data for such populations is being confused with the relevance of multiple conditions for those groups.

One can strongly argue that multiple causes are meaningful for every category of the population. In the USA in 1969, age variations in the mean number of medical conditions reported on the death certificate were small, ranging from 2.22 for infants to 2.85 for persons 85 and over. Time changes in multiple-cause reports seem to be more a function of past reporting deficiencies than of trends in multiple-cause incidence. There is no inherent reason for assuming that people in less-developed regions of the world are less likely to be exposed to morbid conditions than those in more-developed regions. Moreover, although infections and parasitic diseases have a more obvious and sudden effect on survival than most chronic diseases, why should we believe that populations highly subject to infectious and parasitic diseases do not have other morbid conditions that affect not only the incidence of death but the timing of it as well?

3. Multiple-Cause-of-Death Measurement

There are two basic ways, and some variations of each, in which multiple-cause information can be converted into numerator data for mortality analysis. These are, first, a count of total mentions of a cause and, secondly, combinations of reported causes (Guralnick 1966; Krueger 1966; Manton and Stallard 1984).

The *total-mentions approach* is a means of assessing the presence of a condition at the time of death without regard for whether or not the condition was classified as an underlying cause. Because the accuracy of physician determination of the underlying cause is often questioned, this approach identifies conditions that have the potential for being either an underlying or contributing or associated cause of death. On the other hand, one may view it as a count of the incidence of particular morbid states which may have a greater or lesser impact on the incidence or timing of death.

Certain causes of death are more likely to be cited as an underlying cause than others. Conversely, some causes are rarely specified as an underlying cause. Several forms of cancer are most apt to be the only condition reported on the certificate and most probably given as the underlying cause when other conditions are also provided. Diabetes is a condition that is rarely the only cause reported and least likely to be cited as an underlying cause when other conditions are mentioned too. In the case of cancers, the frequency of total mentions exceeds underlying mentions by only about 15% in the USA today. Where it is regarded as underlying, other conditions may hasten the malignancy, lower the resistance to the cancer, or introduce complications in the aetiology of the disease. Where it is not given as underlying, is it conceivable that it is not an effective contributor to the cause of death? In the case of diabetes, in less than one-fifth of the mentions is it cited as an underlying cause. When it is so cited, the probability is high that several other conditions are also present. When it is not cited, it is probable that it is a crucial component of the causal chain or at least a complicating factor.

To be sure, one cannot accept all conditions mentioned as being of equal importance. However, the total-mentions approach calls our attention to conditions that may be underestimated in their impact on the risks of death.

The *cause-combination approach* provides greater depth of comprehension of the process leading to death. All reported conditions are treated as part of a configuration of morbid states that, in combination, determine the incidence and timing of death.

Under this approach, the various conditions may be regarded as having no necessary causal links and thus contributing to the death or they may be examined as relating to each other in a patterned way, possibly in a causal sequence. Whereas the total-mentions approach focuses on causes and the number of decedents who were associated with them, the cause-combination approach focuses on decedents and the number and nature of the cause patterns that were associated with them. Thus, the number of total mentions will exceed the number of deaths, whereas the frequency of cause combinations will equal the number of deaths.

In constructing the cause combinations, one can retain the underlying-cause designation for one of the causes, thereby treating the combination as an elaboration of the underlying cause, or the underlying designation can be ignored, thereby presenting the configuration as a set of associated conditions at the time of death. Both approaches can be found in the limited research literature on this topic (Guralnick 1966; Krueger 1966; Manton and Stallard 1984; Wrigley 1986).

Manton and Stallard (1984) grouped all causes of death into an abbreviated list and assigned binary codes to each. A count of the coexistence of conditions was made by tabulating binary-code patterns. Frequent disease patterns can easily be detected this way, and comparisons of frequencies by characteristics of the decedents can also be made. Going beyond descriptive analysis of the full range of cause combinations is difficult; however, when specified combinations are of analytical interest, one can proceed with chi-square analysis of 2 × 2 tables (Guralnick 1966) or one can work with other statistical association models (Manton and Stallard 1984).

Demographers will usually want to go beyond numerator analysis to assess the importance of multiple-cause indicators, either total mentions or cause combinations. The conventional approaches would be to calculate ratios of total mentions to underlying-cause reports (deaths) for particular cause-of-death categories, and to use populations at risk to compute rates for total mentions and cause combinations. Cause-specific mortality rates can thus be decomposed into cause-combination components.

4. Multiple-Cause Analysis of Differential Mortality

The application of some of these approaches can be illustrated with regard to differential mortality. In a study of deaths in the state of Florida, USA, in 1980, the research was focused on the consequences for cancer mortality differentials by age, sex, and race of using multiple-cause measures instead of underlying-cause designations. The analysis was restricted to the population 45 years and older.

Multiple-cause-of-death data from death certificates in the USA have been coded and entered on to a special computer file since 1968. The file contains all deaths recorded in each calendar year by state of residence. The system for creating these data, called ACME (Automated Classification of Medical Entities), led to a complex file structure but one that permitted comparisons of underlying and multiple-cause classifications for individual death records (Israel *et al.* 1972; Chamblee *et al.* 1979; National Center for Health Statistics 1980; Chamblee and Evans 1982).

The ACME file reproduces coded information on the death certificate in addition to recoding cause-of-death data into two forms—entity axis and record axis. Entity axis coding preserves the information provided originally by the certifying physician. All of the conditions cited on the death record are thus made available for multiple-cause analysis. Record axis coding involves an 'axis translation' of the original cause data to create a set of codes which are free of contradictions among causes reported, devoid of repetitive codes, and the most precise within the ICD-9 classification structure and medical information on the record (Chamblee *et al.* 1979).

The focus of the analysis was on cancer but comparisons with non-cancer causes were desirable. For purposes of this study, causes were grouped into 20 categories covering ten cancer groups and ten non-cancer groups (Table 17.1). Our examples will be confined to ratios of multiple to underlying cancer causes by age, and by sex and race, for specific site groups, and decomposition of certain cancer sites by multiple-cause combinations. These are illustrative of the kinds of information data one can extract from multiple-cause data files.

Since cancer is the major cause of death most likely to be classified as an underlying cause, we would expect minimal effect on cancer rates of using total mentions instead of underlying causes. However, it has been shown that misdiagnoses of cancer and the lethal effects of other causes when the survival time of cancer patients is extended influences age-specific cancer rates. In one study cited, only 10% of cancer patients were found to have

Table 17.1. *Classification scheme of causes of death*

Code	Cause of death	ICD code (Ninth Revision)
1	Cancer of the respiratory system, not secondary	160–169
2	Cancer of the digestive system, not secondary	150–159
3	Cancer of the genito-urinary organs	175–187
4	Cancer of the urinary organs	188–189
5	Cancer of bone, connective tissue, and skin	170–173
6	Female breast cancer	174
7	Lymphatic and haematopoietic cancer	200–203
8	Leukaemia	204–208
9	Cancer of other and unspecified sites	140–149 190–199
10	Benign and unspecified cancers	210–239
11	Chronic ischaemic heart disease	412–414
12	Acute myocardial infarction	410–411
13	Cerebrovascular disease	430–438
14	Other forms of heart disease	390–409 415–429
15	All accidents	800–949
16	Chronic obstructive respiratory disease	490–496
17	Influenza, pneumonia	480–487
18	Cirrhosis of liver	571
19	Diabetes mellitus	250
20	Atherosclerosis	440

died without major complicating factors (Manton and Stallard 1984: 31–7). Thus, even in the case of cancer deaths, a multiple-cause approach is efficient (Wrigley and Nam 1986).

As seen in Table 17.2, among those 45 and older in Florida in 1980, respiratory cancer was cited on the death certificates of 7505 persons. For 6838 of these, cancer of the respiratory system was given as the underlying cause. Hence, using the total-mentions approach identifies 10% more deaths involving that type of cancer. In those cases where cancer of the respiratory system is not diagnosed as the underlying cause, both other forms of cancer and non-cancer causes may be given as the underlying cause.

The ratio of total mentions to underlying causes was less for cancer of the respiratory system than for other cancers. The ratios for other specific cancers ranged from 1.12 for the digestive system to 1.35 for leukaemia. For other and unspecified sites, which for lack of specificity is much less likely to be given as an underlying cause, the ratio rose to 4.37. Thus, the presence of cancer at the time of death was understated by 10 to 337%, dependent on the particular cancer site.

Among the non-cancer causes (not shown in the table), the ratios ranged from 1.1 for acute myocardial infarction and 1.49 for cirrhosis of the liver,

Table 17.2. *Ratios of deaths based on total mentions to those based on underlying cause for nine primary cancer sites, at age 45 years and older, Florida, 1980*

Cancer site	Deaths		Ratio TM/UC
	Based on underlying cause (UC)	Based on total mentions (TM)	
Respiratory	6838	7505	1.10
Digestive	6319	7073	1.12
Genito-urinary	2751	3431	1.25
Urinary	1041	1294	1.24
Bone, tissue, skin	459	554	1.21
Female breast	1739	2002	1.15
Lymphatic and haematopoietic	1091	1334	1.22
Leukaemia	764	1032	1.35
Other and unspecified	2557	11167	4.37

through 1.62 for cerebrovascular disease, 2.25 for chronic obstructive respiratory disease, 2.52 for accidents, and 3.08 for influenza and pneumonia, to 4.17 for diabetes mellitus, 4.71 for other forms of heart disease, and 5.19 for atherosclerosis. Thus, among non-cancer deaths, the extent of understatement of conditions when underlying causes were given varied from 10 to 419%.

These total mentions to underlying cause ratios vary considerably by age. For every major cancer site, the ratio of total-mentions to underlying-cause cases increases with age (Table 17.3). This is partly due to a slightly greater number of reported conditions at the time of death for older than younger adults, but the persistent pattern by age for all cancer sites suggests that there is no consistent tendency for one type of cancer to be reported as underlying when two or more types of cancer are identified as present.

Table 17.3. *Ratios of deaths based on total mentions to those based on underlying cause for nine primary cancer sites, at age 45 years and older, by age, Florida, 1980*

Cancer sites	Age					
	Total	45–54	55–64	65–74	75–84	85+
Respiratory	1.10	1.04	1.08	1.10	1.12	1.20
Digestive	1.12	1.04	1.07	1.09	1.17	1.21
Genito-urinary	1.25	1.05	1.11	1.19	1.31	1.48
Urinary	1.24	1.10	1.15	1.23	1.27	1.40
Bone, tissue, skin	1.21	1.02	1.17	1.17	1.26	1.54
Female breast	1.15	1.05	1.07	1.15	1.24	1.42
Lymphatic and haematopoietic	1.22	1.06	1.12	1.24	1.27	1.44
Leukaemia	1.35	1.00	1.14	1.37	1.45	1.58
Other and unspecified	4.37	4.10	4.03	4.42	4.72	4.35

Ratios of total mentions to underlying causes also varied considerably when sex and race differences were examined; however, these differences were not as systematic as those for age.

The ultimate purpose of our analysis was to find out what impact alternative measures had on differential mortality. Are the magnitudes of the differentials affected? Are reversals of differentials observed? Table 17.4 presents the results for differentials by sex and race. Since all deaths in the population being studied are included in the analysis (that is, we are not dealing with a sample), tests of statistical significance are not necessary. In order to identify large variations in order of magnitude of the differentials when the total-mentions and underlying-cause approaches were being compared, disparities of 10% or greater were highlighted.

For four of the cancer sites among whites, the sex differential (measured as the ratio of male to female death *rates*) is substantially greater when total mentions are used. As an illustration use of the underlying-cause approach shows that the excess of male over female death rates is about 60%. For instance, when the sex differential is examined within the non-white category, we find some cases where there is an increase and some where there is a decrease. For other and unspecified cancer sites, excess male mortality rates among non-whites drops from 125% when an underlying-cause approach is used to 82% when the method of total mentions is applied (Wrigley and Nam 1986). Similarly, race differentials (measured as the ratio of non-white to white death rates) may vary considerably when observed

Table 17.4. *Mortality differentials based on underlying cause and on total mentions for nine primary cancer sites, at age 45 years and older, by sex and race, Florida, 1980*

Cancer site	Ratio of death rates							
	Male/female				Non-white/white			
	White		Non-white		Male		Female	
	UC	TM	UC	TM	UC	TM	UC	TM
Respiratory	3.19	3.22	5.41	5.64[a]	1.00	1.01	0.59	0.57
Digestive	1.50	1.52	1.50	1.44	0.94	0.87	0.94	0.93
Genito-urinary	1.60	2.00[a]	1.66	1.85	1.62	1.42[b]	1.56	1.54
Urinary	2.89	3.30[a]	2.45	2.73	0.72	0.69	0.84	0.84
Bone, tissue, skin	2.18	2.24	2.92	3.02[a]	1.04[c]	0.87[bc]	0.78	0.64
Female breast	n/a	n/a	n/a	n/a	n/a	n/a	0.69	0.70
Lymphatic and haematopoietic	1.30	1.42[a]	1.43	1.51	0.95	0.89	0.86	0.84
Leukaemia	1.77	1.92[a]	2.12	2.19	0.51	0.46	0.42	0.40
Other and unspecified	1.60	1.52	2.25	1.82[b]	1.14	1.06[b]	0.81	0.89

Notes: UC = underlying cause, TM = total mentions.

[a] Higher by 10% or more.

[b] Lower by 10% or more.

[c] Differential direction reversed.

within categories of sex. For cancer of the bone, connective tissue, and skin, a differential male mortality rate that is 4% higher for non-whites based on the underlying-cause approach changes to a differential that is 13% higher for whites when the total-mentions approach is adopted. In this case, an actual reversal in race differentials occurs.

Table 17.5. *Multiple-cause-of-death combinations for cancer of the digestive system and cancer of other sites,* at age 45 years and older, Florida, 1980*

	No.	%
Cancer of the digestive system		
No other conditions reported	1653	23.4
Other cancer*	1223	17.3
Circulatory	704	10.0
Other cancer*/circulatory	433	6.1
Other diseases	367	5.2
Other cancer*/other diseases	357	5.0
Circulatory/other diseases	244	3.4
Other digestive	239	3.4
Other cancer*/other digestive	191	2.7
Other cancer*/circulatory/other diseases	148	2.1
Other digestive/other diseases	107	1.5
Other respiratory	88	1.2
Other respiratory/other diseases	85	1.2
Circulatory/other respiratory	83	1.2
Circulatory/other digestive	76	1.1
Other combinations	1075	15.2
Total mentions	7073	100.0
Cancer of other sites*		
No other conditions	2806	19.4
Circulatory	1372	9.5
Digestive cancer	1223	8.4
Respiratory cancer	1082	7.5
Other diseases	1024	7.1
Genito-urinary/urinary cancer	738	5.1
Circulatory/other diseases	586	4.0
Respiratory cancer/other diseases	437	3.0
Digestive cancer/circulatory	433	3.0
Respiratory cancer/circulatory	375	2.6
Digestive cancer/other diseases	357	2.5
Genito-urinary/urinary cancer/circulatory	308	2.1
Other respiratory/other diseases	291	2.0
Other respiratory	284	2.0
Genito-urinary/urinary cancer/other diseases	236	1.6
Digestive cancer/other digestive	191	1.3
Other digestive	163	1.1
Circulatory/other respiratory	150	1.0
Digestive cancer/circulatory/other diseases	148	1.0
Other combinations	2274	15.7
Total mentions	14478	100.0

* Cancer of other sites includes all except respiratory, digestive, and genito-urinary and urinary combined.

Table 17.5 provides some illustrations of the decomposition of total mentions of a cause. Using two general cancer sites as examples, the distribution of all combinations of conditions for persons having the specified cancer is shown. To simplify the tabulation of combinations, the cause list was reduced to four cancer groups and six non-cancer groups. The categories can obviously be expanded or narrowed to select the cause combinations of particular interest.

For decedents with cancer of the digestive system, 23% had no other conditions reported. This does not necessarily mean that there were no other complicating conditions, only that the physician did not certify any others. In 17% of the cases, the digestive cancer was identified along with one or more other cancers besides respiratory, genito-urinary, or urinary. In 10% of the cases, digestive cancer was the only cancer detected but circulatory disease was also reported. Numerous other combinations of digestive cancer and the remaining nine cause categories were recorded. Not all of the theoretical combinations of the ten causes were given (Wrigley 1986).

By specifying the coexistence of certain cause combinations involving two or more categories and detailing the category of other combinations, one can mark the incidence of joint occurrence of categories of interest. For instance, one can measure the frequency of multiple cancers of different sites as well as the presence of circulatory ailments when a cancer is also present.

Similar findings result when cancers of other sites are studied (Table 17.5). In this situation, there is a different pattern of multiple combinations from the case of digestive cancer. When these combinations of causes are examined by age, sex, and race (not shown in this table), significant differentials in cause patterns appear.

5. Future Research Agenda and Strategies

If my assumption that multiple conditions at the time of death describe all categories of decedents—in times past as at present, in developing as in developed countries, at all age levels, and for infectious and parasitic as well as the chronic diseases—is correct, then it behoves us to devote greater attention to multiple-cause-of-death analysis and its many variations. The World Health Organization (WHO), in its 1975 Revision of the International Classification of Diseases (WHO 1977), acknowledged progress in multiple-cause coding and analysis but did not incorporate such developments in ICD-9. In several developed nations, coding and tabulation of multiple-cause-of-death information is proceeding. Likewise, academic research has been increasingly focused on this area of analysis (Bradshaw and Frisbie 1986; Crews 1986; Manton 1986; Nam and Eberstein 1986).

The need for extending multiple-cause research to developing countries, and comparing nations, is paramount. We must certainly be cautious about the comparability of data from different countries. Long ago, Janssen (1940) pointed out some limitations of cross-national comparisons of cause-of-death data. First, the habits of physicians in reporting causes of death vary from country to country. Secondly, questions on death records for obtaining cause-of-death information are different. Thirdly, there is variation in responsibility (medical or otherwise) for reporting cause of death. Finally, the rules for assigning cause of death may vary from country to country. Brouard and Lopez (1985) add that the comparability of diagnosis is affected by religious and social constraints on the reporting of some diseases or injuries as well as national variations in the fashionableness of particular diseases among the medical profession.

Before any analysis of this topic can be advanced, it will be necessary to assemble appropriate data sets. A first step will be to locate existing data sources which already provide some multiple-cause-of-death information. Extensive evaluation of this information will be necessary before analysis is undertaken, but useful analysis from such sources is possible. The next step will be to provide for obtaining multiple-cause data in connection with ongoing record systems and surveys. Our experience in this area is growing (Vallin *et al.* 1984) and the inclusion of new items of information should be possible.

The testing of multiple-cause-of-death data collection in field pilot studies is a further necessary step. Some headway is being made in better training of lay medical personnel in reporting causes of death (WHO 1978), and this training must be extended to acquisition of medical histories and/or reporting current complicating morbid conditions. Also, we are seeing more collaborative efforts between survey and record analysts (demographers, epidemiologists, and others) and medical practitioners in amassing data bases. New field-based medical technologies (such as battery-operated echocardiography equipment) lend themselves to incorporation within the context of field interview studies.

Some could argue, however, that we still have considerable room for improving the sets of data on primary causes of death and that we should focus our attention on that before turning to other developments such as multiple causes of death. Others could counter persuasively by arguing that we need meaningful data by cause and that to achieve it we must work to build up such data bases.

Golini and Egidi (1984) have detailed some useful models for linking morbidity and mortality. Their discussion emphasizes the complexity of the network and serves to strengthen the case for development of multiple-cause-of-death information. Newer classification systems for primary causes of death, such as the distinction between aetiological and anatomical axes (Meslé and Vallin 1984), can be merged into multiple-cause classifications.

Once multiple-cause data become a more general reality, there are several areas where they might be used profitably.

(*a*) The principal contribution will be in a more thorough understanding of cause-of-death structures and a refined expression of the relative risks of death by cause.

(*b*) Having multiple-cause data should make it easier to test models of competing causes of death and selection in survival.

(*c*) The fuller cause data base should enable more meaningful analysis of trends and differentials and produce a stronger foundation for mortality projections.

(*d*) Finally, the notion of stages of an epidemiological transition (Omran 1971) may be over-simplified, based as it is on primary causes of death. When examined within a multiple-cause-of-death framework, the transition may appear much more complex and its various stages less distinct than they do now.

References

Bradshaw, B.S., and Frisbie, W.P. (1986), 'Mortality, Migration, and Multiple Causes', Paper presented to *Annual Meeting of the Population Association of America*, San Francisco, Apr.

Brouard, N., and Lopez, A. (1985), 'Causes of Death Patterns in Low Mortality Countries: A Classification Analysis', *International Population Conference*, Florence, i (Liège: IUSSP, Ordina Éditions).

Chamblee, R.F., and Evans, M.C. (1982), 'New Dimensions in Cause of Death Statistics', *American Journal of Public Health*, 72, 11.

—— —— and Rosenberg, H.M. (1979), 'A National Multiple Cause of Death Statistical System', *Proceedings of the Social Statistics Section, American Statistical Association*.

Crews, D.E. (1986), 'Predictors of Mortality in the Old and Very Old: An Application of Multiple-Cause-of-Death Statistics to the Epidemiology of Aging', unpublished manuscript, Florida State University.

Dorn, H.F., and Moriyama, I.M. (1964), 'Uses and Significance of Multiple Cause Tabulations for Mortality Statistics', *American Journal of Public Health*, 54, 3.

Dublin, L.I., and Van Buren, G.H, (1924), 'Contributory Causes of Death: Their Importance and Suggestions for Their Classification', *American Journal of Public Health*, vol. 14.

Erhardt, C.L. (1958), 'What Is the "Cause of Death?"', *Journal of the American Medical Association*, 168, 2.

Fales, W.T., and Moriyama, I.M. (1949), 'International Adoption of Principles of Morbidity and Mortality Classification', *American Journal of Public Health*, 39.

Golini, A., and Egidi, V. (1984), 'Effect of Morbidity Changes on Mortality and Population Size and Structure', in Vallin *et al.* (1984).

Guralnick, L. (1966), 'Some Problems in the Use of Multiple Causes of Death', *Journal of Chronic Diseases*, vol. 19.

Israel, R.A., Templeton, M.C., and Evans, M.C. (1972), 'New Approaches to Coding and Analyzing Mortality Data', *Proceedings of the Social Statistics Section, American Statistical Association*.

Janssen, T.A. (1940), 'Importance of Tabulating Multiple Causes of Death', *American Journal of Public Health*, vol. 30.

Krueger, D.E. (1966), *New Numerators for Old Denominators: Multiple Causes of Death*, National Cancer Institute Monograph, no. 19.

Manton , K.G. (1986), 'Cause Specific Mortality Patterns among the Oldest Old: Multiple Cause of Death Trends 1968 to 1980', *Journal of Gerontology*, 41, 2.

—— and Stallard, E. (1984), *Recent Trends in Mortality Analysis* (Orlando, Fla.: Academic Press).

Meslé, F., and Vallin, J. (1984), 'The Problem of Studying Mortality Patterns by Cause over a Long Period of Time: An Example from France, 1925 to 1978', in Vallin *et al.* (1984).

Moriyama, I.M. (1956), 'Development of the Present Concept of Cause of Death', *American Journal of Public Health*, vol. 46.

Nam, C.B., and Eberstein, I.W, (1986), 'Prior and Proximate Factors in Infant Mortality by Cause', NICHD research project in progress.

National Center for Health Statistics (1980), *Multiple Causes of Death Public Use Tape: ICD-9* (Hyattsville, Md.: US Department of Commerce).

—— (1982), *Annotated Bibliography of Cause-of-Death Validation Studies, 1958–80*, Vital and Health Statistics, series 2, no. 89 (Washington, DC: US Government Printing Office).

Omran, A, (1971), 'The Epidemiological Transition: A Theory of the Epidemiology of Population Change', *Milbank Memorial Fund Quarterly*, vol. 49.

Vallin, J., Pollard, J.H., and Heligman, L. (eds.) (1984), *Methodologies for the Collection and Analysis of Mortality Data* (Liège: Ordina Éditions).

WHO (World Health Organization) (1977), *International Classification of Diseases*, 1975 Revision, i (Geneva: WHO).

—— (1978), *Lay Reporting of Health Information* (Geneva: WHO).

Wrigley, J.M. (1986), 'A Multiple-Cause-of-Death Approach to Cancer Mortality Differentials at the Older Ages', Doctoral dissertation, Florida State University.

Wrigley, J.M., and Nam, C.B. (1986), 'Differential Cancer Mortality at the Older Ages: Comparison of Rates Based on Underlying and Multiple-Cause Designations', Paper presented at the *Annual Meeting of the Population Association of America*, San Francisco, Apr.

18 Studies of Causes of Death in Latin America: The Current Situation and Future Perspectives

JUAN CHACKIEL

Centro Latinoamericano de Demografia (CELADE), Santiago, Chile

1. Introduction

Studies of mortality according to cause of death are still at a very early stage in Latin America and many Third World countries. What are the main reasons for this deficiency? The absence of such studies can be traced first and foremost to the paucity of information and to well-known deficiencies of data available from vital-statistics registers. Even though mortality information can, under certain conditions, be obtained by means of indirect procedures, it seems a much more difficult task to retrieve data on mortality by cause of death without relying on vital statistics.

There are, however, other important factors responsible for the lack of studies on mortality by cause.

(*a*) There is little communication between demographers and health professionals, physicians in particular. Although the discipline of demography in developing countries is well known for its progress in technical aspects of estimation of mortality and fertility trends and for the attempts to broaden its focus by increasing participation in interdisciplinary studies with social scientists, the level of interaction with professionals in the medical area is quite low. In addition, many demographers have had their basic training in the social sciences and very few of them in medicine. As a consequence, demographers experience difficulties in understanding, let alone mastering, the nomenclature of diseases, their aetiology, and related knowledge.

(*b*) Demographers have assigned high priority to the study of *total* levels and trends of vital phenomena. Due to the highly doubtful quality of sources of information available, a great deal of effort has been invested to get basic understanding of the crudest trends and less effort has been allocated to the study of differentials or cause of specific processes. Thus, although countless procedures for data collection have been developed and time and material resources have been invested on the development of indirect estimation techniques and models (model life tables, fertility models, stable populations, etc.), very little progress has been made in the data collection and analysis of mortality or fertility differentials in general, and mortality by cause in particular.

(*c*) Research that relies on deaths classified by causes runs into practical problems due to the great quantity of data that must be handled. For example, consideration of 1000 causes, for a ten-year period, for both sexes separately and 18 age groups, leads to working with a matrix of 360 000 cells. Simplification usually requires a tedious exercise of regrouping the original cause of deaths into a smaller number of categories. Further recategorization of causes of death may lead to problems of comparability across time and regions as well as to the inefficient treatment of some problems in order to deal more efficiently with others. Admittedly, however, current advances in the use of microcomputers should attenuate the effects of this problem.

(*d*) In contrast to other types of demographic research in developing countries, it is very difficult to obtain information about causes of death using indirect methods. Further, there are few examples of successful new data collection on causes of death, and those which are available deal with small, selected communities, thus preventing the generalization of conclusions.

(*e*) The absence of studies in this area has paralysed all research efforts and resulted in a lack of understanding of how to use the information that is available. New ideas about proper use of available data on causes of death are not only needed but, more importantly, feasible.

In this chapter I analyse several of the issues mentioned above, evaluate their current position, and speculate on their future situation in Latin America. I also propose a number of areas of study which appear to be in greatest need of development. In particular, I will consider: (*a*) the quality of death registers by causes and other general aspects related to the treatment of basic information, (*b*) the use of data on causes of death for determining their contribution to the change of general mortality, and (*c*) possible improvements in projections of mortality that rely on information on causes of death.

1. Death Registration in Latin America

The first point of importance regards the heterogeneity across countries with respect to the quality of vital statistics. As in other aspects of the region, there are countries whose register system is satisfactory, others where registers are practically absent, and some which fall between the two. Table 18.1 (columns 2 and 3) displays the percentages of under-registration of total deaths for the periods 1960–5 and 1975–80 for Latin American countries. By and large, these were obtained by the comparison of actually registered data with estimated deaths implied by the population projections elaborated by CELADE jointly with the corresponding national organizations. In most cases, the corrected figures are derived from evaluations

Table 18.1. *Under-registration of the registered deaths of the group of ill-defined causes (B45) and deaths with medical certification, Latin America, between 1960 and 1980* (%)

Country	Under-registration		Ill-defined causes		Total without information 1978	Medical certification	
	1960–65	1975–80	1965	1978		1965	1978
(1)	(2)	(3)	(4)	(5)	(6)	(7)	(8)
Argentina	5.8	1.4	12.4	4.3	5.6	—	—
Bolivia	64.1	66.5	23.2	—	—	—	—
Brazil	34.4	19.2	39.7	—	—	—	—
Chile	1.3	2.5	7.7	10.8	13.0	80.1	—
Colombia	13.3	24.9	13.5	8.6	31.4	59.8	—
Costa Rica	11.9	6.7	9.2	8.0	14.2	55.4	64.0
Cuba	26.4	2.7	1.6	0.03	2.7	98.6	—
Dominican Republic	53.4	45.2	32.1	31.9	62.7	20.8	—
Ecuador	12.8	19.5	21.7	16.5	32.8	39.4	—
El Salvador	31.3	24.3	34.4	28.5	45.9	—	—
Guatemala	12.7	4.0	16.0	18.2	21.5	18.0	—
Haiti	—	79.5	—	—	—	—	—
Honduras	49.8	53.7	41.0	34.0	69.4	—	—
Mexico	11.3	14.9	18.6	8.9	22.5	68.9	—
Nicaragua	56.4	59.6	19.9	27.0	70.5	—	—
Panama	25.2	24.5	18.6	10.8	32.7	54.9	75.0
Paraguay	58.5	38.5	25.8	19.4	50.4	38.2	—
Peru	42.6	35.7	13.7	8.4	41.1	—	67.1
Uruguay	6.0	2.6	6.7	6.9	9.3	—	99.8
Venezuela	24.8	9.8	25.0	14.2	22.6	72.8	83.1

Notes: Cols. 2 and 3, the under-registration was calculated on corrected deaths; 4 and 5, % of group B45 on registered deaths; 6, % without information about corrected deaths: $6 = (3) + (5)*(100 - (3))/100$.

carried out using indirect estimation methods based either on retrospective information from censuses and surveys or on a combination of census and vital-statistics information with the growth balance equation.

Under-registration is fundamental to determining the possibility of working with data on mortality classified by cause of death. In those countries where an important number of deaths escape registration, one must exercise a great deal of caution to interpret results and conclusions, be they cause-specific or not. It would be illusory to believe that non-registered deaths are distributed in the same way as reported deaths. If we were to apply a very demanding criterion, for example using only those countries where under-registration is less than or equal to 10% (for the period 1975–80), we would be able to work with a certain degree of confidence in a handful of countries in the region: Argentina, Chile, Costa Rica, Cuba, Guatemala, Uruguay, and Venezuela. At the other extreme we have countries with an under-registration higher than 35%: Bolivia, the

Dominican Republic, Haiti, Honduras, Nicaragua, Paraguay, and Peru. Although this is a fair assessment of the quality of data at the national level, it may not represent regional situations. Thus, for example, there are cases where the national register may represent a very low percentage of total deaths in the country but may also be of a very good quality for some selected geographical divisions or localities. This is the case of the state of São Paulo in Brazil, where one finds unusually reliable data from vital statistics.

If one focuses on time trends, it is difficult to miss the remarkable case of Venezuela where from a level of under-registration of about 25% in the mid sixties there has been a decline to less than 10%. Cuba is a country that traditionally has had very complete information and the high percentage of under-registration for the period 1960–5 is exceptional, the likely outcome of the disruption caused by the revolution of 1959.

This is a general situation which requires further analysis to uncover important details. Note, for example, that in some cases the omission of deaths is concentrated at early ages; this would still allow the study of youth and adult mortality. Evaluation of male deaths in Venezuela around 1981 showed an under-registration of 30% for those under 1 year of age, 10% in the group 1–4, and 10% in the group 5 and over. In Costa Rica, omission was estimated at 10% for the age group 0–4 and 6% for 5 and over for 1973.

Table 18.1 also displays indicators directly connected with the quality of data on causes of death, namely, the percentage of deaths attributable to ill-defined or unspecified causes (columns 4 and 5) and the percentage of deaths that had medical certification (columns 7 and 8). The frequency of category B45 (ill-defined symptoms and conditions) pertaining to the Abbreviated List of 50 causes of death proposed in the ICD of the World Health Organization prior to the Ninth Revision, reflects very strongly the difficulties of establishing a clear diagnosis of the disease producing death. A high percentage of deaths classified in this group aggravates the problem created by under-registration, since deaths whose cause is unknown are, from the point of view of analysis by cause, essentially equivalent to non-registered deaths.

The high positive correlation between the percentage of under-registration of deaths and the relative importance of group B45 (Table 18.1) is a cursory indicator of the gravity of the situation; in some countries the number of deaths where the cause is known is very small. For example, for the period 1975–80 Honduras had an estimated under-registration of 53.7%; of the registered deaths the cause is unknown for only 34%, which implies that there is no information of causes of death for 69.4% of all deaths. In other words, any analysis by causes of death for Honduras would be based on only 30% of all deaths which actually occurred in the country. There are only three countries for which the deaths with unknown causes account for less than 10% of all deaths (Table 18.1, column 6): Argentina, Cuba, and Uruguay.

It is known that deaths from ill-defined causes cannot be reallocated in the same proportion to those already known and procedures considering different weights have been proposed (Ledermann 1955). It would be useful to compare the results obtained from techniques to reallocate the ill-defined causes using a proportional assumption with those based on regression models. Until now, little work has been done in this area of mortality analysis in Latin America.

By and large, the time trend of the percentage represented by group B45, has steadily decreased in almost all countries, probably due to better disease diagnosis. Venezuela's case is remarkable again, although it still maintained in 1978 a level of 14% of ill-defined causes. Some countries (e.g. Peru) show a high degree of under-registration but also relatively low percentages of ill-defined causes of death. This happens because registered deaths, although subject to omission, correspond to urban areas when a great fraction of them have medical certification.

Another aspect to be considered when analysing the proportion of ill-defined causes is their age pattern. Table 18.2 displays the age distribution for the category B45 for selected countries of the region. Several patterns can be observed. The most general one shows that the proportion of ill-defined causes is higher for persons over 65 years of age and for children (in

Table 18.2. *Relative distribution of the B45 group (ill-defined causes) in the total of deaths, by age, for selected countries, between 1965 and 1978* (%)

Country	Year	Age						
		Total	0	1–4	5–14	15–44	45–64	65+
Chile	1965	7.7	5.5	9.2	4.8	3.8	4.7	13.9
	1978	10.8	14.3	15.1	10.2	6.6	6.9	13.0
Colombia	1968	8.6	7.4	8.4	9.6	5.7	7.5	12.4
	1975	9.4	7.7	12.4	12.3	5.7	7.1	12.4
Costa Rica	1967	7.5	7.4	7.0	6.7	3.3	4.6	10.5
	1980	7.8	8.8	11.0	6.1	4.0	5.8	9.4
Cuba	1968	1.3	0.9	1.6	1.9	1.5	0.6	1.6
	1977	0.3	0.2	0.6	0.7	0.8	0.2	0.2
Dominican Republic	1965	47.2	46.8	45.3	41.8	33.8	44.1	60.9
	1976	35.1	30.0	35.6	31.7	23.6	31.7	47.3
Guatemala	1965	16.0	6.1	12.9	17.9	18.5	17.8	39.6
	1982	12.4	2.5	13.8	14.6	9.3	9.6	26.1
Honduras	1978	31.8	27.5	33.7	35.0	19.1	23.8	30.6
Paraguay	1967	23.9	6.0	24.5	23.3	18.3	22.9	38.5
	1980	19.4	17.7	16.4	18.5	14.3	17.0	23.9
Peru	1968	13.7	10.4	12.3	13.1	11.1	12.3	20.7
	1978	8.4	5.7	8.1	9.0	9.8	6.5	1.1
Uruguay	1966	6.7	5.5	10.0	4.9	7.1	7.5	6.4
	1975	6.4	7.2	9.3	6.8	6.9	7.3	5.7
Venezuela	1968	24.8	25.3	36.5	27.1	16.5	20.3	27.0
	1978	14.2	13.3	22.0	13.0	7.5	12.6	18.2

particular, those in the age group 1–4) than in other age groups. It follows that any attempt to reallocate these ill-defined causes must be age-specific. A similar qualification applies to the treatment of sex-specific mortality.

The percentage of deaths that are certified by a physician is another indicator of the quality of the information about cause of death. It is likely that in those cases the diagnosis of the disease which led to death will be more precise. For many countries of the region under study this information is not available, but what is known leads us to believe that there have been recent small improvements which are correlated with a decrease in the proportion of deaths allocated to ill-defined causes. There are some inconsistencies, however, which are not easy to explain. For example, although in 1978 Costa Rica had a lower percentage of medical certification than other countries, the overall quality of information on mortality is very good. Similarly, although in 1965 Guatemala had only 18% of deaths certified by a physician, it is a country with relatively complete statistics. This is somewhat unexpected since Guatemala exhibits one of the lowest levels of socio-economic development in Latin America. The exceptional quality of registration may be related to a strong tradition of record-keeping developed during the Spanish colonial period, when she acted as the capital of the Capitanía General of Guatemala. This interpretation may explain in part the rather surprising fact that with a 67% rural population and a 42% native population, a low percentage of deaths certified by a physician occurs jointly with relatively high levels of registration. It may also be the case that the fraction of registered deaths in countries with low levels of socio-economic development corresponds to urban areas with a high probability of medical certification and good diagnosis of the cause of death. This could explain why in cases like Peru, for example, there is only an 8% rate of ill-defined causes, about 70% of deaths certified by a physician, but somewhat deficient registration.

We could classify Latin American countries according to the quality of available data, by considering the percentage of under-registration of deaths as well as the proportion of ill-defined causes (Table 18.1, column 6):

- Countries with *very good information* on causes of death (less than 15%): Argentina, Chile, Costa Rica, Cuba, and Uruguay;
- Countries with *relatively good information* on causes of death (between 15% and 25%): Guatemala, Mexico, and Venezuela;
- Countries with *less reliable data* (between 25% and 40%): Colombia, Ecuador, and Panama.
- Countries with *deficient information* (over 40%): Bolivia, Brazil, the Dominican Republic, El Salvador, Haiti, Honduras, Nicaragua, Paraguay, and Peru.

In the first two groups of countries, it would be possible to use the data to carry out studies about mortality patterns and to inform the design of health policies. In the other countries more caution is required and the utilization of the data should be subjected to careful scrutiny. On this matter there are divergent opinions with some researchers dismissing the possibility of making any analysis when the information is very deficient taking one extreme, and with others who think that it is always possible to make corrections to bring the quality up to standard taking the opposite stand. Although no information can be considered as absolutely useless, any analysis should be carried out by first setting boundaries about what can be extracted from the data and when it is unreasonable to expect any useful results. If the basic data is of very poor quality, no degree of sophistication in correction and estimation methods will lead to improvements in the results.

2. Grouping of Causes of Death

All countries use the ICD of the WHO, which is revised every ten years. In 1975 the Ninth Revision was carried out. Vallin and Nizard (1978) and WHO (1978) date the emergence of the first classifications to the eighteenth century. In Chicago in 1893, at a meeting of the ISI, J. Bertillon proposed what was the beginning of the current classification which, although modified by several changes, maintains its general structure. Bertillon's classification was applied in Latin America for the first time in San Luis Potosí, Mexico.

The ICD presents a detailed three-digit classification with four-digit subcategories, which allows specific regrouping according to criteria more appropriate to the concrete objectives of the research. To date there have been many proposals to group the causes of death into a smaller number of categories with the purpose of facilitating the handling of information as well as to make it more consistent with specific research objectives. In Latin America, in particular, it was common until recently to utilize the five groups proposed by the United Nations (1963) in its *Population Bulletin*, no. 6. These groups were created according to their sensitivity to medical improvements and health programmes, using as a baseline the ICD's Abbreviated List of 50 causes. Subsequently, several other groupings have been proposed using aetiological or anatomic criteria or both in combination. Thus, for example, Baloche and Nizard (1973) use a categorization consisting of six groups and originally attributed to J.N. Biraben and P.E. Vincent, which is based on disease aetiology. The aetiological criterion is useful since it is more appropriate for the stochastic independence invoked by many of the techniques used to analyse these data.

Preston *et al.* (1972) have proposed 12 groups mostly based on an aetiological criterion. The most elaborate work has been done by Vallin and Nizard (1978), who proposed the use of one aetiological and one anatomical axis, using the ICD list of four digits.

While important progress in the definition of pure criteria for grouping causes is important, in developing countries with a relatively high mortality it is also worth while to group causes according to the likelihood of curbing the incidence of the diseases using different kinds of health programme. This approach is certainly more attractive in studies aimed at the formulation of health policies. Taucher (1978) has worked in this direction in CELADE, proposing a two-tiered classification of causes according to their degree of avoidability with *current* medical technology (avoidable and non-avoidable). Avoidable deaths are further classified into four groups according to the type of action that can be taken to prevent them:

(*a*) avoidable by vaccination or preventive treatment;
(*b*) avoidable by early diagnosis and medical treatment;
(*c*) avoidable by improvement of environmental sanitary conditions;
(*d*) avoidable by mixed actions.

The remaining deaths fall into the following groups:

(*e*) difficult to avoid with the present state of knowledge and current technological development;
(*f*) ill-defined causes of death;
(*g*) residual group.

Due to limitations of data in Latin America, Taucher proposed to group causes of death from the intermediate list of 150 causes called List A in the Seventh Revision. For the Eighth and Ninth Revisions, a correspondence with the three-digit classification was established. Table 18.3 shows for Chile, Uruguay, and Venezuela the relative distribution of deaths according to the grouping of causes proposed by Taucher.

Uruguay shows a different structure by cause of death compared to the other two countries. In Uruguay the prevalence of non-avoidable deaths is twice that of the other two countries. This is in part due to the older age structure of the population, and in part due to the fact that its mortality level is lower, which is generally linked with a relatively lower proportion of

Table 18.3. *Relative distribution of deaths by groups of causes proposed by Taucher, Chile, 1974–1975, Uruguay, 1974–1975, and Venezuela, 1978*

Grouping	Chile	Uruguay	Venezuela
Avoidable			
Vaccination or preventive treatment	1.3	0.8	1.1
Early diagnosis or treatment	4.3	5.2	3.4
Sanitary environment	3.9	1.5	5.5
Mixed measures	34.2	13.5	30.4
Non-avoidable	28.9	52.5	26.6
Ill-defined	9.8	6.4	14.2
Others	17.6	20.1	18.9
TOTAL	100.0	100.0	100.0
Life expectancy at birth, 1975–80	67.6	69.6	66.2

Sources: Taucher (1978); Damonte (1983); Schkolnik (1986).

Table 18.4. *Relative distribution of deaths grouped by avoidability considering the aetiologial axis of Biraben and Vincent, Venezuela, 1978*

Aetiological axis	Classification by avoidance degree (%)							
	Total	(*a*)	(*b*)	(*c*)	(*d*)	(*e*)	(*f*)	(*g*)
Infectious and parasitic	23.0	100.0	5.7	98.0	26.9	—	—	43.6
Intoxication and allergies	3.4	—	0.1	1.0	4.4	—	—	10.5
Accidents	15.6	—	—	—	50.9	0.6	—	—
Tumours	10.1	—	65.8	—	—	27.9	—	2.3
Malformations	7.1	—	14.6	—	14.2	8.0	—	0.7
Degenerative	27.0	—	13.8	1.0	3.6	63.5	2.6	42.9
Ill-defined	13.8	—	—	—	—	—	97.4	—
TOTAL	100.0	100.0	100.0	100.0	100.0	100.0	100.0	100.0

Source: Schkolnick (1986).

avoidable deaths. In addition there may be differences in life-style, particularly in dietary habits.

The proposed classification leaves a residual with a high percentage of deaths which are not classified according to their degree of avoidability. The 'ill-defined' and the 'others' account for approximately a quarter of the deaths in the three countries. Further refinements of the classification should reduce this residual considerably. Adoption of the classification would require periodic revisions to ensure that it is consistent with medical advances that make it necessary to redefine the criteria of non-avoidability of some causes of death.

Additional improvements obtain if one combines the two-way classification proposed by Vallin and Nizard with a grouping of causes of death by the degree of avoidability along an aetiological axis. This allows us to see the nature of the diseases which make up each group. This example of Venezuela is presented in Table 18.4 using an aetiological grouping proposed by Biraben and Vincent. One can verify in the table that all deaths due to causes avoidable by 'vaccination and preventive treatment' correspond to 'infectious and parasitic diseases', that 80% of those due to causes avoidable by 'diagnosis or early medical treatment' are 'tumours' and 'malformations', and almost all those avoidable by 'environmental sanitation' correspond to 'infectious and parasitic diseases' and 'accidents'. On the other hand, most non-avoidable deaths are of degenerative origin (63.5%), followed by tumours (27.9%). This information can be very useful for decision-making about health programmes, since it makes it possible to evaluate the possible effects of measures according to the incidence of different types of disease.

3. Causes of Death as Factors of Mortality Change

In this section, we examine the contribution of different causes of death to total mortality change in a given period, discuss the role of causes of death as immediate factors affecting mortality, and the geographic distribution of mortality by causes of death. These three aspects are essential for a fruitful use of cause-of-death information to improve knowledge of mortality change and to supply vital information to planners for the development of health policies.

3.1. The Contribution of Causes of Death to Mortality Change

This is perhaps one of the most frequently studied subjects in research on causes of death. The goal is to investigate the relative contribution of each cause of death to the decline observed in the total deaths in a given period of time. To avoid the effects of the age structure of the population, most work has been done with age-standardized death rates or with single (multiple) probabilities of dying by age from specific causes and the corresponding effects on life expectancy at birth. The analysis of the relation between age-standardized rates by causes and age-standardized mortality rates was first carried out in UN *Population Bulletin*, no. 6 (UN 1963), where the existence of a strong relation between the general level of mortality and the causal structure of mortality was established. Preston (1976) extended the model with the aim of determining typical patterns of mortality change and explaining regional and temporal differences in the causal structure of mortality. The author indicates that the model does not attempt to substitute for studies using data from each country, but he considers it useful as a general framework. With 11 groups of causes of death (Preston *et al.* 1972) in 165 populations he proposed the linear model

$$M(i) = a(i) + b(i).M,$$

where $M(i)$ is the standardized death rate from group of causes i, M is the standardized death rate from all causes combined, and $a(i)$ and $b(i)$ are the regression parameters. The parameter $b(i)$, which represents the slope of the regression line, has the property of expressing the change that occurs in the mortality rate from each group of causes per unit of change in the total rate M. Thus, the sum of the $b(i)$ is equal to 1 (Espenshade 1973). This procedure reveals that, leaving aside the residual group, the group of causes classified under 'influenza, pneumonia, and bronchitis' contributes about 25% of the total change, that the group under 'other infectious and parasitic diseases' accounts for 14% of the total change, and that 10% is contributed by causes under 'respiratory and tuberculosis' and 'diarrhoeal' diseases. The groups 'neoplasms' in both sexes and 'cardiovascular' for men have a negative $b(i)$, suggesting that mortality is increasing by these causes as the general level declines. While the model proposed by Preston may be valid for the time periods covered by his data, it is possible that the negative slope he found for some of the groups of causes does not represent current relations for many of the populations included in his sample due to alteration of medical knowledge. This linear model can be used to evaluate information for a given country by comparing standardized observed rates with predicted rates from all causes combined. Deviations from the predicted structure may signal peculiarities in the mortality patterns but may also result from idiosyncrasies of the classification procedures or from defective data, or both.

Since the parameters of the model are relatively straightforward to estimate, it is very tempting to apply it to populations having data for a time series or for geographical subdivisions of a single country. This exercise should yield *b(i)* values to analyse variability over time in the same country or country-specific regional patterns.

An application of this type was made for Latin America by Palloni and Wyrick (1981), who considered data from 11 countries for several years (49 cases) and the same cause-of-death grouping utilized by Preston. Values of the regression coefficient *b(i)* are very similar. The differences are greater in 'other infectious and parasitic diseases', which in Preston's model contribute to about 14% of total changes, whereas Palloni and Wyrick's Latin American model suggests a contribution of the order of 25%. The other group which appears to have greater importance in Latin American countries is 'diarrhoea' (13% versus 10%, according to Preston's). Reasons for these differences have been already mentioned in both studies: 'diarrhoea' and 'other infectious and parasitic diseases' can be increased by bad living conditions of the population, malnutrition, ignorance of necessary sanitary measures, etc. The high percentage of the contribution of infectious and parasitic diseases may also be due to the presence of malaria. In the same study, Palloni and Wyrick show that the countries with the highest death rates due to malaria are also those where infectious and parasitic diseases exhibit a larger contribution to total changes. Although this is not the appropriate place for a discussion of this point, it has been mentioned to demonstrate how necessary it is to carry out in-depth research for countries and for certain regions within them. Lastly, it is important to remember the warnings issued by Palloni and Wyrick about the limitations of this kind of exercise: first, the slope of the regression line only considers the influence of a basic cause of death; it does not show the magnitude of its real contribution through its dependency on other causes of death. Secondly, only an estimate of the mean level of the contribution of one cause of death is obtained, but the heterogeneity of this process is left obscure.

A characteristic of the methodology examined above is that it determines the contribution of groups of causes from an analysis where time is not explicitly considered. Instead, regression equations are estimated according to variability related to general mortality levels. An alternative way of working would be to compare the contribution of particular causes of death to total mortality at the beginning and at the end of a given period for a country or region. Taucher (1978) uses a grouping of causes of death by degree of avoidability to investigate the contribution of each group in the period 1955–6 to 1974–5 in the case of Chile. She proposed the following comparisons:

- mortality rates observed in 1974–5 compared with expected mortality rates taking the standard population from 1974 to 1975 and age-specific death rates by group of causes from 1955 to 1956. This comparison shows changes in risk of death for each group of causes;
- deaths observed from 1974 to 1975 compared with expected death according to age-standardized rates from 1955 to 1956. Differences between these represent avoided deaths for 1974–5 for each group of causes. The relative distribution of the differences shows the contribution of each group of causes to the absolute number of avoided deaths.

It should be noted that these results depend strongly on the standard population used. Table 18.5 presents results of the above-mentioned exercise. The rates showing the largest decline are those of the group avoidable by 'prevention', followed by causes avoidable by 'mixed measures'. Deaths avoidable by 'early diagnosis and medical treatment' and non-avoidable causes experience a small increase. This increase is probably due to the contribution of neoplasms in the first group and to certain types of cancer, cardiovascular diseases, and congenital malformation among the non-avoidable causes of death. The highest contribution to avoided deaths corresponds to causes avoidable by mixed measures, whereas the lowest contribution corresponds to deaths avoidable by preventive measures (4.9%). This is a somewhat awkward result since these deaths also appear to have experienced a higher relative decline. It is likely that these results are the outcome of the level of the rates and of the age structure of the standard population.

One can also analyse the contribution of each group of causes to the increase in life expectancy at birth. For this purpose, the mortality tables by cause of death may be used to calculate the expectation of life at birth for all causes combined from the probabilities of dying by age for each group of causes. Baloche and Nizard (1973) have proposed the following formula:

$$e(o)=0.5+\sum_{a=0}^{a=104}\prod_{x=0}^{a}\left[\prod_{i=1}^{i=m}\{1-q(x,i)\}\right],$$

where $q(x, i)$ is the probability of dying from cause i between ages x and $x+1$.

If in the expression for the beginning of the time period under study we use the probability of dying from a group of causes observed at the end of the period, we obtain the life expectancy that should result after the change in the probability of death due only to causes in that group. It is then possible to obtain the magnitude of increase in the life expectancy at birth that results from decreases in the probability of dying for each separate cause and assess its relative contribution to the total increase in life expectancy experienced during the period. The results of this exercise are, of

Table 18.5. *Rates and deaths by expected and observed causes of death, 1974–1975, if the observed rates, 1955–1956, remain constant, Chile*

Grouping of causes	Rates × 100 000		Change (%)	Deaths			Decline (%)
	Observed	Expected[a]		Observed	Expected	Avoidable	
Avoidable							
Prevention	9.6	33.6	−71.4	976	3413	−2437	4.9
Diagnosis and treatment	32.3	31.7	1.9	3287	3226	61	—
Sanitation	29.0	37.4	−22.5	2944	3797	−853	1.7
Mixed measures	256.5	552.8	−53.6	26059	56170	−30111	60.0
Non-avoidable	217.1	216.9	0.1	22059	22039	20	—
Ill-defined	73.8	154.5	−52.2	7504	15697	−8193	16.3
Others	131.9	216.5	−39.1	13404	22001	−8597	17.1
TOTAL	750.3	1243.5	−39.7	76233	126343	−50191[b]	100.0

[a] Standardized age rates, 1955–6, with standard population, 1974–5.
[b] Totals are the sum of negative differences: observed–expected.

Source: Taucher (1978).

course, limited by the assumption of independence among the various groups of causes of death (for a different method that side-steps this difficulty, see Chapter 15). An attractive aspect of this kind of procedure, however, is its usefulness for projecting future trends in life expectancy at birth. This will be explored later in the chapter.

3.2. Causes of Death as Immediate Factors Affecting Mortality

During the last decade there have been considerable advances in the study of mortality differentials by geographical and socio-economic characteristics. Among these are the studies that resulted from the programme of Infant Mortality Research in Latin America (IMIAL), carried out by CELADE with information about children ever born and children surviving, obtained from population censuses and surveys. Although these studies play an important exploratory role for understanding the determinants of early childhood mortality and may be a useful guide to health planners, they are not refined enough to permit a full understanding of the complexity of the processes and mechanisms determining the levels of early child mortality. Some of this complexity is revealed in recently developed conceptual models about the factors affecting early child mortality. These models include socio-economic factors as 'background' determinants of mortality but acting through 'intermediate variables' which affect directly the prevalence of disease and the levels of mortality risks. Mosley (1983) identifies 15 intermediate variables pertaining to five great groups: (*a*) maternal fertility factors; (*b*) environmental contamination with infectious agents; (*c*) availability of nutrients to the foetus and infant; (*d*) injuries; (*e*) personal disease control factors.

Mosley states that cause of death must not be considered as a biological *determinant* of mortality, but as the final link of the chain connecting socio-economic with intermediate factors. To emphasize this idea he uses two Latin American experiences as examples. One is based on a study in a rural area in Guatemala, where it is reported that childhood death results from the combination of poor diet and multiple recurrent infections. The other is a study of 13 areas in different countries revealing the difficulties of classifying diarrhoeal disease as a cause of death because of the existence of a synergistic relation between diarrhoea and inadequate nutrition.

These difficulties notwithstanding, the cause of death (if accurately reported) is still the final step of a morbid process and provides information that could be useful as an indicator of the intermediate factors which produce death. It would be interesting to classify deaths by cause in order to infer, within certain limits, the role played by each one of the biological and intermediate factors.

Some studies have attempted to explain how living conditions affect mortality. By and large, inferences depend on statistical analyses using data on cause of death. For Latin American countries, for example, it was established that on average, about 45% of mortality change was the result of an improvement in living conditions (Palloni and Wyrick 1981) as measured by GNP, the percentage of illiterate adults, and the availability of potable water. Further, the same authors found that *exogenous* factors which do not depend directly on improvements in living conditions are more important in countries having a strong presence of malaria and other infectious diseases the incidence of which can be checked with medical technology not directly linked with economic development.

3.3. Analysis of Mortality by Cause of Death according to Geographical Area

One of the main objectives of mortality analysis according to cause of death is the production of knowledge for health policies to reduce mortality in those population groups at greater risk. To identify these target groups it is necessary to measure mortality levels with a certain level of geographical disaggregation. There is an increasing demand for data for small areas and small communities, which imposes strong requirements on available information. This has contributed to bolstering the development of procedures to improve data quality and analysis procedures.

For an agenda of investigation of mortality variation, high priority should be given to the creation of an atlas of mortality by cause of death within sufficiently small administrative units. This type of exercise could be done both with causes of death and morbidity statistics, showing the incidence of the most important illnesses in each geographical area. In some countries of Latin America collections of maps have already been elaborated with this objective.

4. Mortality Projection

Procedures for mortality projections used in Latin American countries do not take into account the incidence of causes of death and the probable reduction (increases) in the prevalence of certain diseases. Generally, the expected evolution of life expectancy at birth of each sex is projected according to some fixed empirical rules:

(*a*) the gains in years of life expectancy at birth projected to occur in the future approximate to other countries' experiences;
(*b*) the gains decrease as mortality declines;
(*c*) the sex differentials in life expectancy at birth between sexes is defined as an increasing function of life expectancy. The age pattern of mortality is either directly taken from model life tables or is obtained by interpolation of probabilities of death by age from a life table for the country at the beginning of the projection period and another one considered to be an upper boundary. That is not to say

that what is being done is incorrect, it is probably the most reasonable way of working with available information for most countries.

When available, reliable information on cause of death should be used to improve the quality of the projection and to provide a better justification of its prospective evolution by anchoring it to the current incidence of dominant diseases. The focus can be on the identification of causes of death which will contribute to increases in life expectancy, or on those which explain male over-mortality and which will produce increasing sex differentials or, finally, on those which will determine the age pattern of mortality and its variations. This is not an easy subject and for many countries its application can be regarded as irrelevant since it poses strong demands on the quality of the information. Yet, these are the same countries that stand to gain the most from such applications: in countries where a sharp mortality decline is likely, efforts must be made to improve its forecasts, because in most of these cases future mortality trends will have important effects on the growth and perhaps the structure of the population.

The difficulties resulting from non-comparability of data over time, changes in the quality of data collection, improvements in diagnosis, and variation in the classification of diseases all require a parsimonious yet flexible classification of deaths. The projection of mortality by cause and by sex and age groups, should be based on life tables by cause of death. Although life tables by cause of death are well known, rarely have they been used for analyses of Latin American mortality or for its projection into the future. Keyfitz (1977) proposed a procedure to project mortality from life tables by cause of death. The procedure relies on the assumption that the mortality rate from all causes of death combined is equal to the sum of the cause-specific rates as if each one of them operated alone:

$$\mu(a) = \sum_{i=1}^{n} \mu(a, i),$$

where $u(a)$ is the force of mortality at age a and $u(a, i)$ is the force of mortality for age a and cause i. This implies that

$$l(x) = \prod_{i=0}^{n} \ell(x,i)$$

and therefore that

$$e(o) = \int_{0}^{w} \prod_{i=1}^{n} \ell(x,i)dx.$$

To extrapolate life expectancy at birth, values of $l(x, i)$ must be projected and Keyfitz suggests using Brass's (1974) logit relation:

$$Y(x) = a + b \cdot YS(x),$$

where

$$Y(x) = 0.5 \cdot 1n[1 - 1(x)]/1(x),$$

which when applied to life tables by cause of death leads to:

$$Y(x, i) = a(i) + b(i) \cdot YS(x, i).$$

Trends over time in the parameters $a(i)$ and $b(i)$ could be estimated and then projected to get projected values of $Y(x, i)$, the corresponding values of $1(x, i)$ and, finally $1(x)$.

Using data from Uruguay, Ana Maria Damonte (1983) has carried out interesting work starting from formulae for life expectancy at birth which are defined as functions of probabilities of dying by cause of death as proposed by Baloche and Nizard (1973). She sets as a target for the future the evolution of certain groups where the greatest benefits in current advances in medicine can be found. This idea could be the basis for projecting mortality by cause of death taking as a limit some model given by lower mortality countries. The projected path of mortality by cause, however, ought to be consistent with the specific conditions of each country.

The projection may be simplified by working directly with mortality rates by cause of death and by age group. Since

$$m(x) = \sum_{i=1}^{n} m(x, i),$$

the total mortality rate could be obtained as the sum of the separately *projected* rates for each cause of death. This procedure was utilized for Canada's mortality projections, which were based on simple estimates of trends observed in previous periods (Statistics Canada 1975).

All procedures that rely on extrapolation of mortality trends by cause, independently for each sex and age group, require that we evaluate very carefully the consistency of the resulting total life tables for each projection period not only from the point of view of their overall levels of mortality but also with regard to age patterns.

5. Summary and Conclusions

Several Latin American countries are in a position to carry out research with data on cause of death obtained from vital statistics. In an analysis of the quality of registration as measured by levels of under-registration and some indicators of the declaration of the cause of death, five countries were found to have very good information. These are Argentina, Chile, Costa Rica, Cuba, and Uruguay. Three other countries were found to have relatively good information: Guatemala, Mexico, and Venezuela. It is thus possible to work with a good representation of mortality levels and geographical locations.

Due to the massive data handling required, it is convenient to group causes of death into a smaller number of categories according to criteria that are specially tailored for the proposed research. Generally, aetiological and anatomical criteria or a combination of both have provided acceptable results. In Latin America an attempt has been made to establish a grouping of causes of death according to their degree of avoidability given extant medical technology. This type of grouping facilitates the task of evaluating the effect of different collective or individual measures to check the incidence of diseases with available medical knowledge. Interesting possibilities are offered by a classification simultaneously using degree-of-avoidability and aetiological criteria, as it provides us with information about the nature of diseases which contribute to mortality change.

The studies of the contribution of causes or groups of causes to mortality change can be carried out using regression techniques as well as standardization of mortality rates by age and cause of death. Applications to Latin America show important effects attributable to the reduction of infectious and parasitic diseases, influenza, pneumonia, bronchitis, and diarrhoea.

The area of determinants of mortality has been little explored in Latin America but deserves more attention. In particular, techniques focusing on cause of death and recently formulated models of relation between morbid process and background and intermediate factors should be utilized more fully.

Finally, we suggested the possibility of improving mortality projections by sex and age using knowledge of the evolution of the causal structure of mortality. Some promising procedures for carrying out the projections, including the use of life tables by cause of death were examined and evaluated.

References

Baloche, C., and Nizard, A. (1973), 'Évolution de la mortalité par type de causes de décès en France, 1950–1967: Essai d'une nouvelle mesure de la mortalité', *International Population Conference*, Florence, iii (Liège: IUSSP, Ordina Éditions).

Brass, W. (1974), *Métodos para Estimar la Fecundidad y la Mortalidad en Poblaciones con Datos Limitados: Selección de Trabajos*, series E, no. 14 (Santiago: CELADE).

Damonte, A.M. (1983), *Uruguay: La Mortalidad por Causas en 1975 y sus Perspectivas Futuras*, series C, no. 162 (Santiago: CELADE).

Espenshade (1973), *The Cost of Children in the Urban United States*, Population Monograph, no. 14 (Berkeley: University of California, Institute of International Studies).

Keyfitz, N. (1977), 'Proyecciones sobre Mortalidad', *Investigación Demográfica en México*, (México: Consejo Nacional de Ciencia y Tecnología).

Ledermann, S. (1955), 'La répartition des décès de causes "indéterminées"', *Revue de l'Institut International de Statistique*, 23(1–3).

Mosley, H. (1983), *Will Primary Health Care Reduce Infant and Child Mortality? A Critique of Some Current Strategies: With Special Reference to Africa and Asia* (Paris: INED, IUSSP).

Palloni, A., and Wyrick, R. (1981), 'Mortality Decline in Latin America: Changes in the Structure and Causes of Deaths, 1950–1975', *Social Biology*, 28(3–4).

Preston, S. (1976), *Mortality Patterns in National Populations* (New York, San Francisco, London: Academic Press).

—— Keyfitz, N., and Schoen, R. (1972), *Causes of Death: Life Tables for National Populations* (New York and London: Seminar Press).

Schkolnik, S. (1986), 'Consideraciones Preliminares para el Estudio de las Causas de Muerte en Venezuela', CELADE, Santiago (unpublished).

Statistics Canada (1975), *Technical Report on Population Projections for Canada and the Provinces* (Ottawa: Statistics Canada).

Taucher, E. (1978), *Chile: Mortalidad desde 1955 a 1975: Tendencias y Causas*, series A, no. 162 (Santiago: CELADE).

UN (United Nations) (1963), *United Nations Population Bulletin*, no. 6 (New York: UN).

Vallin, J., and Nizard, A. (1978), 'Les Causes de décès en France, I: Pour une typologie simple et homogène: Application à la période 1968–1974', *Population*, 33(3), 547–608.

WHO (World Health Organization) (1978), *Clasificación Internacional de Enfermedades: Revisión 1975* (Washington: WHO).

19 Socio-economic Careers and Differential Mortality among Older Men in the United States

ROBERT D. MARE

Center for Demography and Ecology, University of Wisconsin, Madison, Wisconsin, USA

---

1. Introduction

A main focus of mortality analysis for developed nations is differentials in levels and patterns of mortality across socio-economic groups and the social processes that give rise to those differences. Stemming from the strong regularity that mortality rates vary inversely with positions in socio-economic hierarchies, these investigations document an important effect of social inequality within societies and assess the effects of socio-economic change of mortality trends. To the extent that death rates vary across social groups, such analyses suggest how to improve health and prolong life for large segments of the population and aid in understanding long-run mortality trends.

The goals of this chapter are: (*a*) to describe socio-economic mortality differences among adult males in the United States; (*b*) to examine the effects of socio-economic differentiation early in life on the timing of mortality in later adulthood; (*c*) to obtain estimates of socio-economic effects that are relatively free from biases due to health-related selection into socio-economic positions; (*d*) to illustrate a strategy for estimating the effects of several dimensions of socio-economic statuses at multiple periods of life; (*e*) to show the value to mortality analysis of longitudinal data that are gathered for other purposes; and (*f*) to illustrate modern methods of survival analysis in the study of social differences in the timing of death.

Sections 2–5 of this chapter briefly review studies of socio-economic mortality differences and place the present investigation within the context of prior research (2). They discuss the effects of health-related selection into socio-economic statuses on estimates of social differences in mortality (3–4), and describe a rudimentary model of the effects of socio-economic changes over a man's lifetime on his mortality risks (5). Section 6 discusses the data, and Section 7 the analytic methods used in this chapter. Section 8 presents empirical results. Finally, Section 9 includes a summary of main findings, conclusions, and future research.

2. The Study of Socio-economic Effects

The investigation of socio-economic mortality differences in developed nations generally and the United States in particular has been limited by the availability of good data on the socio-economic characteristics of decedents and comparable surviving populations. (See Rosenberg and McMillen 1983 for a recent review of US data.) In the United States, the analysis of adult socio-economic differences on national populations has been largely confined to the analysis of aggregate 'ecological' associations between social characteristics of areas (states, counties, census tracts) and mortality rates; and to a single cross-sectional study of individuals, the 1960 Matched Records Study, a linkage of 1960 census records to certificates of deaths in the four-month period following the 1960 enumeration (Kitagawa and Hauser 1973). The Matched Records Study documented large cross-sectional differentials in mortality rates among levels of educational attainment, and provided some evidence of differentials by occupation and income as well. Although the Kitagawa and Hauser results are valuable, the individual-level data from which they derive have not been preserved, thus precluding further analysis. Moreover, given their cross-sectional nature, the data provide no information on longer-run effects of age-dependent socio-economic conditions. More recently, researchers have attempted to exploit data obtained for other purposes to provide fuller analyses of socio-economic effects (Rosenberg and McMillen 1983). Such studies advance previous work by including multivariate analyses of socio-economic effects. They remain, however, largely cross-sectional, and thus often control poorly for health-related selection into social positions (see below).[1]

In contrast to the largely cross-sectional US research, major prospective and longitudinal studies have been carried out in several European nations (see Fox (1984) for a review). Some of these studies consist of cross-sectional censuses (or samples) combined with mortality records in later years. Others obtain longitudinal information on social characteristics as well as on the timing of death. These data largely focus on differences in mortality among broad socio-economic categories defined at an initial observation point, rather than on multivariate analysis of the effects of socio-economic standing at several stages of life (e.g. Fox *et al.* 1985). The European studies, however, generally rest on data superior to those used in US studies. Because the European studies are longitudinal, moreover, they can separate socio-economic effects from the contaminating effects of health-related selection.

Since high-quality longitudinal data on mortality are not generally available for the USA, mortality studies must exploit data obtained for other purposes. This chapter uses the National Longitudinal Survey of

[1] For the most part, moreover, the recent studies have not used modern methods for the multivariate analysis of survival data.

Labor Market Experience of Mature Men (NLS), a panel survey of 5 020 men in the United States aged approximately 45–59 in 1966 (Center for Human Resources Research 1985). Designed as a source of information about work careers and retirement, these data are a unique resource for studying mortality among older men in the USA.[2] The data are unique in that they include detailed information on respondents' work, educational, and family histories, including wealth, income, and socio-economic characteristics of respondents' parents. Information about experience up to 1966 is obtained retrospectively, whereas information after that year is obtained through a sequence of interviews up to 1983. Although the data come from a sample that is modest in size relative to major mortality studies in developed countries, by application of suitable methods one can use the data to investigate socio-economic effects on mortality. The analysis reported here shows that retrospective socio-economic histories of workers are a valuable and relatively inexpensive way to study socio-economic effects. The data are described in further detail below.

3. Selection into Socio-economic Positions and Mortality Differentials

As is well known, estimated mortality differences across social characteristics such as marital status, occupation, and income can be distorted by the tendency for persons in poor health to enter disproportionately statuses and positions that are observed to have higher mortality rates. Conversely persons in good health tend to attain or remain in higher-status positions. Observed differentials in death rates among social categories, therefore, result from both the effect of those categories on mortality and also health-related selection of persons into the categories (e.g. Kitagawa and Hauser 1973; Fox 1979, 1984; Fox *et al.* 1982; Fox *et al.* 1985). Several strategies are available to neutralize these selection biases.

One strategy is to measure socio-economic conditions several years prior to the period during which deaths are observed. This strategy minimizes the likelihood that the health conditions that lead to death or survival during the observation period also determine socio-economic position. A second strategy is to measure health conditions directly. If valid measures of morbidity are available, then they can serve as statistical controls in the assessment of socio-economic effects. A third strategy is to assess the effects of socio-economic standing just prior to the period during which deaths are observed, but to control for prior socio-economic standing. This last strategy indirectly controls selection biases in so far as prior socio-economic standing affects subsequent health. As is discussed further below, the first and third of these strategies are used in the analyses reported here.

[2] One previous analysis of the data (Mott and Haurin 1985) examined socio-economic effects on mortality. The present paper considers additional aspects of socio-economic status not included by Mott and Haurin and applies more appropriate statistical methods to the data.

4. Differential Mortality and the Socio-economic Career

The effects of social conditions on the timing of death cumulate through life. Mortality in later adulthood is conditional on experiences in childhood, adolescence, and young adulthood. Social and economic conditions and experiences are a sequence of opportunities for and constraints upon styles of life (diet, recreation, places of residence, consumption of preventive medical care, etc.) that may affect long-run morbidity and mortality risks. The risks at any moment are a weighted function of all past activities. The weights, however, are difficult to specify because so many hypotheses are possible. For example, past experiences may cumulate so that individuals can be described by an age-dependent index that is their average socio-economic level over all previous socio-economic levels that they have experienced. Alternatively, experiences in the remote past may decline in importance as time passes and thus have less weight. Another possibility is that very recent conditions and experiences have not had enough time to affect mortality and must be discounted relative to the more remote past. Yet another possibility is that conditions at specific stages of life (e.g. late childhood or early adulthood) have a disproportionate impact because individuals are particularly vulnerable to environmental deprivation at those stages. Finally, the rate and direction of change in socio-economic conditions, rather than their average level, may affect mortality. Of course, data that reveal these processes are scarce. More importantly, we lack specific theories upon which to base the function relating current and past activities and statuses to present mortality risks.

5. A Simple Approach to the Effects of Socio-economic Careers

The complex relationship between cumulative life experience and mortality on the one hand, and the contaminating influence of health-related selection into socio-economic positions on the other, imply that our understanding of mortality can benefit from an unstructured investigation of the combined effects of social conditions at various stages of life. That is, we can view mortality risks in mid to late adulthood as a function of current and past socio-economic conditions, where the weights are determined empirically. Such an approach is preliminary to more structured investigations in which the weights are constrained to follow specific hypotheses about the cumulative impact of experiences. This approach, moreover, enables us to assess socio-economic mortality differences and also take account of the effects of health-related selectivity on gross associations between socio-economic standing and mortality.

This chapter examines the effects of socio-economic statuses that are measurable with the available data and have been investigated in the study of phenomena other than mortality. Numerous studies have examined the effects of the social standing of the family of orientation, schooling, and early labour force achievements on income, and occupational mobility and attainment at maturity (e.g. Duncan *et al.* 1972; Sewell and Hauser 1975). These variables index potentially important dimensions of socio-economic variation in mortality and provide an outline of individuals' socio-economic careers.

More specifically, we view mortality risks as a function of socio-economic conditions in childhood, schooling, early occupational experience, occupation at maturity, and financial assets. Each of these factors affects mortality both directly and also indirectly through later socio-economic statuses. We index family socio-economic background by a classification of respondents' fathers' occupations during the respondents' teenage years. Socio-economic background affects adult mortality both directly and through its effect on socio-economic statuses established later in life, especially educational attainment. Educational attainment, measured by formal schooling completed, also affects mortality directly and through its impact on occupational achievement and on the accumulation of wealth. Men's occupations on entry into the labour force are a function of their educational and social backgrounds and may affect mortality rates later in adulthood. Occupational positions and financial assets at maturity are affected by family background, schooling, and occupation at labour force entry, and themselves may affect rates of mortality in late adulthood.

Although many mechanisms govern the associations between early and later socio-economic standing, a key relationship from the standpoint of the present study is that early socio-economic conditions may affect health and morbidity, which, in turn, affect occupational and financial standing in later adulthood. As discussed above, occupational and financial standing may be associated with subsequent mortality rates because of both socio-economic effects and health-related selection into occupational and financial levels. However, the net associations between occupational and financial status on the one hand and mortality on the other, once earlier socio-economic statuses are controlled, provide estimates that are relatively free of the contaminating effects of selection.

To summarize, then, we examine the effects of socio-economic experiences, positions, and experiences over a broad span of life—childhood, youth, young adulthood, and maturity—on mortality rates from mid-life onwards. This enables us to achieve two goals: (*a*) to describe the effects of additional dimensions of social standing to those that have been examined in previous research, including essential aspects of individuals' socio-economic careers, and (*b*) to obtain estimates of the effects of later socio-economic standing on mortality that are purged of some of the contaminating effects of health on socio-economic standing.

6. Data

6.1. Sample Design and Outline

The 5020 NLS respondents were sampled from the US civilian non-institutional population and were first interviewed in 1966 and on 11 subsequent occasions between 1967 and 1983 via face-to-face or telephone interviews. The analysis, therefore, is of survival patterns over the 17-year period from 1966 to 1983; that is, the age spans 45–62 for the youngest birth cohort up to 59–76 for the oldest cohort. The NLS provides retrospective data on date of birth, race and ethnicity, educational attainment and training, military service, labour force experiences up to 1966, and parental socio-economic characteristics when respondents were teenagers. It also includes prospective data on health, occupational, and other labour market experiences, income and wealth, and family circumstances from 1966 onwards. Although the survey provides no direct information on mortality, deaths are recorded as reasons for permanent attrition from the panel. Respondents' exact ages and causes of death are unknown.

6.2. Survival and Attrition

Table 19.1 summarizes the mortality information provided by the NLS. The first panel of the table classifies NLS respondents by their age in 1966 and whether they died or survived the observation period. The analyses reported here exclude 14 of the 5020 respondents because the survey provides unreliable information on their ages. (Although the target population for the survey was persons aged 45–59 in 1966, some respondents fall outside this interval, as indicated by the 44–61 age range in the table.) The survey provides information on 1495 decedents and 3511 survivors. The latter include not only persons who survive until 1983, but also persons who drop out of the panel prior to 1983 for reasons other than death (refusal, unreported change of address, etc.). The second panel of Table 19.1 classifies the observations into survivors, deaths, and withdrawals for reasons other than death for each of the 11 intervals between panel waves. Year-to-year variation in the relative importance of deaths and other withdrawals is most likely the result of variation over time in survey methods, especially between telephone and face-to-face interviewing. The increasing relative importance of deaths arises from both accelerating mortality with age and the increasingly selective group of 'loyal' respondents who remain in the survey. Such variations are unlikely to affect the analysis of socio-economic effects on mortality.

The analyses reported in this chapter assume that the NLS provides unbiased estimates of the mortality experience of the US male population.[3]

[3] The NLS is confined to the 1966 civilian non-institutional population. Less than 2% of 45- to 59-year-old men were inmates of institutions in the 1960s (Land and Hough 1986: 64).

Table 19.1. *Survival status of men in National Longitudinal Survey of mature men, by age in 1966 and by observation period, 1966–1983*

Age at 3.1.66	Total	Dead	Other
44–47	807	152	655
47–50	1088	240	848
50–53	1053	323	730
53–56	971	337	634
56–59	833	332	501
59–61	254	111	143
TOTAL	5006	1495	3511
Interval between observations	**Survive**	**Die**	**Lost**
1966–67	4813	60	133
1967–68	4669	72	72
1968–69	4449	102	118
1969–71	4207	165	77
1971–73	3983	168	56
1973–75	3754	174	55
1975–76	3509	100	145
1976–78	3258	161	90
1978–80	3031	202	25
1980–81	2892	101	38
1981–83	2623	190	79
1983–	0	0	2623

More specifically, they assume that interviewers correctly report deaths and that probabilities of attrition from the NLS due to causes other than death are independent of probabilities of death. Deaths may be misclassified as attrition from other causes if interviewers cannot verify their occurrence. Withdrawal from the panel may result from serious illnesses that closely precede death, implying a lower death rate among respondents than among drop-outs. Multivariate analyses of the effects of socio-economic factors on mortality may also be affected by biases in mortality rates, although these distortions are reduced to the extent that factors predicting attrition from other causes are also controlled in the models. As shown below, however, life table functions estimated from NLS data agree rather closely with those based on official vital statistics.

6.3. Estimating Age of Death

The multivariate methods used in this chapter (see below) assume that one can rank all decedents by their age at death with few ties. If the ages at death are grouped or many ties are observed, then the computational burden of the multivariate analysis greatly increases. Ages at death of NLS respondents, however, must be inferred from the intersurvey interval in which they die because exact age or date of death is not reported. As indicated in Table 19.1, this interval is either one or two years wide. The NLS does report date of birth and, for completed interviews, date of interview to the exact day. The interview dates, date of birth, and interval of death enable us to estimate an exact age of death. For the ith individual let t_{0i} denote exact date of birth, t_{ki} denote exact date of the kth interview ($k = 1, \ldots, K$), and $\bar{t}_{k+1}$ denote the average date of interview for persons who respond to the $(k+1)$ th interview. Then, for an individual who dies between the kth and $(k+1)$ th interview, estimated age at death is

$$a_{ki} = t_{ki} - t_{0i} + [(\bar{t}_{k+1} - t_{ki})/2].$$

That is, we assume that a decedent lives for exactly one-half of the interval between his last interview and the average interview date for men who survive and respond to the next interview. By incorporating information on exact days of birth and last interview, we virtually eliminate ties in estimated ages of death.

6.4. Measurement of Independent Variables

All independent variables are measured categorically and, in the multivariate analyses, are represented by sets of dichotomous variables. The variables and their categories are as follows: (*a*) race (black versus non-black); (*b*) years of formal schooling ($<$ 8 years, 8–11 years, 12 years, $\geq$ 13 years); (*c*) father's occupation when respondent was age 15 (professional or managerial, clerical or sales, craftsman, operative, service or private household, farm labourer, non-farm labourer, farm owner or manager (including tenant), member of the armed forces); (*d*) respondent's first occupation after leaving school (same categories as for father's occupation); (*e*) respondent's occupation in 1966 (same categories as for father's occupation, but excluding armed forces); and (*f*) total family assets in 1966 (in quartiles).

The occupational categories are a grouping of 1960 US census major occupation categories. Respondent's occupation in 1966 excludes the armed forces because the survey is restricted to persons who were civilians in 1966. For persons who were not employed in 1966, 1966 occupation corresponds to their most recent job. Family assets include the sum of the values of house and other real estate, family business, automobiles, stocks and bonds, and savings accounts, minus debts. All of the independent variables except for race have missing data for at least some respondents. Life table functions specific to levels of the socio-economic variables (see below) are estimated over all persons for whom the data are present on the variable. In the multivariate analyses, each classification is augmented with an additional dichotomous variable that equals one if data are missing on that variable for a respondent and zero otherwise.

7. Models and Methods

The analysis of mortality differentials reported in this chapter applies two general methods: (*a*) life table analyses of average years lived over the observation period for selected socio-economic groups, and (*b*) regression models for the partial effects of socio-economic factors on (instantaneous) mortality rates. The former analyses focus explicitly on differences in durations of life. Whereas regression models for duration of life are available, provided one assumes a particular parametric distribution of survival times (e.g. Kalbfleisch and Prentice 1980: ch. 3), this analysis uses regression methods that do not assume a particular distribution of lifetimes and that estimate differences in instantaneous rates of death. The life table functions, therefore, reveal the quantitative importance in years of life of the differences in rates detected by the regression models.

7.1. Life Table Estimation

We report estimates of average years lived during the observation period within cohorts of men of approximately equal ages in 1966. These estimates derive from typical life table formulae except that they are based on the experience of actual rather than synthetic cohorts and they take account of the data lost during the observation period. Let l_x denote the number of survivors of a cohort to exact age m, T_x denote the years of life remaining beyond x for that cohort, n_x denote the number of men who enter the age interval x to $x+1$, c_x denote the number of men who are surveyed between ages x and $x+1$, and d_x denote the number of men who die between ages x and $x+1$. Then, for example, for men aged 50 at the 1966 survey, we estimate average years lived during the 17-year observation period as

$$(T_{50}-T_{67})/l_{50}=\left[\sum_{x=46}^{x=63}(l_x+l_{x+1})/2\right]/l_{50}, \tag{1}$$

where $1_{50}=n_{50}$ is estimated directly from sample observations in 1966; for $x>50$,

$$l_x=l_{x-1}\{1-[d_{x-1}/(n_{x-1}-c_{x-1}/2)]\};$$

and

$$n_x=n_{x-1}-c_{x-1}-d_{x-1}.$$

These calculations assume that, within single-year intervals, mortality follows a uniform distribution and data lost for attrition are observed for an average of half a year.

7.2. Proportional Hazards Model

To examine the net effects of the independent variables on survival we use a modified version of the proportional hazards regression model (e.g. Cox 1972; Kalbfleisch and Prentice 1980). Under the proportional hazards model, changes in the covariates (independent variables) adjust the instantaneous rate of mortality proportionately. A key feature of this model is that the baseline hazard function is unspecified; that is, the age path of mortality is arbitrary and only the proportionality of the effects of the covariates is assumed. Let t denote age, X_{ki} denote the value on the kth covariate for the ith individual $(i=1,\ldots,I;\ k=1,\ldots,K)$, $\lambda_i(t)$ denote the mortality hazard for the ith individual, $\lambda_0(t)$ denote the (unspecified) baseline mortality hazard, and β_k denote a parameter to be estimated for the effect of the kth covariate on the hazard. Then the proportional hazards model is

$$\lambda_i(t \mid X_{1i},\ldots,X_{ki}) = \lambda_0(t)\left[\exp\left(\sum_k \beta_k X_{ki}\right)\right]. \qquad (2)$$

The parameters β_k denote the effects of unit change in the covariates X_k on the log of the hazard rate within categories of (i.e. 'holding constant') the other independent variables in the model. For categorical independent variables, which are used in the present analysis, β_k denotes the deviation in the log hazard rate for the group for which $X_k=1$ from the baseline group (for whom all $X_k=0$). The quantity $\exp(\beta_k)$ for categorical X_k expresses the hazard of the group for which $X_k=1$ as a proportion of the baseline hazard. For example, if $\exp(\beta_k)=1.25$, then mortality in the group for which $X_k=1$ is 25% higher than for the baseline group.

The covariates X_{ki} are specified in equation (2) to be time invariant, a suitable specification for variables which may be regarded as fixed as of the beginning of the observation period, such as family background, formal educational attainment, and prior labour force experiences. Some covariates, however, such as occupational position or family assets after 1966, may vary with time. To include the latter covariates in the model it would be necessary to generalize equation (2) by replacing X_{ki} with $X_{ki}(t)$. In practice, models with time-varying covariates require substantially more computation to estimate than models with only fixed covariates. Analyses reported in this chapter are confined to examining the effects of fixed covariates. In assessing the effects of occupations and family assets at mid life, therefore, the present analysis treats these variables as fixed at their 1966 levels.

7.3. Stratification for Cohort Differences

Because the NLS data represent men aged 44–60 in 1966, they include a number of birth cohorts. The broad range of cohorts may make equation (2) unsuitable for analysis of socio-economic effects because of intercohort changes in mortality, cohort differences in selection of healthy individuals, and a 'built-in' non-proportionality of cohort mortality differences after age 44. Intercohort trends in mortality may alter the level or shape of λ_0, invalidating the assumption of a fixed baseline hazard. In addition, the sample design may create heterogeneity among birth cohorts in proportions surviving to selected ages. For example, 100% of sample men aged 55 in 1966 survive to age 55, but less than 100% of men aged 45 in 1966 survive to age 55. As of 1966, therefore, older cohorts are more selective than younger cohorts of men with good survival chances. The selectivity may bias estimates of age and cohort patterns of mortality, as well as the estimated effects of covariates (Vaupel *et al.* 1979; Manton and Stallard 1984; Vaupel and Yashin 1985). Finally, even in the absence of differential selectivity or intercohort change, the design of the NLS dictates that cohort-specific hazards are non-proportional. By design, the mortality hazard is zero for sample observations until 1966. After 1966, the hazard follows the appropriate age pattern of mortality. In the absence of cohort trends or selectivity effects, different cohorts will have different and severely non-parallel hazard functions from age 45 onward.

One strategy for taking account of cohort effects is simply to include indicators of cohort membership as covariates in equation (2). This controls for proportional upward or downward shifts in mortality rates over cohorts and for correlations between cohort membership and other covariates in the model. This strategy fails, however, to control for non-proportional cohort differences. Thus, we stratify the data by birth cohort, and estimate equation (2) within cohorts, albeit with common effects β_k of the covariates across cohorts (Kalbfleisch and Prentice 1980: 87–8; Breslow *et al.* 1983). For the jth birth cohort ($j = 1, \ldots, J$), the model is

$$\lambda_{ij}(t \mid X_{1ij}, \ldots, X_{kij}) = \lambda_{0j}(t) \left[\exp \left(\sum_k \beta_k X_{kij} \right) \right], \tag{3}$$

where all other notation is as defined above. Unlike equation (2), which assumes a single unobserved baseline hazard $\lambda_0(t)$, this model allows for J baseline hazards $\lambda_{0j}(t)$, one for each cohort. The β_k denote deviations from cohort-specific baseline hazards resulting from variation in the X_k.

In practice, we stratify the observations into six cohorts that are exact ages 44–7, 47–50, 50–3, 53–6, 56–9, and 59–61 in 1966. We estimate equation (3) for selected combinations of independent variables using the partial likelihood method (e.g. Kalbfleisch and Prentice 1980), which is implemented with the BMDP computer package (Dixon 1983).

8. Empirical Results

This section first summarizes the socio-economic characteristics of the NLS sample and their implications for differential mortality. Secondly, it describes differentials in average years of life over the 1966–83 period for the NLS respondents. Thirdly, it discusses the estimates of proportional hazards models that show the gross effects of the socio-economic variables. Finally, it presents multivariate results on the net effects of socio-economic factors on mortality.

8.1. Socio-economic Distributions

Table 19.2 presents the distributions of the NLS sample by categories of the independent variables. Because the survey over-represents blacks, distributions by socio-economic status are more heavily weighted toward lower socio-economic positions relative to the population. The 1966 family assets measure, however, is based on sample quartiles. Twenty per cent of sample respondents are missing data on assets in 1966.

The occupation distributions illustrate well-known patterns of inter- and intragenerational changes in occupation distributions in the USA (e.g.

Table 19.2. *Proportions of men aged 45–59 in selected socio-economic categories, National Longitudinal Survey, 1966* (no. = 5006)

Race	%			
Non-black	0.72			
Black	0.28			
Schooling (grades)	**%**			
< 8	0.29			
8–11	0.36			
12	0.21			
13 +	0.15			
Occupation		**Father's (%)***	**First (%)***	**1966 (%)***
Professional, managerial		0.13	0.08	0.21
Clerical, sales		0.04	0.12	0.09
Craftsmen		0.11	0.06	0.22
Operatives		0.12	0.23	0.20
Service, private household		0.06	0.06	0.08
Farm labourers		0.02	0.23	0.03
Non-farm labourers		0.07	0.14	0.10
Farm owners, managers, tenants		0.36	0.04	0.06
Armed forces		0.00	0.01	—
Family assets (A) (1966) (by quartile)	**%***		**$**	
1st	0.20		− 32 304 < A < 720	
2nd	0.20		720 < A < 8 243	
3rd	0.20		8 243 < A < 21 778	
4th	0.20		21 778 < A < 1028 805	

* Proportions may not add up to 1 because all classifications include a category for missing data.

Featherman and Hauser 1978). In particular, men aged 45–59 in 1966 were much more likely than their fathers to be employed in 'white-collar' (professional, managerial, clerical, and sales) occupations, and much less likely to be the owners, managers, or tenants of farms. Within the NLS cohort, 37% of men begin their work lives as farm or non-farm labourers but only 13% occupy these positions by age 45–59. In addition, from first to 1966 jobs, the occupation distributions show relative increases in numbers of men in upper white-collar and upper blue-collar positions. These intra-generational differences result from both typical age patterns of employment and also historical trends in the occupation composition of the labour force.

8.2. Differences in Average Years Lived

Life table analyses illustrate patterns of socio-economic differentials by duration of life. Such analyses are ill suited to multivariate investigations, but they describe basic socio-economic differences. Table 19.3 reports

Table 19.3. *Estimated average years lived between 1966 and 1983, by cohort and selected socio-economic category*

Socio-economic category	Approximate age span observed 1966–83					
	46–63	49–66	52–69	55–72	58–75	60–77
All (NLS)	15.4	15.0	14.4	13.9	13.5	12.9
(No. of observations)	(807)	(1088)	(1053)	(971)	(833)	(254)
All (1975 US male life table)	15.6	15.2	14.7	14.2	13.5	13.1
Schooling (by grade)						
< 8	14.4	14.8	13.9	14.1	12.6	12.0
8–11	15.4	14.9	14.1	13.8	13.7	12.7
12	15.8	15.2	14.7	13.8	14.6	13.0
13+	16.2	15.5	15.3	14.3	14.9	14.5
Occupation, 1966						
Professional, managerial	16.1	15.6	14.5	14.1	14.4	13.6
Clerical, sales	15.1	15.1	14.8	13.8	13.8	12.9
Craftsmen	15.6	15.3	14.6	14.3	13.9	12.5
Operatives	15.0	14.7	14.4	13.8	13.8	14.5
Service, private household	14.5	14.2	13.5	13.6	12.0	11.2
Farm labourers	13.8	13.8	13.5	14.2	11.9	12.6
Non-farm labourers	14.9	14.8	13.9	12.8	11.8	9.6
Farm owners, managers	16.4	15.5	14.8	15.3	14.7	13.5
Family assets, 1966 (by quartile)						
1st	14.3	13.5	13.0	13.3	12.0	11.4
2nd	15.4	15.0	14.7	13.8	12.9	12.8
3rd	15.7	15.6	14.3	14.3	13.7	12.5
4th	16.3	15.6	15.4	14.4	15.0	13.5

expected years of life between selected ages of late adulthood for the NLS men. The age spans correspond to three-year birth cohorts. The first panel of the table compares estimates of average years lived in a 17-year interval for NLS men to corresponding life table functions based on US published vital statistics (US Department of Health, Education, and Welfare 1979) for all socio-economic groups combined. The vital statistics are based on 1975 deaths to US males, events that occur approximately midway through the NLS observation period. As these results indicate, NLS and vital-statistics estimates of average years lived agree closely. At most the two estimates differ by 0.3 of a year. Like the vital-statistics estimates, the NLS averages trace the age pattern of mortality, which implies that average years lived vary inversely with initial age of the intervals.

The second panel of Table 19.3 shows substantial differences in average years lived among men with varying amounts of formal schooling. For five of the six cohorts, years lived varies directly with educational status, producing a difference of approximately two years between men with only elementary school (< 8 grades) and those with at least some post-secondary schooling (13+ grades). These results are highly consistent with those obtained from the 1960 Matched Records Study. Kitagawa and Hauser (1973: 17) report a difference between elementary- and college-educated men in expected years of life remaining at age 45 of 1.7 to 2.2 years, which they attribute mainly to mortality from age 45 to 65 (rather than after 65). The estimates in Table 19.3 indicate similarly large differences in years lived by educational status. This pattern is unlikely to result from adverse selection of persons in poor health into lower-status groups. Rather, differentials in schooling are established early in life and index differences in style of life and socio-economic status that persist into middle age.

Table 19.3 also reports differences in average years lived by 1966 occupation and family assets. These estimates are more likely to result from both genuine socio-economic effects on mortality and adverse selection of persons in poor health into the lower-status groups. The differentials are, none the less, broadly consistent with those across schooling groups. In most cohorts men in the most advantaged occupational groups—professionals, managers, farm owners, and farm managers—average 1.5 to 2 years of life more than those in the least advantaged groups—labourers, service, and private-household workers. The farm occupations, including both owners and labourers, are a dwindling proportion of the labour force, suggesting that both highly unfavourable and highly favourable occupations have been reduced in number as the working population has become more urbanized. Table 19.3 also shows that average years of life vary directly with family assets, resulting in a difference of approximately two years between the poorest and wealthiest quartiles.

8.3. Gross Effects of Socio-economic Characteristics

Whereas the life table analyses reveal socio-economic variation in durations of life from 1966 to 1983, proportional hazards models show the effects of socio-economic characteristics on instantaneous mortality rates during the observation period. Each model controls for within-sample cohort differences by allowing, through stratification, for cohort-specific baseline hazard functions. Table 19.4 reports the socio-economic effects for some elementary models. The first three columns of the table present models for the effects of each socio-economic characteristics considered alone. The first column reports the βs for the difference in the log hazard rate [$\log\{\lambda_i(t)\}$] between each social category and the baseline category (in parentheses). The second reports the asymptotic normal statistic $Z(\beta)$ for each parameter. Under simple random sampling, values of these statistics in excess of ±1.96 indicate significance beyond the 0.05 level for a two-tailed test.[4] The third reports the quantities $\exp(\beta)$, which denote proportionate differences in mortality rates between each category and the baseline category.

Table 19.4. *Socio-economic effects on mortality, 1966–1983, men aged 45–59 in 1966, selected proportional hazard models*[a]

Variable[b]	Gross effects			Net effect after controlling for race, education, father's occupation		
	β	$[\beta/SE(\beta)]$	$\exp(\beta)$	β	$[\beta/SE(\beta)]$	$\exp(\beta)$
Race						
(Non-black)						
Black	0.373	(6.9)	1.45	0.288	(4.7)	1.33
		(−2 log L = 19122)				
Schooling (by grade)						
(< 8)						
8–11	−0.192	(−3.2)	0.83	−0.145	(−2.2)	0.87
12	−0.386	(−5.0)	0.68	−0.329	(−3.8)	0.72
13+	−0.608	(−6.6)	0.54	−0.542	(−5.2)	0.58
		(−2 log L = 19108)				
Father's occupation						
(Professional, managerial)						
Clerical, sales	0.119	(0.7)	1.13	0.100	(0.6)	1.11
Craftsmen	0.188	(1.7)	1.21	0.075	(0.6)	1.08
Operatives	0.365	(3.4)	1.44	0.171	(1.5)	1.19

[4] The NLS is a multi-stage probability sample, rather than a random sample. The exact effect of this on statistical tests is unknown. A conservative estimate of statistical significance at the 0.05 level, however, may require that $Z(\beta)$ be as large as 3 for such a complex sample. Strictly speaking, therefore, the reported Z are of descriptive value only.

Table 19.4. *Continued*

Variable[b]	Gross effects			Net effect after controlling for race, education, father's occupation		
	β	$[\beta/SE(\beta)]$	$\exp(\beta)$	β	$[\beta/SE(\beta)]$	$\exp(\beta)$
Service, private household	0.170	(1.3)	1.19	−0.123	(−0.9)	0.88
Farm labourers	0.550	(3.2)	1.73	0.107	(0.6)	1.11
Non-farm labourers	0.285	(2.3)	1.33	−0.012	(−0.1)	0.99
Farm owners, managers, tenants	0.161	(1.8)	1.17	−0.158	(−1.6)	0.85
Armed forces	0.089	(0.1)	1.09	0.118	(0.1)	1.13
	(−2 log L = 19143)			(−2 log L = 19065)		
First occupation						
(Professional, managerial)						
Clerical, sales	0.400	(2.8)	1.49			
Craftsmen	0.550	(3.5)	1.73			
Operatives	0.571	(4.4)	1.77			
Service, private household	0.668	(4.2)	1.95			
Farm labourers	0.638	(4.9)	1.89			
Non-farm labourers	0.658	(4.9)	1.93			
Farm owners, managers, tenants	0.295	(1.7)	1.34			
Armed forces	0.790	(2.3)	2.20			
	(−2 log L = 19126)					
Occupation, 1966						
(Professional, managerial)						
Clerical, sales	0.169	(1.5)	1.18			
Craftsmen	0.132	(1.5)	1.14			
Operatives	0.278	(3.3)	1.32			
Service, private household	0.592	(5.9)	1.81			
Farm labourers	0.630	(4.8)	1.88			
Non-farm labourers	0.621	(6.6)	1.86			
Farm owners, managers, tenants	−0.121	(−0.9)	0.87			
	(−2 log L = 19081)					
Family assets, 1966 (by quartile)						
(1st)						
2nd	−0.404	(−5.4)	0.67			
3rd	−0.546	(−6.9)	0.58			
4th	−0.782	(−9.6)	0.46			
	(−2 log L = 19059)					

[a] All models are estimated within strata defined by three-year year-of-birth categories (see text). (No. = 5006.)

[b] All classifications include a category for missing data.

8.3.1 Race and schooling The gross effects of socio-economic factors on the mortality hazard are largely what one would expect from prior research and from the life table estimates discussed above. The mortality hazard for blacks exceeds that of whites (non-blacks) by approximately 45%, a difference similar to that between men with high-school degrees and those with only elementary-school educations (1.0/0.68 ≐ 1.47). Mortality varies inversely with schooling level. Even at the post-secondary-schooling level, marginal increases in schooling continue to dampen the mortality hazard.

8.3.2. Father's occupation Occupational variation in the mortality hazard has a similar pattern to that for schooling but varies in strength across measure of occupation (father's, first, and 1966). Father's occupation affects the survival chances of sons in middle age. The mortality hazards of sons of operatives, farm labourers, and non-farm labourers exceed those for sons of professionals and managers by 44%, 73%, and 33% respectively. More generally, men from blue-collar backgrounds experience higher death rates during middle age than men from white-collar backgrounds, although many of the specific contrasts among occupations are not statistically significant. Whether the differentials by father's occupation indicate specific life-long effects of social origins or of socio-economic circumstances that are correlated from one generation to the next will be discussed further below.

8.3.3. First occupation The effects of first occupation on the mortality hazard are generally stronger than for father's occupation. Mortality hazards for men who enter the labour force in lower blue-collar occupations are nearly twice those of men who enter as professionals or managers. As Table 19.2 shows, however, less than 10% of NLS men had first jobs as professionals or managers. Although differences in mortality hazards among blue-collar occupations and between blue-collar and lower white-collar occupations for first job follow an approximate socio-economic gradient, these differences are small relative to the contrast with the upper white-collar occupations. Finally, while the mortality hazard for men who start their careers in the armed forces is the highest for any occupation, possibly indicating the poor health prospects of World War II veterans, this estimate is based on only 1% of the NLS sample (see Table 19.2).

8.3.4. 1966 occupation The gross differences in mortality hazards among occupations at the onset of the study in 1966 resemble those among first occupations. Professionals and managers have superior survival chances to all other groups, although the difference between them and the small-farm owner and manager category is trivial. Relative to the first-job differentials, however, 1966 occupation differentials show a clearer socio-economic gradient. For example, non-farm labourers in 1966 have approximately a

40% higher hazard than operatives (1.86/1.32) and a 60% higher hazard than craftsmen. The hazard for clerical and sales workers is only about 15–20% higher than for the highest white-collar group, but well below that of all blue-collar occupations except craftsmen.

8.3.5. Family assets The mortality hazard has a strong inverse relationship with family assets. The largest contrast occurs between the lowest quarter of the asset distribution and the remainder of the asset distribution. Men in the second quartile of the asset distribution experience mortality at approximately two-thirds the rate of men in the first quartile, whereas men in the highest quartile have one-half the mortality hazard of men in the poorest families.

8.4. Net Effects of Socio-economic Characteristics

The gross effects of socio-economic characteristics reported above establish the size of mortality differentials and imply that socio-economic processes deserve further investigation and interpretation. They do not, of course, indicate causal relationships between socio-economic factors and survival. The effects of 'early' socio-economic status, indexed by father's occupation, schooling, and first occupation, are unlikely to result from adverse selection into lower-status positions of persons in poor health because these statuses apply to circumstances 25 or more years prior to the observation period. Whether their effects persist when later events and statuses are controlled, however, remains to be investigated. Conversely, the gross effects of 'late' socio-economic status, while statistically strong, may result from patterns of health-related selection.

The models discussed in this section introduce the effects of socio-economic positions in their approximate temporal order. The last three columns of Table 19.4 report the net effects of race, schooling, and father's occupation to show whether the gross father's-occupation effects persist when schooling and race are controlled. As the results indicate, the gross effects of father's occupation virtually disappear when race and schooling are included in the model.[5] The net effect of schooling on the mortality hazard, in contrast, is only slightly smaller than its gross effect. Evidently, the gross effects of father's occupation are almost entirely due to the ability of sons from higher socio-economic origins to acquire more schooling, rather than from persistent effects of socio-economic origins through life. This result agrees with those for the effects of social origins on outcomes other than mortality. Most of the association between father's occupation and son's occupation at maturity, for example, is attributable to education differentials by social origins (for example, Featherman and Hauser 1978).

[5] Estimates not shown here reveal that the decline mainly results from the control for schooling rather than for race.

Table 19.5. *Socio-economic effects on mortality, 1966–1983, men aged 45–59 in 1966, selected proportional hazard models*[a]

Variable	Race, schooling, first occupation			Race, schooling, first and 1966 occupations			Race, schooling, occupations, assets		
	β	$[Z(\beta)]^b$	$\exp(\beta)$	β	$[(Z(\beta)]^b$	$\exp(\beta)$	β	$[Z(\beta)]^b$	$\exp(\beta)$
Race									
(Non-black)									
Black	0.253	(4.1)	1.29	0.167	(2.6)	1.18	0.082	(1.3)	1.09
Schooling (by grade)									
(< 8)									
8–11	−0.117	(−1.8)	0.89	−0.077	(−1.1)	0.93	−0.019	(−0.3)	0.98
12	−0.294	(−3.4)	0.75	−0.251	(−2.8)	0.78	−0.167	(−1.8)	0.85
13 +	−0.411	(−3.5)	0.66	−0.365	(−3.0)	0.69	−0.278	(−2.2)	0.76
First occupation									
(Professional, managerial)									
Clerical, sales	0.294	(1.9)	1.34	0.273	(1.8)	1.31	0.287	(1.9)	1.33
Craftsmen	0.363	(2.1)	1.44	0.387	(2.3)	1.47	0.402	(2.4)	1.49
Operatives	0.312	(2.1)	1.37	0.300	(2.0)	1.35	0.299	(2.0)	1.35
Service, private household	0.330	(1.9)	1.39	0.294	(1.7)	1.34	0.290	(1.6)	1.34
Farm labourers	0.253	(1.7)	1.29	0.246	(1.6)	1.28	0.247	(1.6)	1.28
Non-farm labourers	0.335	(2.2)	1.40	0.318	(2.0)	1.37	0.304	(1.9)	1.35
Farm owners, managers	−0.033	(−0.2)	0.97	0.042	(0.2)	1.04	0.036	(0.2)	1.04
Armed Forces	0.724	(2.1)	2.06	0.703	(2.1)	2.02	0.736	(2.2)	2.09

Occupation, 1966									
(Professional, managerial)									
Clerical, sales				0.075	(0.6)	1.08	0.036	(0.3)	1.04
Craftsmen				−0.081	(−0.9)	0.92	−0.123	(−1.3)	0.88
Operatives				0.002	(0.0)	1.00	−0.068	(−0.7)	0.93
Service, private household				0.320	(2.9)	1.38	0.219	(1.9)	1.24
Farm labourers				0.305	(2.1)	1.36	0.108	(0.7)	1.11
Non-farm labourers				0.285	(2.6)	1.33	0.171	(1.5)	1.19
Farm owners, managers				−0.274	(−1.9)	0.76	−0.240	(1.7)	0.79
Family assets, 1966 (by quartile)									
(1st)									
2nd							−0.307	(−3.9)	0.74
3rd							−0.398	(−4.6)	0.67
4th							−0.527	(−5.5)	0.59
−2 log L		19075			19038			19002	

[a] All models are estimated within strata defined by three-year year-of-birth categories. (No. = 5006.) All classifications include a category for missing data.

[b] $Z(\beta)$ denotes the asymptotic normal statistic associated with each estimated β, that is, $\beta/SE(\beta)$.

In view of these results, the remaining models include race and schooling, but not father's occupation.

Table 19.5 reports estimates of models that include the effects of NLS respondents' own occupations, controlling for race and schooling. The first three columns report a model that includes race, schooling, and first occupation. They show that approximately one-third of the gross mortality differences by race and by educational level are transmitted to middle-aged men through their initial occupations. (Compare, for example, net βs for schooling of -0.117, -0.294, and -0.411 to gross βs of -0.192, -0.386, and -0.608 (see Table 19.4)). Within first-occupation categories men with some post-secondary schooling have a mortality hazard that is about two-thirds of the hazard for men with only elementary schooling, in contrast to a gross differential of almost 50%. Conversely, much of the gross mortality differential among first occupations results from the association between educational attainment and status of first job. Once race and schooling are controlled, the only important contrast in hazards that remains is between professionals or managers and all other white- and blue-collar occupations for which the hazard is 30–40% higher. This contrast is roughly half the size of the gross contrast reported in Table 19.4. (Differences between farm owners and members of the armed forces and their first occupations are large, but, as noted above, farm ownership and the military are uncommon first occupations.)

The middle three columns of Table 19.5 report estimates for a model that includes race, school, and both first and 1966 occupations. The net effects of 1966 occupation are substantially smaller than the gross effects. Mortality hazards for lower white-collar and upper blue-collar workers are essentially the same as for professionals and managers once the other variables are controlled. However, the hazards for lower blue-collar occupations are 35% higher than for these groups. Among the small fraction of men who are farm owners and managers in 1966, the mortality hazard is lowest of all occupations, approximately 75% of that for professionals and managers. This contrast, however, is barely statistically significant.

Taking account of 1966 occupation reduces the size of contrasts among schooling groups by approximately three-quarters compared to the model that includes only first occupation. A clear educational gradient none the less remains, indicating that the effects of schooling persist, even when occupations are taken into account. Additionally, the net effects of first occupation are virtually unchanged once 1966 occupation is taken into account. That is, the mortality hazards of lower white-collar and blue-collar first occupations are all approximately 30% higher than for professional and managerial first occupations. In contrast to occupation at maturity, where lower blue-collar workers experience higher mortality hazards than other groups, for first occupation all blue-collar and lower white-collar groups are disadvantaged compared to upper white-collar occupations (and farm owners).

The last model, reported in the final three columns of Table 19.5, includes the effects of family assets in 1966. Although the gross effect of assets is reduced by approximately one-fifth when race, education, and first and 1966 occupations are controlled, the net effects are substantial. Controlling all other variables, the mortality hazard for men in the top quartile of the asset distribution is only 60% of that for men in the bottom quartile. Differences in wealth also account for substantial portions of the effects of 1966 occupations, which still indicate higher net mortality for blue-collar than white-collar positions but with a much smaller differential than is observed when assets are not taken into account. Similarly, the effects of educational attainment are further reduced once wealth is controlled. Their final net effects continue to show a clear inverse relationship between school attainment and the mortality hazard, but the net differentials are approximately 40% of the gross differentials. In contrast, the effects of first occupation are unchanged once family assets are taken into account. A lifelong benefit to entering the labour force in professional or managerial occupations remains, even taking account of the later occupational and monetary benefits that may accrue to workers.

8.5. The Effects of 'Occupational Careers'

First and 1966 occupations are not a full summary of occupational experience, but they provide a considerably fuller picture of men's careers than a single occupational measure.[6] Another way of illustrating the combined effects of early and later occupations on mortality is to examine the relative mortality risks of men in given combinations of first and 1966 occupations.

Table 19.6 re-expresses the estimates of the second model in Table 19.5 by reporting the proportionate difference in mortality for men in each combination of occupations compared to men who were in professional or managerial positions for both their first and 1966 occupations. These calculations assume that the effects of the two occupational statuses are multiplicative on the hazard (additive on the log hazard), and that no more complex interactions are present. They show that, controlling for educational and racial differences among occupations, men who are in labouring occupations at both the beginning and toward the ends of their working lives have mortality risks approximately 80% higher than men who are professionals or managers both early and late in their careers. (See diagonal elements in Table 19.6 of 1.74 for farm labourers and 1.82 for non-farm

[6] Only 25% of NLS men occupied the same first and 1966 major occupational category.

Table 19.6. *Relative hazards for combinations of first and 1966 occupations**

First occupation	1966 occupation							
	PM	CS	C	O	SP	FL	NL	FO
Professional, managerial (PM)	1.00	1.08	0.92	1.00	1.38	1.36	1.33	0.76
Clerical, sales (CS)	1.31	1.41	1.20	1.31	1.81	1.78	1.74	1.00
Craftsmen (C)	1.47	1.59	1.35	1.47	2.03	2.00	1.96	1.12
Operatives (O)	1.35	1.46	1.24	1.35	1.86	1.84	1.80	1.03
Service, private households (SP)	1.34	1.45	1.24	1.34	1.85	1.82	1.78	1.02
Farm labourers (FL)	1.28	1.38	1.18	1.28	1.77	1.74	1.70	0.97
Non-farm labourers (NL)	1.37	1.48	1.26	1.37	1.89	1.86	1.82	1.04
Farm owners, managers (FO)	1.04	1.12	0.96	1.04	1.44	1.41	1.38	0.79

* Relative hazards are computed from model 2 in Table 19.5 and assume a constant race–education composition across occupations.

labourers). A man who enters the labour force as a labourer but enjoys mobility into upper blue-collar or white-collar occupations can reduce approximately half of the disadvantage associated with a lifelong career as a labourer.

9. Summary and Conclusion

9.1. Summary of Findings

This chapter has documented the large socio-economic differences in mortality in the USA that have been reported by other researchers and, in addition, has provided more refined estimates of the way in which socio-economic conditions affect the timing of death in later adulthood. Although men's mortality hazards in late adulthood vary inversely with the socio-economic status of their parents—as indicated by fathers' occupations, this effect is transmitted through later experiences, especially educational attainment. Men with at least some post-secondary schooling enjoy two more years of life and a mortality hazard 50% lower than men with only elementary schooling. About half of this differential, however, results from the occupational and financial advantages enjoyed by men with more schooling. Men who enter the labour force as professionals or managers enjoy a 30% lower mortality hazard during the late adult years, a differential that persists when the later socio-economic advantages enjoyed by such men are taken into account. Mortality hazards vary inversely with occupation at maturity (e.g. in 1966) along a more continuous socio-economic gradient on which the hazards for upper blue-collar and lower white-collar workers fall between those of labourers and professionals or managers. Most of the later occupation effect, however, results from differences in the wealth associated with occupation, which negatively affects the mortality hazard.

9.2. Conclusions

The analyses reported here provide several estimates of socio-economic effects that are relatively free from the biases from health-related selection into socio-economic positions. First, differentials in schooling and first occupation are associated with variation in mortality rates measured at least 25 years after these statuses are required. Except among a small number of men who endure ill health from their childhood on, schooling and initial occupational status are not determined by health conditions that also determine the timing of death. Rather, these statuses themselves may determine the style and quality of life, which in turn govern the timing of life-threatening illnesses. These results also suggest that recent and future gains in life expectancy may be realized as a result of increases in levels of educational attainment. Almost 30% of NLS respondents fell into the lowest educational level, a much higher proportion than in cohorts born since 1950, for whom proportions with only elementary school education fell well below 10% (e.g. Mare 1981).

Secondly, in so far as early statuses do affect long-run health, the net effects of later occupational attainment and family wealth reported here are purged of selection biases to a degree. From such analyses as these, of course, one cannot rule out contaminating influences of those aspects of ill health that are unrelated to the early socio-economic conditions controlled in the models.

The analyses reported here illustrate the value of relatively small-scale surveys for the understanding of differential mortality. No doubt, sampling errors for estimated effects are larger than desirable. But powerful statistical methods, appropriately applied, enable one to obtain valid estimates of socio-economic effects that are sufficiently precise to be compared to those derived from large-scale data sources that are dedicated to mortality analysis.

9.3. Work in Progress

The NLS not only provides the basis for estimating basic socio-economic differences in mortality among older men in the USA, but will also support several more refined investigations that are now under way (Mare and Palloni 1984). These include (*a*) the incorporation into basic models of differential mortality short-term variation in socio-economic characteristics, such as assets, employment status, and occupational standing; (*b*) analysis of the effects of self-reports of health on mortality for the purpose of further controlling for health-related selection into occupations as a possible source of bias in estimates of socio-economic mortality differences; and (*c*) joint analysis of the survival of NLS men and their wives. The last of these topics includes investigation of the effects of changes in marital status on mortality and exploration of possible 'bereavement effects' on the mortality hazard of the surviving spouse. It also enables us to examine the relative importance of common family experiences (life-style, place of residence, wealth, etc.) and specific aspects of work and socio-economic background that differ between spouses in determining mortality chances.

References

Breslow, N.E., Lubin, J.H., Marek, P., and Langholz, B. (1983), 'Multiplicative Models and Cohort Analysis', *Journal of the American Statistical Association*, 78, 1–12.

Center for Human Resource Research (1985), *The National Longitudinal Surveys Handbook 1985* (Columbus, Ohio: Center for Human Resource Research, College of Administrative Science, Ohio State University).

Cox, D.R. (1972), 'Regression Models and Life Tables (with Discussion)', *Journal of the Royal Statistical Society*, series B, 34, 187–220.

Dixon, W.J. (1983), *BMDP Statistical Software, 1983 Printing with Additions* (Berkeley: University of California Press).

Duncan, O.D., Featherman, D.L., and Duncan, B. (1972), *Socioeconomic Background and Achievement* (New York: Seminar Press).

Featherman, D.L., and Hauser, R.M. (1978), *Opportunity and Change* (New York: Academic Press).

Fox, A.J. (1979), 'Prospects for Change in Differential Mortality', in *Proceedings of the Meeting on Socioeconomic Determinants and Consequences of Mortality*, Mexico City, pp. 515–61.

—— (1984), 'Design Problems and Data Collection Strategies in Studies of Mortality Differentials: Developed Countries', in J. Vallin, J.H. Pollard, and L. Heligman (eds.), *Methodologies for the Collection and Analysis of Mortality Data* (Liège: Ordina Éditions).

—— Goldblatt, P.O., and Adelstein, A.M. (1982), 'Selection and Mortality Differentials', *Journal of Epidemiology and Community Health*, 36, 69–79.

—— and Jones, D.R. (1985), 'Social Class Mortality Differentials: Artefact, Selection, or Life Circumstances?', *Journal of Epidemiology and Community Health*, 39, 1–8.

Kalbfleisch, J.D., and Prentice, R.L. (1980), *The Statistical Analysis of Failure Time Data* (New York: John Wiley and Sons).

Kitagawa, E.M., and Hauser, P.M. (1973), *Differential Mortality in the United States* (Cambridge, Mass: Harvard University Press).

Land, K.C., and Hough, G.C., jun. (1986), 'Improving the Accuracy of Intercensal Estimates and Postcensal Projections of the Civilian Noninstitutional Population', *Journal of the American Statistical Association*, 81, 62–74.

Manton, K.G., and Stallard, E. (1984), 'Heterogeneity and Its Effects on Mortality Measurement', in J. Vallin, J.H. Pollard, and L. Heligman (eds.), *Methodologies for the Collection and Analysis of Mortality Data* (Liège, Belgium: Ordina Éditions).

Mare, R.D. (1981), 'Trends in Schooling: Demography, Performance, and Organization', *The Annals of the American Academy of Political and Social Science*, 453, 96–122.

—— and Palloni, A. (1984), *Socioeconomic Effects on Mortality for Older*

American Men: Proposal to National Institute of Aging (Washington: US National Institutes of Health).

Mott, F.L., and Haurin, R.J. (1985), 'Factors Affecting Mortality in the Years Surrounding Retirement', in H.S. Parnes, J.E. Crowley, R.J. Hauring, W.R. Morgan, F.L. Mott, and G. Nestel (eds.), *Retirement among American Men* (Lexington, Mass.: Lexington Books).

Rosenberg, H.M., and McMillen, M.M. (1983), 'New Research Directions on Socioeconomic Differentials in Mortality: United States of America', Paper presented to the *UN/WHO/CICRED Network on Socio-Economic Differential Mortality in the Industrialized Societies, Third Meeting*, Rome, 24–7 May.

Sewell, W.H., and Hauser, R.M. (1975), *Education, Occupation, and Earnings: Achievement in the Early Career* (New York: Academic Press).

US Department of Health, Education, and Welfare (1979), *Vital Statistics of the United States 1975*, ii: *Mortality*, part A (Hyattsville, Md.: US Department of Health, Education, and Welfare, Public Health Service, National Center for Health Statistics).

Vaupel, J.W., and Yashin, A.I. (1985), 'Heterogeneity's Ruses: Some Surprising Effects of Selection on Population Dynamics', *American Statistician*, 39, 176–85.

—— Manton, K.G., and Stallard, E. (1979), 'The Impact of Heterogeneity in Individual Frailty on the Dynamics of Mortality', *Demography*, 16, 439–54.

20 Assessing Morbidity Levels from Mortality Measurement: The Case of Cancer in Italy, 1955–1979

VIVIANA EGIDI

Dipartimento di Scienze Demografiche, Università di Roma 'La Sapienza', Rome, Italy

ANTONIO GOLINI

Istituto di Ricerche sulla Popolazione (IRP), Rome, Italy

and RICCARDO CAPOCACCIA and ARDUINO VERDECCHIA

Istituto Superiore di Sanità (ISS), Rome, Italy

The sharp fall in mortality experienced in Italy and other developed countries since the war has radically altered the cause-structure of mortality. The incidence of acute illness has gradually declined, while there has been a continuing increase in chronic and degenerative illness, namely, long-term illness which can temporarily or permanently disable the individual, often curbing his activities to a considerable degree. It is therefore important to appraise not only levels and trends of mortality but also the levels and trends of morbidity and health status in the population. A clearer understanding of these factors in demographic and social terms opens up considerable possibilities for further research and action (Golini and Egidi 1981). In fact, knowledge of morbid processes would enable us to have a theoretical apparatus and quantitative data to be used for the study of population differentials, for predicting future mortality trends, and for designing public health policies. The search for exogenous and endogenous mortality determinants could be greatly enhanced if potential risk exposure were linked to the onset of illness (the primary event) rather than death (the final event). It would also be possible to analyse relations eliminating the confounding effects of other factors that might affect the duration and final outcome of the morbid process.

Many countries have begun collecting information on the incidence and spread of the most common morbid processes, especially by instituting registers (for tumours and, in some cases, for heart diesease), or by running health surveys on selected samples of the population. Yet these efforts are insufficient: there is a serious shortage of data, and there are many countries with no reliable national estimates of morbidity. As a consequence, many researchers have focused their efforts on formulating procedures for obtaining indirect estimates using mortality data (Klementiev 1977; Kitsul 1980; Manton and Stallard 1982*a*,*b*,1984; Verdecchia *et al.* 1984, 1985).

This chapter attempts to assess the morbidity levels (incidence and prevalence) of tumours in the Italian population, and the changes that have taken place in these levels over the period 1955–79, using the data on tumour deaths in Italy and, where these do not exist, using tumour survival curves based on appropriately adapted US data. For this purpose, we have used a statistical model designed and tested on real data with satisfactory results (Capocaccia *et al.* 1985; Verdecchia *et al.* 1989). An attempt is made to interpret mortality trends for all tumours in the light of (*a*) the estimated tumour occurrence variation, (*b*) the lethality rate variation, and (*c*) the variation of the average survival period after onset.

1. The Essential Features of the Model Used

1.1. Definitions

The following symbols and definitions will be used:

- c_i^a crude probability of dying from the specific cause between exact i and $i+1$ for cohort a;
- d_{ij}^a crude probability of dying from the specific cause between age i and $i+1$ for sick people in cohort a who contracted the disease between age j and $j+1$;
- N_i^a number of people in cohort a still alive at exact age i;
- H_i^a number of people in cohort a still alive and free from the specific disease at exact age i;
- S_{ij}^a number of people in cohort a who contracted the specific disease between exact age j and $j+1$, and who survive at exact age i;
- μ_i^a crude probability for an individual belonging to cohort a and still alive at exact age i to be observed as a new case of the specific disease between exact ages i and $i+1$ (hereafter referred to as 'incidence');
- ν_i^a proportion of people in cohort a that have the specific disease at exact age i;
- $g^a(x, y)$ differential force of elimination for cohort a (or 'relative risk') at age x for individuals who contracted the specific disease at exact age y.

1.2. Model Development

The individuals of a cohort a at exact age i can be divided into two groups according to their health status. The number of sick people and their distribution by time of onset of the disease will be determined by the prior action of two forces on the whole cohort, i.e. morbidity and attrition by death or migration. In symbols:

$$N_i^a = H_i^a + \sum_{j=0}^{i} S_{ij}^a. \tag{1}$$

Similarly, the number of deaths from the specific disease considered can be expressed as:

$$N_i^a c_i^a = N_i^a (1 - \nu_i^a) \mu_i^a d_{ii}^a + \sum_{j=0}^{i-1} S_{ij}^a d_{ij}^a. \tag{2}$$

If we make the assumption that the specific force of attrition associated with the disease is independent of and combines additively with the general force of attrition, equation (1) and (2) can be ultimately written as:

$$\nu_i^a = \sum_{j=0}^{i-1} (1 - \nu_j^a) \mu_j^a \exp\left[- \int_{j+\frac{1}{2}}^{i} g^a(x, j + 1/2) dx, \right] \tag{3}$$

$$c_i^a = (1 - \nu_i^a) \mu_i^a d_{ii}^a + \sum_{j=0}^{i-1} (1 - \nu_j^a) \mu_j^a d_{ij}^a \exp\left[- \int_{j+\frac{1}{2}}^{i} g^a(x, j + 1/2) dx. \right] \tag{4}$$

Equations (3) and (4) describe morbidity and mortality probabilities by age in each cohort a. Notice that all terms represent only proportions or probabilities referring either to the specific disease or to the excess mortality which sick people are subjected to. In other words, these equations only include terms that discriminate between sick and healthy populations. Both the population attrition for any and all causes until age i, and the initial size of the cohort, are not in equation (3) and (4) as they are common to both sick and healthy people. As a result, the model rests neither on assumptions about population structure, i.e. stability or stationarity, nor about past mortality and migration. Only differential effect of both these two phenomena on sick people and the general population is relevant.

1.3. Maximum Likelihood Estimation of Model Parameters

Two different approaches can be used for the estimation of the parameters in the model: direct solution of the equation system (3) and (4) (Capocaccia *et al.* 1985) and maximum likelihood estimation. The latter approach has been preferred here both to minimize the effect of random fluctuations of mortality data and to facilitate the study of the dynamics of the disease. In what follows, we sketch only the main features of the procedure. A more detailed description is reported in Verdecchia *et al.* (1989).

Suppose that we have available mortality and population data on an age-per-period rectangular array. We assume the incidence of the disease to be a logistic function of age, period, and birth cohort of the type:

$$\text{Logit } \mu = \alpha + \sum_{k=0}^{k_1} \beta_k (\text{age})^k + \sum_{k=0}^{k_2} \gamma_k (\text{cohort})^k + \sum_{k=0}^{k_3} \delta_k (\text{period})^k, \qquad (5)$$

where α, β_k, γ_k, and δ_k are parameters to be estimated with the maximum likelihood method that best reproduce observed mortality data. The numbers $k1$, $k2$, and $k3$ have been set equal to 3. If we assume that the observed deaths are independent and follow a Poisson distribution, then the maximum likelihood estimates of the parameters can be obtained by means of an iteratively weighted least square method (Frome 1983). The following procedure has been adopted here:

(*a*) we choose starting values for the parameters and evaluate the incidence function by means of expression (5);
(*b*) expected mortality is computed from the values attained by the incidence function by means of the equations (3) and (4);
(*c*) a weighted sum of squares of observed versus expected deaths is calculated;
(*d*) new values of the model parameters are computed in order to minimize the sum of squares using Rao scoring method (a modification of the Newton-Raphson algorithm);
(*e*) steps (*b*) to (*d*) are repeated until the sum of squares convergence criterion is satisfied.

The final sets of parameters determines in a unique way the incidence and prevalence of the disease. The entire procedure has been carried out using BMDP 3R non-linear regression program.

2. A Central Issue: Choosing the Survival Function

As stated earlier, there are no reliable estimates on the survival of cancer patients in Italy. It will be several years before the longitudinal records that have been created produce useful survival estimates, and the data from many clinical studies conducted are not suitable for generalization to cover the whole population. We have therefore used the results obtained in countries where records have been held for a long time on the beginning and end of illness from tumours. The main source has been the survival curves of cancer patients produced in the United States under the Program for Cancer Surveillance Epidemiology and End Results (SEER Program) covering the years 1950–4 (Axtell *et al.* 1972, 1976) and 1973–9 (Ries *et al.* 1983). These curves have been assumed to reflect the survival rate of cancer patients in

Italy in 1955 (the initial year of the estimate) and 1979 (final year) respectively.

Admittedly the model's capacity to describe the real situation of morbidity in Italy is hampered by the fact that curves referring to another country have been used. A number of preliminary tests were carried out using mortality data on one Italian region which does have a tumour register (Lombardy), and the results were then compared with the actual data taken from the tumour register. The result showed that site-specific survival were appropriate (Capocaccia *et al.* 1985; Verdecchia *et al.* 1989).

2.1. Survival Function Adjustment for Site Distribution

To minimize the distortions in the survival curve that could be produced by a different distribution of patients by primary-tumour site in Italy and the United States, we reconstructed a survival curve for all cancer sites combined. This was done assuming that the specific survival curves for the main sites were the same in Italy and the United States during equivalent time periods, but applying the Italian distribution of tumour location of newly diagnosed tumour patients. The method used for this reconstruction is based on the assumption that once one knows the survival function for each type of tumour and the corresponding number of new cases as a proportion of the total number of tumour cases in both populations, the overall survival function can be calculated using the following expression:

$$\sigma(t) = \sum_{l=1}^{n} r_l \sigma_l(t),$$

$$\sigma'(t) = \sum_{l=1}^{n} r'_l \sigma_l(t) \qquad (l = 1, \ldots, n),$$

where t indicates time from diagnosis, $\sigma(t)$ and $\sigma'(t)$ indicate the overall functions for all the tumours in the American and the Italian populations respectively, r_l and r'_l are the corresponding proportions of each type of tumour, and $\sigma_l(t)$ are the specific survival curves per site, which, because of lack of Italian data, are assumed to be the same for both populations.

In practice, this formula cannot be applied because of the large variety of tumours for many of which the survival curves are either not available at all or are based on too few cases to be reliable. The problem can be approximately solved if we consider only the data on the most common tumours or those which are unevenly spread throughout the two populations. We have taken $\sigma_l(t)$ $(l = 1, \ldots, n_l < n)$ to be the survival rate for the types of tumour selected, and R and R' to be the overall proportions of those selected tumours in the American and Italian populations respectively. Assuming that the relative frequency of all types of tumour that have not been considered are the same in both populations, one can show that $\sigma'(t)$ can be calculated with the following expression:

$$\sigma'(t)=\sigma(t)\frac{1-R'}{1-R}+\sum_{l=1}^{n_l}\sigma_l(t)\left(r_l'-r_l\frac{1-R'}{1-R}\right).$$

In this chapter, the overall survival curve has been adjusted taking into account the following cancer localizations: malignant tumours of the stomach (no. 151 of the Ninth Revision of the ICD), colon and rectum (nos. 152–4), pancreas (no. 157), lung and bronchus (no. 162), malignant melanoma of the skin (no. 172), and malignant tumours of the bladder (no. 188), kidney (no. 189), and brain (no. 191).

Using models (3) and (4), we obtain for each of the sites and periods considered an estimate of the number of newly diagnosed cases and an approximate measure of their contribution to the total number of new tumour cases in a year (r_l') in the Italian population. The information about the samples on which the American survival curves are based has been taken directly from the SEER publications.

Table 20.1 shows the survival curves, the proportions used, and the curves for all cancer sites that are obtained after the adjustment. These reconstructed curves predict an improvement in the survival rate of tumour patients, which will lead to a reduction in the lethality rate from 0.78 to 0.63 between 1955 and 1979. These improvements are most marked in the first year of illness (where patient mortality dropped by 25%, and from year four onwards (falling by between 35% in year four and by almost 100% in year nine and over). The timing of the attrition of patients, which can be retrieved from the average duration of the illness of the persons dying of the disease (T_d), fell slightly from 1.3 to 1.2 years. To appraise the overall duration of the illness we should take into account the experience of both the persons who die and the survivors. If we assume that people who have survived for t_m years (maximum duration of the disease) are no longer ill, the average duration of the illness (T) may be expressed by the formula

$$T=[1-\sigma'(t_m)]\ T_d+t_m\sigma'(t_m).$$

The length of the interval during which an individual who has contracted a tumour can be considered to be still suffering from it is of great importance from both the clinical point of view (to assess the effectiveness of the treatment) and from the public health point of view (to assess the magnitude and structure of the population of the tumour sufferers). Apart from the clinical information which identifies the patient in terms of tumour type, site, time of the treatment, permanence of certain symptoms, and other characteristics, reference may also be made to statistical criteria for identifying the members of the population who must continue to be considered as patients as time passes. One criterion might be to exclude the patients whose risk of death is very close to that of the population as a whole, namely, with a relative mortality risk very close to 1 (say 1.001). Using this criterion, the duration of the illness becomes a highly dynamic element. For example, whereas in 1955 an individual with cancer had to survive at least 15 years

Table 20.1. *Estimate of Italian survival curves for all cancers, relative survival and primary sites distribution for new cancer cases in USA and Italy (r_l and r'_l), males, 1955 and 1979*

ICD Code[a]	Primary site	1955						1979					
		Relative survival after				r_l USA	r'_l Italy	Relative survival after				r_l USA	r'_l Italy
		1 year	3 years	5 years	10 years			1 year	3 years	5 years	10 years		
151	Stomach	0.290	0.129	0.112	0.109	6	23	0.381	0.154	0.125	0.121	3	8
152–154	Colon, rectum	0.658	0.446	0.404	0.357	14	9	0.713	0.530	0.478	0.454	15	8
157	Pancreas	0.169	0.020	0.005	0.003	3	2	0.150	0.037	0.025	0.023	3	2
162	Lung, bronchus	0.191	0.061	0.047	0.045	17	11	0.331	0.131	0.108	0.104	21	22
172	Melanoma of the skin	0.740	0.499	0.416	0.376	1	1	0.919	0.783	0.711	0.669	2	3
188	Urinary, bladder	0.733	0.578	0.522	0.473	7	4	0.882	0.767	0.710	0.692	7	9
189	Kidney	0.597	0.368	0.296	0.246	2	1	0.682	0.524	0.495	0.488	2	2
191	Brain, cranial meninges	0.391	0.193	0.170	0.167	3	3	0.452	0.230	0.198	0.192	2	2
	All sites[b]	0.480	0.298	0.251	0.223	100	100	0.612	0.416	0.377	0.365	100	100

[a] Ninth Revision.
[b] Estimated for Italy.

Table 20.2. *Relative survival curves for cancer patients, by age at diagnosis, Italian males, 1955 and 1979*

Years from diagnosis	Age at diagnosis											
	1955						1979					
	30	40	50	60	70	All	30	40	50	60	70	All
0	1.000	1.000	1.000	1.000	1.000	1.000	1.000	1.000	1.000	1.000	1.000	1.000
1	0.633	0.579	0.528	0.480	0.472	0.480	0.822	0.726	0.622	0.612	0.612	0.612
2	0.523	0.460	0.405	0.353	0.345	0.353	0.743	0.615	0.486	0.475	0.475	0.475
3	0.471	0.406	0.349	0.298	0.290	0.298	0.705	0.564	0.429	0.416	0.416	0.416
4	0.441	0.376	0.319	0.269	0.261	0.269	0.687	0.540	0.402	0.390	0.390	0.390
5	0.424	0.358	0.301	0.252	0.244	0.251	0.678	0.529	0.389	0.377	0.377	0.377
6	0.412	0.346	0.290	0.241	0.233	0.240	0.673	0.523	0.383	0.371	0.371	0.371
7	0.404	0.338	0.282	0.234	0.226	0.233	0.671	0.520	0.380	0.368	0.368	0.368
8	0.399	0.333	0.277	0.229	0.221	0.228	0.670	0.519	0.378	0.366	0.366	0.366
9	0.395	0.330	0.274	0.225	0.218	0.225	0.669	0.518	0.378	0.366	0.366	0.365
10	0.393	0.327	0.271	0.223	0.215	0.223	0.669	0.518	0.377	0.365	0.365	0.365
11	0.391	0.325	0.270	0.222	0.214	0.221	0.669	0.518	0.377	0.365	0.365	0.365
12	0.390	0.324	0.269	0.220	0.213	0.220	0.669	0.518	0.377	0.365	0.365	0.365
13	0.389	0.323	0.268	0.220	0.212	0.219	0.669	0.518	0.377	0.365	0.365	0.365
14	0.388	0.322	0.268	0.219	0.212	0.219	0.669	0.518	0.377	0.365	0.365	0.365
15	0.388	0.322	0.267	0.219	0.211	0.218	0.669	0.518	0.377	0.365	0.365	0.365
16	0.388	0.322	0.266	0.219	0.211	0.218	0.669	0.518	0.377	0.365	0.365	0.365

after onset to close the gap between him and the overall population to be able to consider himself 'cured', in 1979 the relative risk fell below one per 1000 by the eighth year after onset (see Table 20.2). As the following examples show, the average durations of the illness differ markedly in the two periods.

$$1955\text{: } T_d = 1.30;\ \sigma'(9) = 0.225;\ \sigma'(15) = 0.218;$$
$$1979\text{: } T_d = 1.21;\ \sigma'(9) = 0.365;\ \sigma'(15) = 0.365.$$

If we take t_m to be constant and equal to 15 years, the duration of the illness will be 4.3 years and 6.2 years in 1955 and 1979, respectively. But if we take t_m to be variable (and equal to 15 years in 1955 and 9 years in 1979), the duration of the illness becomes 4.3 and 4.0 years, respectively (see Table 20.3).

It is interesting to note that the improvements recorded in recent years in the survival rates of tumour patients have been brought about by the simultaneous action of two factors: one is the reduction of the specific risk of death for all the main tumour sites, and the other is a favourable change in the internal site structure of malignant tumours. A simple simulation shows that about 79% of the total reduction in the overall lethality can be attributed to the reduction in the specific death risks of patients, 7% to

Table 20.3. *Some indicators of survivorship of a cohort of cancer patients, by age at diagnosis, estimates for Italian males, 1955 and 1979*

	Age at diagnosis					
	30	40	50	60	70	All
Fatality rate (per 1000)						
1955	612	678	734	782	789	782
1979	331	482	622	635	635	635
Percentage change	−46	−29	−15	−19	−20	−19
Average duration of disease						
Total dead						
1955[a]	1.52	1.44	1.37	1.30	1.29	1.30
1979[a]	1.40	1.32	1.23	1.22	1.22	1.22
Percentage change	−8	−8	−10	−6	−5	−6
1979[b]	1.38	1.31	1.22	1.21	1.21	1.21
Percentage change	−9	−9	−10	−7	−6	−7
Total (dead and surviving)						
1955[a]	6.75	5.81	5.00	4.30	4.18	4.30
1979[a]	10.50	8.40	6.42	6.25	6.25	6.25
Percentage change	+56	+45	+28	+45	+50	+45
1979[b]	6.47	5.29	4.16	4.05	4.05	4.05
Percentage change	−4	−9	−17	−6	−3	−6

[a] Taking into account the first 15 years from diagnosis.
[b] Taking into account the first 9 years from diagnosis.

favourable changes in the internal structure of the malignant tumours, and 14% to the combined effects of these two factors.

2.2. The Relationship between Age and Cancer Survival

To assess survival in terms of age, we assumed that throughout the entire period of survival the instantaneous risk is proportional to an age-dependent factor, and that this is the same in both populations. Taking h and h' to stand for the instantaneous hazard of the American and the Italian populations respectively, this assumption can be expressed by the following relations:

$$h(t, x_i) = b_i h(t, x_0); \; h'(t, x_i) = b_i h'(t, x_0),$$

where x_i indicates the age of onset of the disease, x_0 is a benchmark value (assuming $x_0 =$ age 60), and b represents parameters which determine the dependence of survival on age. The same relations can be written in terms of the survival curves:

$$\sigma(t, x_i) = \sigma(t, x_0)^{b_i}, \tag{6}$$

for the American sample, and

$$\sigma'(t, x_i) = \sigma'(t, x_0)^{b_i}, \tag{7}$$

for the Italian population.

The values of b_i have been estimated from the age curves given in Axtell *et al.* (1976) and Ries *et al.* (1983) using the relation

$$b_i = \frac{\log \sigma(t, x_i)}{\log \sigma(t, x_0)}.$$

To be able to apply the relations (6) and (7) and hence determine the survival rate by age of the Italian population, one needs to know the survival curve $\sigma'(t, x_0)$ for the benchmark age x_0. The latter is easy to estimate if the sum of the specific survival curves per age, approximately weighted by the proportion to the total of cases in each age group, is set equal to the observed value. Thus, for every t, the following equation has been solved in terms of $\sigma'(t, x_0)$:

$$\sigma'(t) = \sum_i \zeta_i \sigma'(t, x_0)^{b_i},$$

where ζ_i indicates the proportion of diagnosed cases which belong to the ith age group. The specific rates per age in the first and last years of the interval considered (1955 and 1979), which are obtained using this procedure, are displayed in Table 20.2. The most important reductions in lethality occur at the lower ages and peak around age 30 (with a reduction of just 50%). At older ages the estimated improvements are much smaller, with minima around age 50 (reductions of about 15%) (see Table 20.3).

2.3. The Time Trend of Survival

The aforementioned procedure was used to estimate the survival trend in terms of age in the first $\{\sigma'_{1955}(t)\}$ and last $\{\sigma'_{1979}(t)\}$ year of the period of observation. For the intervening years, the values σ'_p have been estimated by a linear interpolation. Since no information was available on the years preceding 1955, the survival rate of the initial period was used. In the computations we assumed that the relative survival probability $\sigma'_{a+j}(i-j-1/2, j-1/2)$ given by these curves was an estimate of the term

$$\exp\left[-\int_{j+\frac{1}{2}}^{i} g^a(x, j+\tfrac{1}{2})dx\right]$$

appearing in equations (3) and (4). The crude probability of cancer death for sick people d^a_{ij} has been computed from relative survival $\sigma'_{a+j}(i-j-1/2, j+1/2)$ and the probability of cancer death for the general population.

3. Main Results

The model for estimating the incidence and prevalence of tumours (equations 3 and 4) was applied to the Italian tumour mortality rate in each calendar year from 1955 to 1979, by single years of age. In the first stage, we estimate the parameters of the logistical function by age, cohort, and period (equation 5). This function, which represents the development and age-related mortality rate from tumours, produced very satisfactory results. To increase precision even further a function using a large number of parameters was selected. The model which interpolates the matrix of 80×25 mortality data cells explains 99.6% of the total variance. The results and comments that we present refer only to males in the age groups between 30 and 74 years. At ages younger than 30, tumour mortality rates are too small and hence subject to substantial random noise. At ages older than 74 they are not entirely reliable due to the general low degree of accuracy in the identification of the causes of death.

Figure 20.1 displays mortality and estimated morbidity (incidence) probabilities by age for the two years 1955 and 1979. In addition to the substantial increase in the risk of death from tumours (the average probability in the 30–74 age group rose by 61%, from 23 to 37 per 10 000) there is an even higher risk of illness (the mean probability rose by 87% from 35 to 65 per 10 000). It is to the significant improvement in the survival of patients that one has to attribute effects offsetting the deterioration of mortality due to the increase in the incidence of tumours. The ageing of the population during the time period considered has also played an important role in bringing about the increase: in the risk of death due to tumour the ratio in the age groups 30–49 to those in the age group 50–74 fell from 1.43 to 1.16 between 1955 and 1979. Yet ageing had a much greater effect on mortality

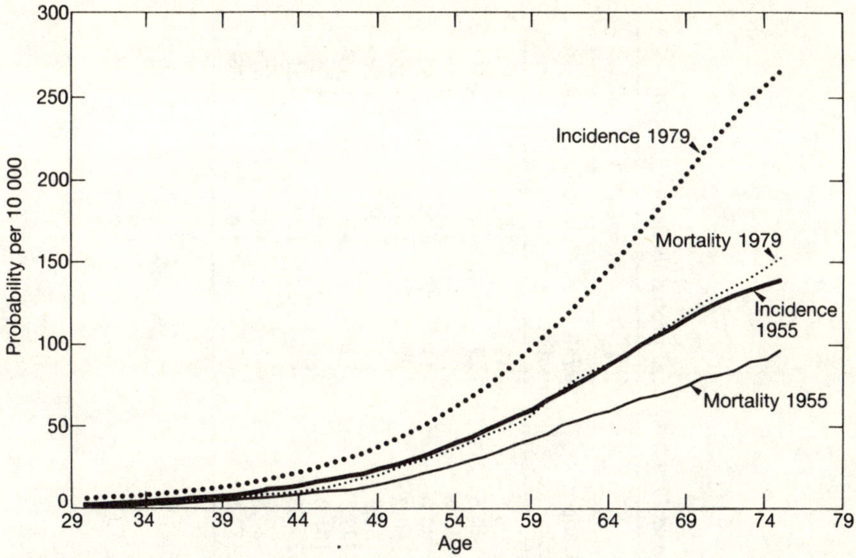

Fig. 20.1. Cancer morbidity and mortality in Italy. Probabilities by age, 1955 and 1979, males

(accounting for 31% of the change) than on the incidence (accounting for 24% of the increase) as a result of the lower survival rate of the older patients.

It is evident that, all the other conditions being equal, at least part of the increase in incidence can be related to improved diagnostic capabilities, which enable physicians to identify more readily this type of disease. On the other hand, heightened awareness about their presence, nature, and consequences has certainly played a role in bringing about changes in the probabilities of proper diagnosis and in a higher proportion of cases being recorded. The older age groups exhibit the most marked deterioration in terms of incidence and mortality (see Table 20.4 and Figure 20.2): for ages above 60, the increase in the risk of dying from tumours was between 40% and 55%, and the increase in the incidence ranged between 65% and 81%. In earlier years the gap between the incidence and mortality trends is comparatively greater: in the age group 30–4, for example, an almost 60% increase incidence led to an increase in mortality of only 5%. Particular attention should be paid to the age group 45–54, where there is a much higher increase in the risk of tumour death than experienced by the population in the immediately following age groups. This relative disadvantage does not stem from any singular increase of risks of illness (see Table 20.4), but rather it is due to a smaller improvement in the survival rate of patients around 50 years of age. At this age, the fall in the lethality rate is about 15%, compared with 19–20% in subsequent years (see Table 20.3).

Table 20.4. *Mortality and incidence of cancer, by age, in Italian males, 1955 and 1979* (net mean probabilities per 10 000)

Age	1955			1979			Change (%)		
	Overall mortality	Cancer mortality	Cancer incidence	Overall mortality	Cancer mortality	Cancer incidence	Overall mortality	Cancer mortality	Cancer incidence
30–34	19.6	2.0	4.1	10.0	2.1	6.5	−49	+5	+59
35–39	23.1	3.3	6.8	16.4	3.8	10.6	−29	+15	+56
40–44	35.4	6.3	11.7	26.4	7.3	17.6	−25	+16	+50
45–49	57.0	11.7	19.6	49.7	16.0	30.0	−13	+37	+53
50–54	92.1	21.7	32.5	85.4	30.8	50.2	−7	+42	+55
55–59	139.5	36.1	51.5	133.1	48.5	81.3	−5	+34	+58
60–64	216.4	54.1	77.2	211.4	75.3	127.3	−2	+39	+65
65–69	319.6	70.5	105.8	320.1	105.0	181.3	..	+49	+71
70–74	505.8	87.2	131.2	486.0	135.0	237.5	−4	+55	+81
30–74	108.6	22.8	34.7	115.7	36.8	64.8	+7	+61	+87
30–74[a]	—	—	—	101.0	32.5	57.6	−7	+43	+66

[a] Age-adjusted.

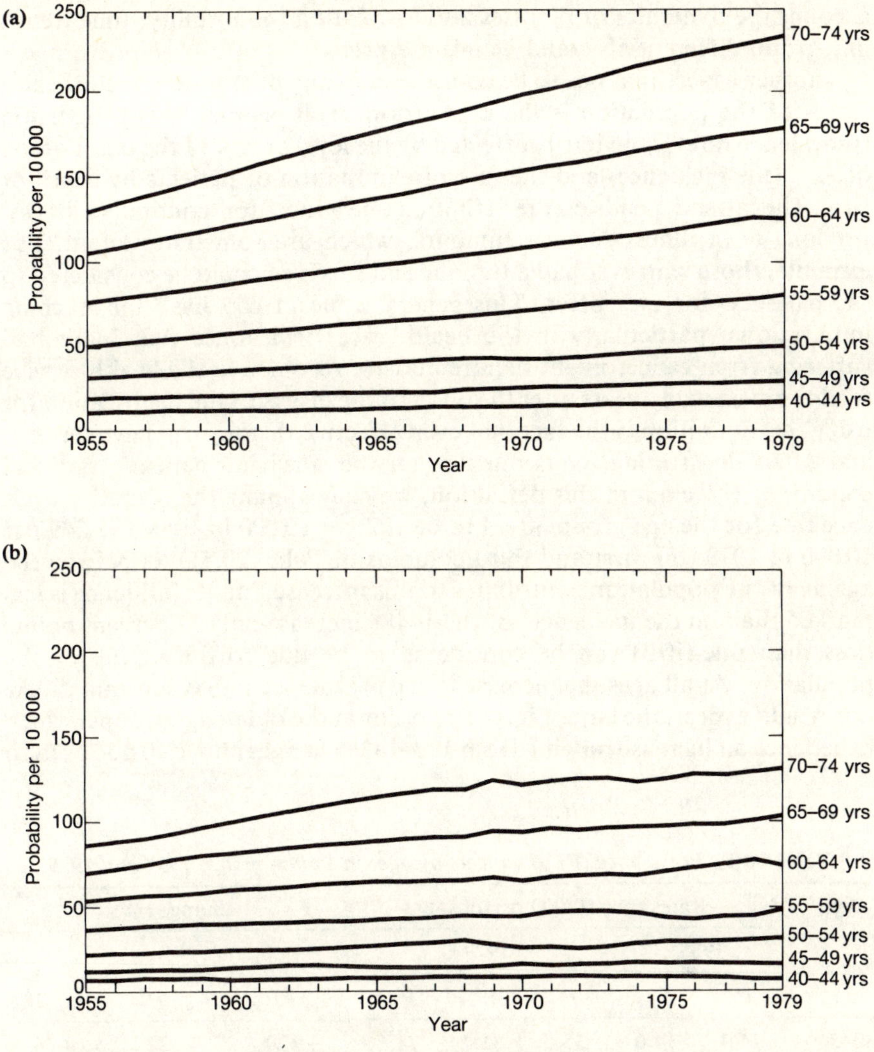

Fig. 20.2. Evolution of cancer morbidity and mortality. Mean incidence and death probabilities (per 10 000), males
(*a*) Incidence
(*b*) Mortality

These observations suggest that proper examination of mortality trends requires taking into account the entire morbidity process, from onset of illness to termination rather than focusing on deaths alone. By introducing this dimension into the mortality analysis, one can get a clearer understanding of the mechanisms underlying time trends and differences between population groups (especially between the two sexes). Without taking into account the dynamics of illnesses and its relation to mortality, time trends and group differentials could be misinterpreted.

Another aspect that has to be considered in any description of the health status of the population is the proportion of ill people. This proportion (prevalence rate) is obviously affected by the level of risk of the onset of the disease (the incidence) and the rate of elimination of patients by death or cure. The latter depends on the definition one adopts for 'continuing illness' for long-term illness such as tumours, which are sometimes latent. For example, those who ever had a tumour and survived could be considered to be 'patients' for ever after. This general concept can have far-reaching implications, particularly in the health care field, since that individual suffering from cancer might require massive recourse to the health service facilities even many years after the onset of the disease. One justification for using this definition is the fact that even effective therapy can have serious, and often devastating, consequences on the surviving patient's physical condition. If we adopt this definition, we can estimate the overall prevalence rate for the age group 30–74 to be 102 per 10 000 in 1955 and 249 per 10 000 in 1979 (the first and third columns in Table 20.5). As before, the ageing of the population contributes to this increase, but its influence is less marked than on the incidence: of the 144% increase only 27 percent points (less than one-fifth) can be considered to be due to the ageing of the population. At all ages the increase in the prevalence rate is substantial. As one would expect, the largest increases occur at the oldest age groups, which experience an increase ranging from 120–142% at ages above 60 due both to

Table 20.5. *Prevalence (P) of cancer, by age, in Italian males, 1955 and 1979*

Age	Rates (per 10 000 persons)					Change (%)		
	1955		1979					
	P	P^a	P	P^a	P^b	$P_{79}-P_{55}$	$P^a_{79}-P^a_{55}$	$P^b_{79}-P^a_{55}$
30–34	16.1	14.0	35.5	31.1	24.6	+120	+122	+76
35–39	26.2	22.9	55.2	48.5	38.1	+110	+112	+66
40–44	40.2	35.0	83.3	72.8	56.8	+107	+108	+62
45–49	62.9	54.6	127.1	110.3	86.6	+102	+102	+59
50–54	97.6	84.9	192.7	166.6	131.4	+97	+96	+55
55–59	146.3	127.1	289.2	250.0	198.7	+98	+97	+56
60–64	212.7	184.0	469.3	405.7	324.4	+120	+120	+76
65–69	293.1	251.2	661.3	568.6	450.0	+126	+126	+79
70–74	379.3	320.4	916.8	777.2	603.0	+142	+143	+88
30–74	102.1	88.0	248.8	214.1	168.7	+144	+143	+92
30–74[c]	—	—	221.5	190.8	150.4	+117	+117	+71

[a] Taking into account the first 15 years from diagnosis.
[b] Taking into account the first 9 years from diagnosis.
[c] Age-adjusted.

the increased risk of illness and the substantial improvements in the survival rates of patients in younger ages. In 1955 in the population 70–4 years old 1 in 26 persons had a present or past history of tumours whereas by 1979 this figure had risen to 1 in 11. Even before age 60 the rates rose considerably, and in 25 years they virtually doubled. This steep increase does not disappear when the definition of 'patient' is restricted to those who contracted the disease in the previous 15 years (second and fourth column in Table 20.5). As we mentioned before when discussing the survival curves, this definition is justified at least for the opening years of the period considered, since beyond the 15-year limit the patient runs just about the same risk of death as the population as a whole and, for all practical puposes, can therefore be considered cured. Using the same reasoning for the final year, the duration of disease that would lead to define an individual as still ill falls to 9 years in 1979. In this instance the comparison is between the proportion of patients whose illness lasted not more than 15 years in 1955 (88 per 10 000 between age 30 and 74) and not more than 9 years in 1979 (169 per 10 000). Even though the 92% increase is still very high, it is far smaller than the result one would obtain using the other two criteria (144% and 143%). This is certainly the best yardstick of the population's health status, which takes fully into account the progress made on the treatment of tumours and the associated increase in the chances of surviving the onset of a tumour. In this instance, the prevalence trend is mainly affected by variations in incidence.

4. Conclusions

In view of the approximations that we had to use to overcome the lack of survival curves for cancer patients in Italy, our conclusions are necessarily tentative. Yet the results do reveal important consequences that require careful consideration, either because they are real, namely, they do reflect the tumour morbidity–death process in Italy, or because they are realistic since the model and its application may be deemed to be at least a simulation of real processes.

From the point of view of health care policy, there is no doubt that during the 25 years studied here, curative medicine has been very effective, managing to reduce tumour lethality by an average of 19% (with peaks at 46% in the case of patients contracting a cancer when aged 30). Of these, 3 out of 10 survived in 1955, compared with 6 out of 10 in 1979. Primary preventive medicine has been less effective: the probability of contracting a tumour has risen drastically for all age groups and the incidence has risen from a minimum of 50% for the age group 40–4 to a maximum of 81% for the age group 70–4. Not only has there been inefficiency in the control of factors that lead to the onset of tumours, but their consequences have actually intensified. Secondary prevention has been effective, however, by encouraging the public to become aware of the symptoms of a tumour in order to have it diagnosed promptly. This may have played a major part in enhancing the effectiveness of treatment and cure, since not only has the number of survivors increased dramatically, but the period after which their excess cancer mortality becomes negligible has been reduced from 15 to 9 years. This is also important from the economic point of view, because it makes it possible to reduce the number of years' sickness (namely, prevalence) which would otherwise occur in a population if the number of deaths alone were to be reduced.

Much more should be done to understand some of the specific regularities, such as the fact that the death rate varies so widely by age group but is minimal at age 50, and the more general ones such as the increase in incidence in the higher age groups which could be an 'inevitable' result of a reduction in other types of illness. From the demographic point of view, the whole vexed question of differential mortality should become clearer once the morbidity–mortality analysis for both sexes and for the population in terms of race, ethnic group, social class, and place of birth is completed. Many demographic patterns such as nuptiality, divorce rate, and fertility experience change whenever there is a substantial change in morbidity rates and in the duration and death rates associated with certain illnesses.

Apart from these more general considerations, the fact remains that studies such as this make it possible to refine the analysis of mortality and to avoid misinterpretations when appraising mortality levels and trends. This type of analysis provides valuable information on the effects of the population's ageing, since it makes it possible to appraise these effects in terms of morbidity (incidence and prevalence). For example, whereas the increased incidence of cancer between 1955 and 1979 by single years of age groups between 30 and 74 years varied from a minimum of 50% to a maximum of 81%, for the age group 30–74 as a whole the increase was 87%. The same applies to prevalence: the increase in individual age groups varied between 55% and 88% but the increase for the age group as a whole was as high as 92%.

Lastly, this type of study is naturally relevant to evaluations of the impact of morbidity levels and trends on health and social security costs as well as their effects on the labour force, productivity, and production systems.

Although in need of refinement, the results obtained here are important since they show that it is necessary to devote increasing attention to the study of sickness and to the health status distribution of the population.

References

Axtell, L. M., Culter, S. J., and Myers, M. H. (1972), *End Result in Cancer*, Report no. 4, NIH Publication no. 73-272.

—— Asire, A. J., and Myers, M. H. (1976), *Cancer Patient Survival*, Report no. 5, NIH Publication no. 81-992.

Bishop, Y. M. M., Fienberg, S. E., and Holland, P. W. (1975), *Discrete Multivariate Analysis: Theory and Practice* (Cambridge, Mass: MIT Press).

Capocaccia, R., Egidi, V., Golini, A., and Verdecchia, A. (1985), 'From Mortality to Morbidity: An Estimation Method', Spontaneous paper presented at the *International Population Conference*, Florence, June (Liège: IUSSP, Ordina Éditions).

Frome, E.L. (1983), 'The Analysis of Rates Using Poisson Regression Models', *Biometrics*, 39, 665–74.

Golini, A., and Egidi, V. (1981), 'Effects of Morbidity Changes on Mortality and Population Size and Structure', *IUSSP Seminar on Methodology and Data Collection in Mortality Studies*, Dakar, Senegal, 7–10 July.

Kitsul, P. (1980), *A Dynamic Approach to the Estimation of Morbidity*, Working Paper of the International Institute for Applied Systems Analysis, WP-80-71 (Luxemburg: IIASA).

Klementiev, A.A. (1977), *On the Estimation of Morbidity*, International Institute for Applied Systems Analysis, RM-77-41 (Luxemburg: IIASA).

Manton, K., and Stallard E. (1982*a*), 'A Cohort Analysis of US Stomach Cancer Mortality 1950–1977', *International Journal of Epidemiology*, 11, 49–61.

—— —— (1982*b*), 'The Use of Mortality Time Series Data to Produce Hypothetical Morbidity Distributions and Project Mortality Trends', *Demography*, 19, 223–40.

—— —— (1984), *Recent Trends in Mortality Analysis* (New York: Academic Press).

Ries, L.G., Pollack, E.S., and Young, J.L., jun. (1983), 'Cancer Patient Survival: Surveillance, Epidemiology and End Results Program 1973–79', *Journal of the National Cancer Institute*, 70(4).

Verdecchia, A., Capocaccia, R., and Mariotti, S. (1984), 'Estimation of Cardiovascular Diseases Morbidity Using Mortality Data', in W. Eimeren, R. Engelbrecht, and C. Flagle (eds.), *Proceedings of the Third International Conference on Systems Science in Health Care* (Berlin, Heidelberg: Springer Verlag), 44–7.

—— —— and Mazzoni, C. (1985), 'Estimation of Cancer Morbidity Using Mortality Data', *Tumori*, 71, 431–9.

—— —— Egidi, V., and Golini, A. (1989), 'A Method for the Estimation of Chronic Disease Morbidity and Trends from Mortality Data, *Statistics in Medicine*, 8, 201–16.

Subject Index

Author Index